ACID PRECIPITATION
Effects on Ecological Systems

ACID PRECIPITATION
Effects on Ecological Systems

Edited by
FRANK M. D'ITRI

ANN ARBOR SCIENCE
THE BUTTERWORTH GROUP

Copyright © 1982 by Ann Arbor Science Publishers
230 Collingwood, P.O. Box 1425, Ann Arbor, Michigan 48106

Library of Congress Catalog Card Number 81-69072
ISBN 0-250-40509-1

0082787

Butterworths, Ltd., Borough Green, Sevenoaks
Kent TN15 8PH, England

120033

5/4.5222

DIT

PREFACE

Along with high living standards in industrial countries
have come increased automobile exhausts, industrial effluents
and the pollutants from electricity generation. Potentially
toxic gases or combinations of sulfur dioxide, ozone, nitrogen oxides, hydrogen fluorides and hydrogen chloride are released into the atmosphere only to return to earth in the form
of acid precipitation. Almost everyone associated with the
problem agrees that the quantity, quality and impact of these
emissions on the environment require further study. However,
there is substantial disagreement with respect to their cause
and the extent of damages and real cost to society, both economic and environmental.

This book reflects the efforts of representatives of government, academia and industry to examine the critical issues
and offer suggestions regarding the direction and focus of
future research. The history and current status of the ecological consequences of acid precipitation are reviewed. Data
are presented for contemporary research on sources, monitoring,
transformation, long-range transport, and the effects on
aquatic and terrestrial ecosystems. This includes an historical review of the progress in the scientific and public
understanding of the acid precipitation problem, from 1661 to
the present.

Frank M. D'Itri

ACKNOWLEDGMENTS

The success of the Conference on the Effects of Acid Precipitation on Ecological Systems, convened at Michigan State University, was due to the interest and dedication of the many individuals and organizations that were involved, and it is my pleasure to acknowledge as many of them as possible. First, a grateful acknowledgment is extended to all of the authors whose research as well as contributions of time, effort and counsel made this book possible. Every attempt was made to provide an accurate transcript of the proceedings; however, I accept responsibility for any errors and/or omissions with apologies to the participants. Secondly, I wish to thank the Office of Water Research and Technology, U.S. Department of the Interior for sponsoring the conference and providing the financial support with a grant through the University of North Carolina to Michigan State University. Additional financial support was provided by the Fish and Wildlife Service and the Detroit Edison Company. Special thanks are extended to Mr. Ken Suter of the University of North Carolina, Dr. N. Jay Bassin of the Fish and Wildlife Service, Dr. William P. Kovalak of the Detroit Edison Company, and Mr. Michael D. Marcus, President, Western Aquatics, Inc., Laramie, Wyoming. Thirdly, I especially appreciate the support received within Michigan State University. In the Institute of Water Research I acknowledge and thank Terry Waters for her expert secretarial and typing skills and Lois Wolfson for her editorial assistance in the preparation of this book.

Frank M. D'Itri is Professor of Water Chemistry, Institute of
Water Research and Department of Fisheries and Wildlife,
Michigan State University in East Lansing. With a PhD in
analytical chemistry, his primary research emphasizes the
analytical aspects of water and sediment chemistry, especially the transformation and translocation of phosphorus,
nitrogen, heavy metals and hazardous organic chemicals in the
environment.

Professor D'Itri is listed in *American Men of Science,
Physical and Biological Science*. He is the author of *The
Environmental Mercury Problem*, co-author of *Mercury Contam-
ination: A Human Tragedy* and *An Assessment of Mercury In the
Environment*, and editor of *Wastewater Renovation and Reuse*.
Dr. D'Itri is also the editor of *Land Treatment of Municipal
Wastewater: Vegetation Selection and Management*, published
by Ann Arbor Science. In addition, he is the author of more
than forty scientific articles on a variety of environmental
topics.

Dr. D'Itri has served as chairperson for the National
Research Council's panel reviewing the environmental effects
of mercury and a number of symposia dealing with the analytical problems related to environmental pollution. He has been
the recipient of fellowships from Socony-Mobil, the National
Institutes of Health, the Rockefeller Foundation and the Japan
Society for the Promotion of Science.

LIST OF CONTRIBUTORS

(Numbers in parentheses indicate the pages on which the author's contribution begins.)

*John D. Aber (411), Assistant Professor, Department of Forestry, University of Wisconsin, Madison, WI 53706.

*Jon W. Allan (209), Graduate Research Assistant, Department of Zoology, Michigan State University, East Lansing, MI 48824.

*Leonard A. Barrie (141), Research Scientist, Atmospheric Environment Service, 4905 Dufferin Street, Downsview, Ontario M3H 5T4.

*Leigh M. Blake (251), Regional Fisheries Manager, New York State Department of Environment Conservation, 317 Washington Street, Watertown, NY 13601.

*Daniel B. Botkin (411), Professor and Director, Environmental Studies Program, University of California-Santa Barbara, Santa Barbara, CA 93016.

*Thomas M. Burton (209), Associate Professor, Department of Zoology, Department of Fisheries and Wildlife, and Institute of Water Research, Michigan State University, East Lansing, MI 48824.

*Philip S. S. Chang (297), Aquatic Biologist, Department of Fisheries and Oceans, Freshwater Institute, 501 University Crescent, Winnipeg, Manitoba R3T 2N6, Canada.

*Carl W. Chen (237), Vice President, Tetra Tech, Inc., 3746 Mt. Diablo Boulevard, Lafayette, CA 94549.

*H. Lee Conway (277), Associate Scientist, Terrestrial and Aquatic Ecology Division, Department of Energy and Environment, Brookhaven National Laboratory, Upton, NY 11973.

*Ellis B. Cowling (3, 43), Chairman, National Atmospheric Deposition Program, Associate Dean, School of Forest Resources, Assistant Director, North Carolina Agricultural Research Service, North Carolina State University, Raleigh, NC 27650.

*Christopher S. Cronan (435), Professor, Department of Botany and Plant Pathology, University of Maine, Orono, ME 04469.

*Ronald Dermott (165), Biologist, Department of Fisheries and Oceans, Great Lakes Biolimnology Laboratory,CCIW, 867 Lakeshore Drive, Box 5050, Burlington, Ontario L7R 4A6.

*Frank M. D'Itri (xi), Professor, Institute of Water Research, and Department of Fisheries and Wildlife, Michigan State University, East Lansing, MI 48824.

Elwin D. Evans, Aquatic Biologist, Water Quality Division, Michigan Department of Natural Resources, Lansing, MI 48909.

*David L. Findlay (297), Phytoplankton Technician, Department of Fisheries and Oceans, Freshwater Institute, 501 University Crescent, Winnipeg, Manitoba R3T 2N6, Canada.

*A.J. Francis (411), Microbiologist, Department of Energy and Environment, Brookhaven National Laboratory, Upton, NY 11973.

*Robert A. Goldstein (237), Project Manager, Environmental Assessment Department, EPRI, 3412 Hillview Avenue, P. O. Box 10412, Palo Alto, CA 94303.

Jeremy M. Hales, Manager, Atmospheric Chemistry Section, Battelle Pacific Northwest Laboratories, Richland, WA 99352.

*George R. Hendrey (277, 411), Head, Terrestrial and Aquatic Ecology Division, Department of Energy and Environment, Brookhaven National Laboratory, Upton, NY 11973.

*William A. Hoffman, Jr. (385), Professor, Department of Chemistry, Denison University, Granville, OH 43023.

Raymond Herrmann, Chief, Water Resources Division, U.S. Department of the Interior, National Park Service, Washington, DC 20240.

*Thomas C. Hutchinson (105), Professor and Chairman, Department of Botany, University of Toronto, Toronto, Ontario M5S 1A1.

*Arland H. Johannes (237), Assistant Professor, Department of Chemical and Environmental Engineering, Rensselaer Polytechnic Institute, Troy, NY 12181.

*William L. Keepers (87), Vice President, Power Production, Wisconsin Power and Light, 222 Washington Avenue, Madison, WI 53701.

*John R.M. Kelso (165), Scientist, Department of Fisheries and Oceans, Great Lakes Biolimnology Laboratory, 875 Queen Street East, Sault St. Marie, Ontario P6A 2B3.

William P. Kovalak, Principal Biologist, Detroit Edison Company, 2000 Second Avenue, Detroit, MI 48826.

*Sagar V. Krupa (469), Associate Professor, Department of Plant Pathology, University of Minnesota, St. Paul, MN 55108.

*Jeffrey J. Lee (453), Research Environmental Scientist, U.S. Environmental Protection Agency, Corvallis Environmental Research Laboratory, 200 S.W. 35th Street, Corvallis, OR 97333.

*Steven E. Lindberg (385), Staff Geochemist, Earth Sciences Section, Oak Ridge National Laboratory, P. O. Box X, Oak Ridge, TN 37830.

*James H. Lipsit (165), Technician, Department of Fisheries and Oceans, Great Lakes Biolimnology Laboratory, 875 Queen Street East, Sault St. Marie, Ontario P6A 2B3.

*Orie L. Loucks (21), Science Director, The Institute of Ecology, 4600 Sunset Avenue, Indianapolis, IN 46208.

*Robert J. Love (165), Biologist, Department of Fisheries and Oceans, Great Lakes Biolimnology Laboratory, 875 Queen Street East, Sault St. Marie, Ontario P6A 2B3.

*Diane F. Malley (297), Research Scientist, Limnology Section, Department of Fisheries and Oceans, Freshwater Institute, 501 University Crescent, Winnipeg, Manitoba R3T 2N6, Canada.

*William W. McFee (435), Professor, Agronomy Department, Purdue University, West Lafayette, IN 47907.

*Jerry M. Melillo (411), Assistant Scientist, Ecosystems Center, Marine Biological Laboratory, Woods Hole, MA 02543.

*Kenneth W. Ragland (123), Professor, Department of Mechanical Engineering, University of Wisconsin, Madison, WI 53706.

Thomas H. Roush, Research Aquatic Biologist, U.S. Environmental Protection Agency, Environmental Research Laboratory, 6201 Congdon Boulevard, Duluth, MN 55804.

*William E. Sharpe (365), Assistant Professor of Forest Resources Extension, Institute for Research on Land and Water Resources, Pennsylvania State University, University Park, PA 16802.

Carl L. Schofield, Senior Research Associate, Department of Natural Resources, Cornell University, Ithaca, NY 14853.

R. Kent Schreiber, Team Leader, Eastern Energy and Land Use Team, National Power Development Group, U.S. Department of the Interior, Fish and Wildlife Service, Ann Arbor, MI 48105.

Thomas A. Seliga, Director, Atmospheric Sciences Program, Ohio State University, Columbus, OH 43210.

*David S. Shriner (385), Staff Ecologist, Terrestrial Ecology Section, Oak Ridge National Laboratory, P. O. Box X, Oak Ridge, TN 37830.

*Robert Singer (329), Assistant Professor, Department of Biology, Colgate University, Hamilton, NY 13346.

*Richard M. Stanford (209), Research Associate, Institute of Water Research, Michigan State University, East Lansing, MI 48824.

*J. Robert Stottlemyer (261), Research Scientist, Department of Biological Sciences, Michigan Technological University, Houghton, MI 49931.

*Kenneth E. Wilkening (123), Research Specialist, Department of Mechanical Engineering, University of Wisconsin, Madison, WI 53706.

Sylvan H. Wittwer, Director, Agricultural Experiment Station, Michigan State University, East Lansing, MI 48824.

*Edward S. Young (365), Research Assistant, School of Forest Resources and The Institute for Research on Land and Water Resources, The Pennsylvania State University, University Park, PA 16802.

*Authors of papers published in these proceedings.

INTRODUCTION

The acidity of rain, snow, and atmospheric particulates that fall upon much of the world appears to have increased significantly over the last four decades; and these effects are now having far-reaching environmental, social, and political consequences. In addition to natural sources of acid precipitation that result from geological weathering, volcanic eruption, anaerobic decomposition of organic matter, airborne sea salt sprays, and lightning, most of the increased acid precipitation burden has been attributed to consumption of fossil fuels, especially coal.

As the standard of living has improved in industrially developed societies, an unwelcome side effect of the greater energy consumption has been exhausts from automobiles as well as heavy manufacturing, metal smelting, and the generation of electricity. With the world's oil and gas reserves being rapidly depleted and the development of viable alternative energy sources still in the distant future, the use of more coal appears to be inevitable as well as the consequent greater sulfur and nitrogen oxide emissions, conservatively estimated to increase five-fold by the year 2000. Ironically, the greater dependence on coal comes at a time when the negative environmental impacts of burning it are also being recognized, and the faltering economy makes it less feasible to introduce more expensive treatment of stack gases or to initiate other emission controls.

Increased surface water acidity has been correlated with sharp decreases, in some cases to the point of extinction, in the number of aquatic organisms in many lakes and streams. Acidic fallout may cause demineralization of soils as well as adverse effects on sensitive agricultural crops, native plants, and forests. While the data relating the effects of acid rain on forest and agricultural crop production and yield are limited, certain sensitive crop systems have been shown in experimental studies to be directly affected by deposition on the foliage of toxic gases such as sulfur dioxide, ozone, nitrogen oxides, hydrogen fluoride, and hydrogen chloride.

In addition, atmospheric deposition may be responsible for a number of indirect soil effects which vary greatly depending on the type of vegetation, natural rates of acid formation in the soil, and the physical-chemical properties of the soil. Acidification may also decrease the rates of many soil microbial processes such as nitrogen fixation and decomposition of organic matter. At the same time, lysimeter studies indicate that soil acidification increases the leaching of exchangeable plant nutrients and/or minerals, such as calcium, magnesium, potassium, iron, manganese, and phosphorus. The solubility and mobility of many toxic metal cations are also affected, such as aluminum and zinc in the soil solution. As a result, soil fertility may decrease. This is especially critical in forests which often have less fertile soils than those used for agriculture and where growth may depend more on atmospheric nutrients. In addition, the weathering, corrosion, and deterioration of man-made materials and buildings have also been accelerated. Some researchers have recently expressed concern that acidic aerosols in the atmosphere may constitute an additional subchronic health threat to human beings, especially those who live in or near large industrial centers.

The damage occurs because the depositions of acidic and other air pollutants change the physio-chemical environment, disrupting the dynamic equilibrium among soil, water, air, and living organisms. Under normal circumstances the abiotic and biotic components are closely linked as they perform essential functions in the terrestrial and aquatic ecosystems. They can tolerate some change without serious disruption, but greater disturbances can alter both the structure and function of the ecosystem. Whereas some organisms may benefit from the change, others may be harmed and reflect decreasing growth rates and numbers. The impact of acidic air pollutants may have a diverse or varied, even magnified, impact on the complex natural ecosystems over both the short and the long term.

The regions of greatest susceptibility to the effects of acid precipitation usually are recently glaciated. The large areas of exposed granitic and other non-calcareous bedrock are highly resistant to chemical weathering and only partially covered by generally thin, coarse-grained glacial deposits of similar lithology that are low in buffering and cation exchange capacity. These soils and freshwaters are naturally low in alkalinity and calcium; consequently, they are poorly buffered and vulnerable to inputs of acidic pollutants. Thus, this airborne acid particularly affects the characteristic chemistry of poorly buffered soils, lakes, and streams in Scandinavia, the northeastern United States, and eastern Canada.

In general, surface water sensitivities to acidification vary among watersheds. The sensitivity of aquatic ecosystems depends on buffering factors internal to both watersheds and surface waters. Watershed factors include soil particle size, texture, ionic absorptive capacity, chemistry, depth, bedrock geology, soil drainage, landform relief, and vegetation. Surface water factors include alkalinity levels, sediment bicarbonates, weak organic acids, sulfate reduction, surface area, water volume, flushing rates, respiration levels, and nutrient levels. The relative importance of these factors can differ among watersheds and surface waters.

Scientists have been aware for more than forty years that the rain and snow falling on many areas of the world have been acidic. Swedish researchers observed in 1920 that numerous small, mountain lakes located at high elevations in Scandinavia becoming more acid and, in some cases, devoid of fish. However, not until the 1960s and early 1970s was enough evidence compiled to correlate acid rain with acidification of surface waters. Then it was suggested that the acidity of rain coincided with the enormous industrial expansion and nearly exponentially rising consumption of energy in the industrialized countries, especially after World War II.

Since then, the acidity of rainfall in some regions of the eastern half of the United States and Canada appears to have increased as much as 50-fold. As the areas with highly acid precipitation (pH < 4) became larger from year to year, it was gradually attributed to increasing emissions of sulfur and nitrogen oxides from the combustion of fossil fuels. Especially in the Scandinavian countries, concern intensified that acid precipitation was causing damage to the natural environment. While the consequences of acidic and other air pollutants being deposited in the environment have not been fully evaluated, the impact may be extensive.

Acid rain has been correlated with environmental damage in many parts of the world, including Japan, Norway, Sweden, Canada, northern Europe, and the United States. As the acidity of precipitation in some instances increased 200-fold since 1956, an estimated 10,000 lakes in Scandinavia are now without fish, and another 10,000 are threatened. Moreover, according to current projections, by the year 2000 an additional 20,000 lakes may be threatened. In eastern Canada and the northeastern United States the effect on susceptible lakes has also been extensive. In the Adirondacks the pH values in many lakes are depressed to the point where fish no longer reproduce, and more than 100 lakes have none at all. Ontario has 140 lakes without fish, whereas Nova Scotia has lost salmon populations in 9 rivers, and 11 more are threatened. Canadian researchers predict that 48,000 Ontario lakes will become devoid of fish over the next 20 years. Similar predictions are also being made for watercourses in Quebec, New Brunswick, and Nova Scotia.

Fish are especially vulnerable to changes in acidity because they have several critically sensitive life stages such as spawning, egg development, fry hatch, and early development. The greatest damage appears to occur in the spring during times of spawning or hatching. This period often coincides with snowmelt which may rapidly flush hydrogen ions from the snow pack which had been accumulated from both atmospheric depositions and soil decomposition. The result is a rapid pulse of increased acidity in streams and lakes which can cause an acute stress that sometimes becomes intense enough to kill adult fish, apparently by upsetting the salt balance in their bodies.

Acidification has other serious consequences for the structure and functions of freshwater aquatic ecosystems. Lake survey data indicate that acidification decreases species richness and leads to simplified ecosystems. The data indicate that animal and plant life in thousands of lakes may be eliminated or drastically altered during the next 10 to 25 years if the acidity continues to increase. Thus, long-term as well as short term changes in the chemistry of freshwater ecosystems are predicted for the United States and Canada.

As more information is needed on the nature, distribution, and fate of acidic precursors in the environment, scientists must identify temporal and spatial trends in acidic depositions in the watersheds to correlate changes in the extent of acid precipitation with meteorological factors. The relative importance of external anthropogenic sources versus internal natural sources needs evaluation to establish ecosystem acidification. Related to this, more information is needed on global atmospheric processes and the complex photochemical reactions that transform exhaust products into acids. Although this process is not completely understood, one hypothesis is that hydrogen peroxide vapor is generated in cloud water and acts as the rate-limiting oxidant in the conversion of nitrogen and sulfur oxides to nitric and sulfuric acids. They then return to earth in rain, snow, or dry fallout.

As researchers monitor the impact of acid on the environment, they have also sought ways to rehabilitate areas that have been damaged. Some suggestions include planting fish strains that are more resistant to low pH in acidified waters, liming these waters and taxing sulfur oxide emitters under some kind of worldwide policy whereby the money would be used to restore the affected water and continue research. This

technology is available to reduce emissions from current sources substantially, albeit with considerable financial cost.

Another prospect is to require industries that consume fossil fuels to restrict the amount of sulfur oxides emitted by burning cleaner coal or cleaning it before it is burned or afterwards with smokestack scrubbers. The cheapest and most common locally implemented alternative is to build taller smoke-stacks so the sulfur emissions are released farther up in the atmosphere. Smelters and plants that burn fossil fuels have steadily built higher smokestacks, especially since the 1960s. Consequently, the acid precursors from these elevated sources may be transported for hundreds of kilometers. During this interval, they can be chemically transformed into nitric and sulfuric acids that are deposited on leaves, soils, surface waters, and man-made structures. This long-range transport increases the likelihood that a local pollution problem will be transformed into a regional, national, or international problem.

Whereas severe, even irreversible damage to susceptible ecosystems is now being studied, any resolution of the long-range transport problem is likely to be more difficult where international cooperation is required. In accordance with the Charter of the United Nations and the principles of international law, sovereign nations have the right to exploit their resources pursuant to their own environmental policies. However, they also have the responsibility to ensure that activities within their jurisdiction or control do not cause damage to the environment of other states or of areas beyond the limits of their national jurisdiction. This is set forth in Principle 21 of the Declaration of the United Nations Conference on the Human Environment (Stockholm, June 5-16, 1972).

While international politics and monetary considerations complicate efforts to control acidic emissions, the resolution finally will depend on strict emission laws that are enforced nationally and with international cooperation. For example, as Canada is on the receiving end of many millions of tonnes of sulfur and nitrogen oxide emissions from the United States, this northern neighbor favors having a bilateral air pollution treaty signed expeditiously. The ailing United States economy has, however, decreased the likelihood that the current adminis-tration will even enforce the standards of the Clean Air Act. Although it has been used to protect air quality near major polluting facilities on a state-by-state basis, this 1970 law and its 1977 amendments have not been fully tested to resolve regional pollution problems and probably will have to be sup-plemented by additional legislation to deal with international issues. At the present time, any regulatory action that is proposed to mitigate acid rain is likely to be perceived as potentially having an adverse effect on the coal industry, electrical utilities, and heavy manufacturing all of which would strongly oppose it.

Industrialists suggest that because of the uncertain scientific evidence, stable emission levels and unlikelihood of catastrophic environmental damage, ample time exist to perform research, weigh the costs and benefits, and work together to develop policies that assure the future well-being of our nation and neighbors. Nonetheless, ultimately, either the industries themselves or the governments of countries that emit acid pollutants must bear the cost of implementing controls. Finally, adoption of stricter pollution control measures must be coupled with continued monitoring and research to obtain the data necessary to critically evaluate the effectiveness of present control measures and establish the degree to which emissions must be controlled in the future. In any event, these measures will be reluctantly instituted as they increase costs and diminish profits accordingly.

As scientists await resolution of the political and economic issues, they can identify sensitive areas of the United States and Canada, and more field studies can be initiated to delineate and explain the impact of this pollution. Acid rain presents the classic scientific conflict between the need to respond quickly to ameliorate the problem and, at the same time, to gather enough data to respond correctly. As scientists have the obligation to obtain the most reliable information in the most economical manner and to formulate an accurate perspective, they will be called upon to aid decision-makers who are now being forewarned as the hazards are foreseen.

Frank M. D'Itri

CONTENTS

Part 1

History and Current Status of
Acid Precipitation and Its
Ecological Consequences

Part 2

General Overview: Sources of
Acid Precipitation, Monitoring,
and Related Problems

Part 3

The Effects of Acid Precipitation
on Aquatic Ecosystems

Part 4

Effects of Acid Precipitation
on Terrestrial Systems

PART 1

HISTORY AND CURRENT STATUS OF
ACID PRECIPITATION AND ITS
ECOLOGICAL CONSEQUENCES

CHAPTER 1

A STATUS REPORT ON ACID PRECIPITATION AND ITS
BIOLOGICAL CONSEQUENCES AS OF APRIL 1981

Ellis B. Cowling
School of Forest Resources
North Carolina State University
Raleigh, North Carolina 27650

INTRODUCTION

During the past several years, the phenomenon of acid precipitation and its biological consequences have become a subject of intense public interest. Numerous discussions and debates have been published in the press, in legislative and administrative hearings, and in the board rooms of industry. These discussions often have been based on incomplete awareness of the many atmospheric and biological processes involved.

The purpose of this chapter is to provide a brief, up-to-date, carefully documented, and rigorously peer-reviewed summary of the present status of knowledge about this important environmental issue. The summary statements listed below were developed in consultation with many scientists engaged in research on acid precipitation in Europe, Canada, and the United States. In most cases, these summary statements are a synthesis of observations and inferences derived from many independent investigations published prior to April 1981. It is beyond the scope and purpose of this summary to provide rigorous and complete documentation for every point; thus, only a few key investigations are cited for each statement. Complete documentation can be obtained by consulting the original literature, access to which is provided by the following reviews and documents: Abrahamsen, 1980; Almer et al., 1974; Altshuller and McBean, 1979, 1980; Ambio, 1976; Bolin et al., 1972; Braekke, 1976; Cowling, 1980, 1981; Dochinger and Seliga, 1976; Drabløs and Tollan, 1980; Galloway et al., 1978; Gorham, 1976; Hendrey, 1978; Howells, 1979; Husar et al., 1978; Hutchinson and Havas, 1980; Likens, 1976; Likens et al., 1979; Malmer, 1973; NAS, 1975; Odén, 1968, 1976; Overrein et al., 1980; and Reuss, 1975.

ATMOSPHERIC CHEMISTRY, TRANSPORT, AND DEPOSITION

1. Acid precipitation (rain, snow, hail, dew, fog, frost) is
    occurring in large regions of the eastern United States
    and Canada, Europe, and Japan. It has also been reported
    in certain urban and rural areas in the western United
    States and Canada and may occur in other regions of the
    world (Odén, 1968; McColl and Bush, 1978; Likens et al.,
    1979; Lewis and Grant, 1980).

2. This widespread occurrence of acid precipitation and dry
    deposition results in large part from man-made emissions
    of oxides of sulfur and nitrogen. These substances are
    transformed in the atmosphere to sulfuric and nitric acids,
    transported over great distances, and deposited on
    vegetation, soils, surface water, and materials (Smith,
    1872, Bolin et al., 1972; Likens, 1976).

3. Acid precipitation is only one special feature of the
    general phenomenon of atmospheric deposition, which
    includes three major mechanisms by which substances are
    transferred from the atmosphere into ecosystems:

    (a) absorption and adsorption of gases,
    (b) impaction and gravitational settling of fine aerosols
         and coarse particles, and
    (c) precipitation, which includes both dissolved sub-
         stances and particles that are removed from the
         atmosphere in rain, snow, hail, dew, fog, and frost.

    Processes a and b are components of dry deposition (Junge
    and Werby, 1958; Galloway and Cowling, 1978).

4. The acidity (or alkalinity) of precipitation is measured
    in terms of pH units--the negative logarithm of the
    concentration of hydrogen ions. Carbon dioxide combines
    with water in the air to form a dilute solution of carbonic
    acid at about pH 5.6. For this reason, acid precipitation
    is arbitrarily defined as precipitation with a pH value
    below 5.6. Because pH is a logarithmic scale, each unit
    change in pH represents a ten-fold change in acidity (Odén,
    1968; Likens, 1976).

5. Evidence for the occurrence of long-term changes in the
    acidity of precipitation in North America includes the
    following:

    (a) changes in average annual pH value of precipitation
         in the years between 1955-56 and 1975-76,

(b) changes in the alkalinity of surface waters; an especially valuable data set of this sort is available for the Hinckley Reservoir near Utica, New York, between 1924 and 1980,

(c) changes in the amounts and patterns of fuel use in the United States and Canada since the industrial revolution began about 1850; at present, about 95 percent of sulfur emissions in eastern North America are from man-made sources, and

(d) changes in the sulfate and lead content of Greenland ice also since about 1850 (Beamish and Harvey, 1972; Cogbill and Likens, 1974; Cragin et al., 1975; Cronan and Schofield,1979; Galloway and Whelpdale, 1979).

6. The ratio of sulfate to nitrate ions in precipitation varies from time to time and from place to place. In much of eastern North America the average equivalent ratio of sulfate to nitrate varies from about 3.3:1 in summer to about 0.7:1 in winter. The average annual ratio of sulfuric to nitric acids is currently about 2:1; but nitric acid is becoming progressively more important relative to sulfuric acid (Likens,1976; Likens et al., 1979; Galloway and Likens, in press).

7. Tall stacks at power plants and smelters decrease nearby ground-level concentrations of SO_2 and NO_x, but simultaneously increase the geographical area of acid deposition. Removal of alkaline particulate matter from stack emissions also increases the acidic character of atmospheric pollutants and, therefore, the acidity of atmospheric deposition (NAS, 1975; Braekke, 1976).

8. Analyses of air-mass movements and chemical transformations in the atmosphere indicate that acid precipitation in one state or region results in part from emissions into the atmosphere in other states or regions, often many hundreds of kilometers from the original source of the emissions (Odén, 1968; OECD, 1977; Galloway and Whelpdale, 1979; Altshuller and McBean, 1979, 1980; Husar and Patterson, 1980; Lyons, 1980).

9. Contemporary precipitation in industrial regions of the world is acidified mainly by man-made emissions of gases and particulate matter. Major anthropogenic sources include combustion of fossil fuels (especially coal and oil), certain industrial processes (especially smelting of ores), exhausts from internal combustion engines, and nitrogen fertilization of agricultural and forest land (Bolin et al., 1972; Kovda, 1975; Likens, 1976).

10. Changes in the acidity of precipitation are reflections of changes in the balance between the major cations and anions in precipitation. Thus, to understand the acidity of precipitation, it is essential to measure all the major anions and cations in precipitation especially hydrogen, sulfate, nitrate, and ammonium ions, but also potassium, sodium, calcium, magnesium, chloride, bicarbonate, sulfite, and phosphate ions (Likens, 1976) .

11. All of the major anions and cations transferred into ecosystems when it is raining or snowing (wet deposition) also are contained in the gases and particulate matter that are transferred from the atmosphere into ecosystems when it is not raining or snowing (dry deposition). Thus, in a chemical mass balance sense, it is both impossible and unrealistic to distinguish the biological effects of "acid precipitation" (wet deposition) from the biological effects of dry deposition (Tamm, 1958; Kovda, 1975; Odén, 1976).

12. Some substances (such as ammonium sulfate) which themselves are not acidic can cause the acidification of soils when they are taken up by plants or modified by soil microorganisms. Thus, the concept of "acidifying precipitation" must be added to the concept of "acid precipitation" (Odén, 1976).

13. The potentially injurious substances in dry and wet deposition include not only acidic substances but also certain toxic gases and organic substances (notably pesticides) as well as various other inorganic substances or heavy metals such as manganese, zinc, copper, iron, molybdenum, boron, chlorine, fluorine, bromine, aluminum, lead, iodine, nickel, cadmium, vanadium, mercury, and arsenic.

GENERAL EFFECTS ON ECOSYSTEMS

1. The elements listed above (item 13) are essential micro-nutrients needed by plants in small amounts. At higher concentrations, however, each of these same elements (and all the others) can be toxic to plants and animals. Furthermore, the acidity of precipitation can affect the solubility, mobility, and toxicity of certain of these elements. These substances can affect the following:

   (a) the foliage and roots of plants,

(b)   animals or microorganisms that may ingest or decompose these plants,

(c)   animals (including man) that may drink water containing these elements, and

(d)   animals (especially fish and other aquatic organisms) that live in chemical equilibrium with waters containing these substances (<u>Ambio</u>, 1976; Odén, 1976).

2.   Significant quantities of plant nutrients, including nitrogen and sulfur, are being added to soil in dry deposition and in precipitation. These atmospheric inputs of beneficial nutrient elements are especially important in forests and rangelands where nutrients from other sources are scarce and where fertilization by man is not a normal management procedure (Hooke, 1687; Tamm, 1958; Wittwer and Bukovac, 1969; Cowling, 1980).

3.   Short-term fertilization effects due to atmospheric deposition of ammonia and nitrate tend to offset long-term nutrient leaching and other detrimental effects of acid precipitation on forest ecosystems. Negative effects of atmospheric deposition on forest growth are most likely when nutrient deficiencies or imbalances are increased by acid deposition (Abrahamsen, 1980).

4.   The net effect of atmospheric deposition can be beneficial or injurious to plants and animals depending on the chemical composition of the deposition (wet and dry), the duration and intensity of deposition episodes, the species and genetic characteristics of the organisms on which the substances are deposited, and the physiological condition, structure, and stage of maturity of the organisms (Grahn <u>et al</u>., 1976; Abrahamsen, 1980; Cowling, 1980; Lee <u>et al</u>., 1980).

5.   Damage or injury to aquatic or terrestrial organisms is most likely when a particularly sensitive life form or life stage developing in poorly buffered waters or soils coincides with major episodic inputs of acid precipitation or other injurious substances (Borgstrøm and Hendrey, 1976; Leivestad <u>et al</u>., 1976; Likens <u>et al</u>., 1979; Hall <u>et al</u>., 1980).

6.   Regions of greatest sensitivity to acid precipitation usually are recently glaciated, with large areas of exposed granitic (crystalline) and other non-calcareous bedrock and thin soils that are low in buffering and cation exchange capacity (Reuss, 1975; McFee, 1978; Abrahamsen, 1980).

7. Terrestrial and aquatic ecosystems are so intimately
    <u>linked</u> that it is both impossible and unrealistic to
    consider the effects of acid precipitation in aquatic
    ecosystems without considering its effects on terrestrial
    vegetation, soils, and the geology of the drainage basin
    (Braekke, 1976; Odén, 1976).

EFFECTS ON AQUATIC ECOSYSTEMS

1. Acids and other soluble substances contained in polluted
    snow are released as contaminated meltwater during warm
    periods in winter or in early spring. The resulting
    release of pollutants can cause major and rapid changes
    in the acidity and other chemical properties of stream and
    lake waters. Fish kills are a dramatic consequence of
    such episodic inputs into aquatic ecosystems (Odén and
    Ahl, 1970; Leivestad <u>et al</u>., 1976).

2. Acid precipitation titrates the acidity/alkalinity of lake
    and stream waters from conditions that are favorable for
    fish and other aquatic organisms to conditions that inhibit
    reproduction and/or recruitment of populations of fish and
    fish-food organisms (Hendrey and Wright, 1975; Henriksen,
    1979; Leivestad <u>et al</u>., 1976; Henriksen, 1980).

3. In oligotrophic lakes and streams between pH 6.0 and pH
    5.0, reproduction of certain species of aquatic organisms
    is inhibited; below pH 5.0, populations of many freshwater
    fish will become extinct (Beamish and Harvey, 1972; Dickson
    <u>et al</u>., 1973; Leivestad <u>et al</u>., 1976; Schofield, 1976).

4. Interference with normal reproductive processes in fish
    populations is induced not only by acidity itself but also
    by increased concentrations of certain cations, notably
    aluminum, in acidified lake and stream waters. Aluminum
    compounds vary greatly in toxicity to fish and other
    aquatic life. In general, ionic forms of aluminum are
    more toxic than aluminum complexed with organic matter
    (Hultberg and Grahn, 1976; Cronan and Schofield, 1979;
    Driscoll, 1980).

5. Reproduction of frogs and salamanders also is inhibited
    by atmospheric acidification of surface waters (Pough,
    1976; Hagström, 1977).

6. As of 1980, hundreds of lakes in the Adirondack region of
    New York State, and many hundreds of lakes in various parts

of southern Ontario and Quebec were showing acid stress in the form of diminished populations or extinction of fish populations. Lakes and streams in other regions of the United States and Canada also are vulnerable to stress by acid precipitation. These regions include northern Minnesota, Wisconsin, and Michigan; parts of southern Appalachia and Florida, and large parts of Washington, Oregon, California, and Idaho, and parts of the Canadian Maritime Provinces (Beamish and Harvey, 1972; Schofield, 1976; Galloway and Cowling, 1978; Altshuller and McBean, 1979; Hendrey et al., 1980).

EFFECTS ON AGRICULTURAL CROPS AND FORESTS

1. In certain industrial regions of the world, substantial damage to forests and agricultural crops is caused by the dry deposition of toxic gases. Sulfur dioxide, ozone, oxides of nitrogen, fluoride, and hydrogen chloride cause serious economic damage to crops and forests that must be considered together with the possible effects of acid deposition. Recent experiments have shown that a combination of acid precipitation and ozone caused greater reduction in yield of soybeans than ozone alone (Heck and Brandt, 1977; Shriner et al., 1980; J. Trioiano, personal communication).

2. Direct and indirect damage to crops and forests have been reported in laboratory, greenhouse, and field experiments in which synthetic rain equivalent (in chemical composition and rate of deposition) to natural rains has been applied. These biological effects include the following:

    (a) Induction of necrotic lesions and other morphological changes in foliage (Shriner, 1978; Evans and Curry, 1979).
    (b) Loss of nutrients due to leaching from foliar organs (Wood and Bormann, 1974).
    (c) Predisposition of plants to infection by bacterial and fungal pathogens (Shriner, 1978).
    (d) Accelerated erosion of waxes on leaf surfaces (Shriner, 1976).
    (e) Inhibition of nodulation of legumes leading to decreased nitrogen fixation by symbiotic bacteria (Shriner and Johnston, 1980).
    (f) Reduced rates of decomposition of leaf litter leading to decreased mineralization of organically-bound nutrients (Abrahamsen, 1980).

(g) Inhibited formation of terminal buds and increased mortality of pine seedlings (Matziris and Nakos, 1977).

(h) Both decreased and increased yield of certain agricultural crops (Evans and Lewin, 1980; Evans et al., 1980; Lee et al., 1980).

(i) Interference with normal reproductive processes (Evans and Conway, 1980).

3. Various specific biological effects of simulated acid rain have been demonstrated in controlled field and laboratory experiments (see item 2 above), but reliable evidence of economic damage to agricultural crops, forest, and other vegetation, and to biological processes in soil by naturally occurring precipitation has been reported only rarely (Jonsson and Sundburg, 1972; Galloway et al., 1978; Cowling, 1980; Ulrich and Khanna, 1980; Tomlinson, 1981).

4. Recent experiments in Germany have demonstrated a significant correlation between amount of soluble aluminum in forest soils, death of feeder roots in spruce, fir, and beech forests, and widespread decline in the growth of these forests. Acid precipitation and dry deposition of acid-yielding substances have been postulated as probable causes of these coincident effects (Ulrich et al., 1980; Schuck et al., 1979; Tomlinson, 1981).

5. Mushrooms, mosses, lichens, and other lesser vegetation in forests accumulate heavy metals from the atmosphere, especially lead and cadmium. Wildlife feeding on these plants themselves accumulate the metals, sometimes making both the plants and the wildlife hazardous for human consumption (Munshower, 1972; Huckabee and Blaylock, 1974; Tyler, 1980).

EFFECTS ON SOILS

1. The consequences of acid inputs to soils vary greatly, depending upon the rates and recent history of atmospheric acid inputs, the character of the vegetation, natural rates of acid formation in the soil, and the physical-chemical properties of the soil (Wiklander, 1973; McFee, 1978).

2. Soil acidification increases leaching of exchangeable plant nutrients such as calcium, magnesium, potassium, iron, and manganese, and increases the rate of weathering of most minerals (Overrein, 1972; Reuss, 1975; Odén, 1976).

3. Acidification decreases the rate of many soil microbial processes such as nitrogen fixation and breakdown of organic matter. Various processes important in nutrient cycling, and critical in most ecosystems, are known to be inhibited by lowering soil pH. Included are the following: nitrogen fixation by Rhizobium bacteria on leguminous plants, by actinomycetes on some non-leguminous plants, and by certain blue-green algae and bacteria; mineralization of nitrogen from forest litter; nitrification of ammonium compounds; and overall decay rates of forest floor materials (Reuss, 1975; Braekke, 1976; Francis et al., 1980; Shriner and Johnston, 1980).

4. Atmospheric deposition of heavy metals also inhibits certain microbial processes in forest litter, especially decomposition of organic matter (Braekke, 1976; Tyler, 1976).

5. Acidification of soils reduces the availability of phosphorus to plants and increases the solubility of other elements, some of which may be toxic to plants. Aluminum is the most abundant toxic element in most forest and agricultural soils. Increasing soil acidity leads to greatly increased solubility and toxicity of aluminum to many crop and forest plants (Odén, 1976; Tomlinson, 1981).

6. Soils differ by orders of magnitude in their susceptibility to acidification. Calcareous soils are unlikely to be damaged by acid inputs, but may be affected by metal deposition. Soils with low cation exchange capacity and degree of base saturation are very susceptible to increased acidification (Reuss, 1975; McFee, 1978).

7. Large quantities of hydrogen ions are added to soils as acid precipitation and as a result of cation uptake processes and both soil-amendment and fertilization practices. Acidification by these processes can be controlled by normal management practices in cultivated soils. But large areas of North America and Europe are not cultivated and have soils that are poorly buffered and, therefore, are susceptible to further acidification. Many of these soils occur in forest and wilderness areas (Reuss, 1975; Odén, 1976).

EFFECTS ON WATER QUALITY

1. The chemical composition of lake, stream, and groundwaters is determined in part by the chemical composition of

precipitation and dry deposition. The chemical composition of precipitation is modified by chemical and biological weathering and exchange processes as precipitation: washes over vegetation; percolates through the soil; interacts with the underlying bedrock of the drainage basin in which the precipitation occurs; and runs off over the soil surface (Braekke, 1976; Odén, 1976).

2. Acid precipitation increases the solubility and mobility of many cations in soil. This increases the concentration of toxic metal cations, including aluminum, manganese, and zinc, in the soil solution. It also increases the leaching of nutrient cations, including potassium, calcium, magnesium, etc. These toxic and nutrient ions are transferred from soils into surface and groundwaters (Braekke, 1976; Odén, 1976).

3. Acidification of groundwaters has been found in western Sweden and attributed to long-term changes in the acidity of precipitation. As a result, hundreds of shallow dug wells and deeper drilled wells for home and farm buildings have shown acidities in the range between pH 4.0 and pH 6.0. A strong positive correlation has been found between the acidity of the well water and the content of metal ions, including copper, zinc, lead, manganese, and aluminum. In some cases, accelerated corrosion of copper piping has resulted in the necessity for early repair and/or replacement of copper piping systems (Hultberg and Wenblad, 1980).

LITERATURE CITED

Abrahamsen, G. 1980. Acid Precipitation, Plant Nutrients, and Forest Growth. In: D. Drabløs and A. Tollan, eds., "Ecological Impacts of Acid Precipitation," Proc. Intern. Conf., Sandefjord, Norway. SNSF Project Report, Ås, Norway.

Almer, B., W. Dickson, C. Ekström, E. Hörnströmm, and U. Miller. 1974. Effects of Acidification on Swedish Lakes. *Ambio* 3:30-36.

Altshuller, A.P. and G.A. McBean, Co-chairmen. 1979. The LRTAP Problem in North America: A Preliminary Overview Prepared by the United States-Canada Bilateral Research Consultation Group on the Long-Range Transport of Air Pollutants. Atmospheric Environment Service, Downsview, Ontario. Part I (1979), 48 pp.; Part II (1980), 39 pp.

Ambio. 1976. Report of the International Conference on the Effects of Acid Precipitation in Telemark, Norway. Ambio 5:200-252.

Beamish, R.J. and H.H. Harvey. 1972. Acidification of the LaCloche Mountain Lakes, Ontario, and Resulting Fish Mortalities. J. Fish. Res. Bd. of Can. 29:1131-1143.

Bolin, B., L. Granat, L.Ingelstom, M. Johannesson, E. Mattsson, S. Odén, H. Rodhe, and C.O. Tamm. 1972. Sweden's Case Study for the United Nations Conference on the Human Environment: Air Pollution Across National Boundaries. The Impact on the Environment of Sulfur in Air and Precipitation. Norstadt and Sons, Stockholm. 97 pp.

Borgstrom, R. and G.R. Hendrey. 1976. pH Tolerance of the First Level Stages of Lepidurus arcticus (pallas) and Adult Gammarus lacustris G.O. Sars. Internal Report22/76, SNSF Project, Ås, Norway.

Braekke, F.H. (ed.). 1976. Impact of Acid Precipitation on Forest and Freshwater Ecosystems in Norway. Res. Rep. 6/76, SNSF Project, Ås, Norway. 111 pp.

Cogbill, C.V. and G.E. Likens. 1974. Acid Precipitation in the Northeastern United States. Water Resources Res. 10:1133-1137.

Cowling, E.B. 1980. Acid Precipitation and Its Effects on Terrestrial and Aquatic Ecosystems. In: T.J. Kneip and P.J. Lioy, eds., "Aerosols: Anthropogenic and Natural, Sources and Transport." Annals New York Academy of Science 338:540-555.

Cowling, E.B. 1981. An Historical Résumé of Progress in Scientific and Public Understanding of Acid Precipitation and Its Biological Consequences. Research Report, SNSF Project, Ås, Norway.

Cragin, J.H., M. Herron, and C.C. Langway, Jr. 1975. The Chemistry of 700 Years of Precipitation at Dye-3 Greenland. U.S. Army Cold Regions Research and Engineering Laboratory, Hanover, New Hampshire. 18 pp.

Cronan, C.S. and C.L. Schofield. 1979. Aluminum Leaching Response to Acid Precipitation: Effects on High-level Watersheds in the Northeast. Science 204:304-305.

Dickson, W., C. Ekström, E. Hörnström, and U. Miller. 1973. Forsurningens inverkan på vastkustsjöar. (The Effects of Acidification of West-Coast Lakes.) Swedish National Environmental Protection Board, Stockholm, Publ. No. 7. 97 pp.

Dochinger, L.S. and T.A. Seliga. 1976. Proceedings First International Symposium on Acid Precipitation and the Forest Ecosystem. USDA Forest Serivce, Gen. Tech. Rept. NE-23, Northeast, For. Exp. Sta., Upper Darby, Pennsylvania. 1079 pp.

Drabløs, D. and A. Tollan (eds.). 1980. Ecological Impact of Acid Precipitation. Proc. Intern. Conf., Sandefjord, Norway. SNSF Project, Ås, Norway. 383 pp.

Driscoll, C.T. 1980. Aqueous Speciation of Aluminum in the Adirondack Region of New York State, USA. In: D. Drabløs and A. Tollan, eds., "Ecological Impact of Acid Precipitation," pp. 214-215. Proc. Intern. Conf., Sandefjord, Norway. SNSF Project, Ås, Norway.

Evans, L.S. and C.A. Conway. 1980. Effects of Acidic Solutions on Sexual Reproduction of Pteridium aquilinum. Amer. J. Bot. 67:866-875.

Evans, L.S., C.A. Conway, and K.F. Lewin. 1980. Yield Responses of Field-Grown Soybeans Exposed to Simulated Acid Rain. In: D. Drabløs and A. Tollan, eds., "Ecological Impact of Acid Precipitation," pp. 162-163. Proc. Intern. Conf., Sandefjord, Norway. SNSF Project, Ås, Norway.

Evans, L.S. and T.M. Curry. 1979. Differential Responses of Plant Foliage to Simulated Acid Rain. Amer. J. Bot. 66:953-962.

Evans, L.S. and K.F. Lewin. 1980. Growth, Development, and Yield Responses of Pinto Beans and Soybeans to Hydrogen Ion Concentrations of Simulated Acidic Rain. Environ. and Exp. Bot. 21:103-113.

Francis, A.J., D. Olson, and R. Bernatsky. 1980. Effects of Acidity on Microbial Processes in a Forest Soil. In: D. Drabløs and A. Tollan, eds., "Ecological Impacts of Acid Precipitation," pp. 166-167. Proc. Intern. Conf., Sandefjord, Norway. SNSF Research Report, Ås, Norway.

Galloway, J.N. and E.B. Cowling. 1978. The Effects of Precipitation on Aquatic and Terrestrial Ecosystems: A Proposed Precipitation Chemistry Network. J. Air Pollut. Cont. Assoc. 28:228-235.

Galloway, J.N., E.B. Cowling, E. Gorham, and W.W. McFee. 1978. A National Program for Assessing the Problem of Atmospheric Deposition (Acid Rain). Nat. Atm. Dep. Prog., Nat. Res. Ecol. Lab., Fort Collins, Colorado. 97 pp.

Galloway, J.N., S.F. Eisenreich, and B.C. Scott (eds.). 1980. Toxic Substances in Atmospheric Deposition: A Review and Assessment. NADP Special Report. Nat. Atm. Dept. Prog., Nat. Res. Ecol. Lab., Fort Collins, Colorado. 217 pp.

Galloway, J.N. and G.E. Likens. 1980. Acid Precipitation: The Importance of Nitric Acid. Atmos. Environ. (in press).

Galloway, J.N. and D.M. Whelpdale. 1979. An Atmospheric Sulfur Budget for Eastern North America. Atmos. Environ. 14:409-417.

Gorham, E. 1961. Factors Influencing Supply of Major Ions to Inland Waters, With Special Reference to the Atmosphere. Geol. Soc. Amer. Bull. 72:795-840.

Gorham, E. 1976. Acid Precpitation and Its Influence Upon Aquatic Ecosystems - An Overview. Water Air Soil Pollut. 6:457-481.

Gorham, E. and A.G. Gordon. 1960. The Influence of Smelter Fumes Upon the Chemical Composition of Lake Waters Near Sudbury, Ontario, and Upon the Surrounding Vegetation. Can. J. Bot. 38:477-487.

Grahn, O., H. Hultberg, and L. Landner. 1974. Oligotrophication -- A Self-Accelerating Process in Lakes Subjected to Excessive Supply of Acid Substances. Ambio 3:93-94.

Hagström, T. 1977. Grodornas forsvinnande i en försurad sjö. (Disappearance of Frogs in an Acidified Lake.) Sveriges Natur 11:367-369.

Hall, R.J., G.E. Likens, S.B. Fiance, and G.R. Hendrey. 1980. Experimental Acidification of a Stream in the Hubbard Brook Experimental Forest, New Hampshire. Ecology 61:976-989.

Heck, W.W. and C.S. Brandt. 1977. Effects on Vegetation: Native, Crops, Forests. In: A.C. Stern, ed., "Air Pollution," Vol. II, pp. 157-229. Academic Press, New York.

Hendrey, G.R. 1978. Limnological Aspects of Acid Precipitation. Proceedings International Workshop, Sagamore Lake, New York. Brookhaven National Laboratory, Upton, New York. 66 pp.

Hendrey, G.R., J.N. Galloway, S.A. Norton, C.L. Schofield, D.A. Burns, and P.W. Schaffer. 1980. Sensitivity of the Eastern United States to Acid Precipitation Impacts on Surface Waters. In: D. Drabløs and A. Tollan, eds., "Ecological Impact of Acid Precipitation," pp. 216-217. Proc. Intern. Conf., Sandefjord, Norway. SNSF Project Report, Ås, Norway.

Hendrey, G.R. and R.F. Wright. 1975. Acid Precipitation in Norway: Effects on Aquatic Fauna. J. Great Lakes Res. 2 (Suppl. 1):192-207.

Henriksen, A. 1980. Acidification of Freshwaters - A Large Scale Titration. In: D. Drabløs and A. Tollan, eds., "Ecological Impact of Acid Precipitation," pp. 68-74. Proc. Intern. Conf., Sandefjord, Norway. SNSF Project, Ås, Norway.

Henriksen, A. 1979. A Simple Approach for Identifying and Measuring Acidification of Freshwater. Nature 278:542-545.

Hooke, R. 1687. An Account of Several Curious Observations and Experiments Concerning the Growth of Trees. Phil. Trans. Royal Soc. London 16:207-313.

Howells, G. 1979. Ecological Effects of Acid Precipitation. Proceedings of a Workshop held at Galloway, Scotland, Sept. 1978. Central Electricity Research Laboratories, Leatherhead, Surrey, England.

Huckabee, J.W. and B.G. Blaylock. 1974. Microcosm Studies on the Transfer of Hg, Cd, and Se From Terrestrial to Aquatic Ecosystems. In: D.D. Hemphill, ed., "Proceedings Univ. of Missouri 8th Annual Conference on Trace Substances in Environmental Health," Columbia, Missouri.

Hultberg, H. and O. Grahn. 1976. Proceedings of the First Specialty Symposium on Atmospheric Contributions to the Chemistry of Lake Waters. J. Great Lakes Res. 2 (Suppl. 1), 208 pp.

Hultberg, H. and A. Wenblad. 1980. Acid Ground Water in Southwestern Sweden. In: D. Drabløs and A. Tollan, eds., "Ecological Impact of Acid Precipitation," pp. 220-221. Proc. Intern. Conf., Sandefjord, Norway. SNSF Project, Ås, Norway.

Husar, R.B., J.P. Lodge, Jr., and D.J. Moore (eds.). 1978. Sulfur in the Atmosphere. Atmos. Environ. 12:1-3 (special issue). 796 pp.

Husar, R.B. and D.E. Patterson. 1980. Regional Scale Air Pollution: Sources and Effects. In: T.J. Kneip and P.J. Lioy, eds., "Aerosols: Anthropogenic and Natural, Sources and Transport," pp. 399-417. Annals New York Academy of Sciences 338.

Hutchinson, T.C. and L.M. Whitby. 1974. Heavy Metal Pollution in the Sudbury Mining and Smelting Region of Canada. 1. Soil and Vegetation Contamination by Nickel, Copper, and Other Metals. Environ. Conserv 1(2):123.

Hutchinson, T.C. and M. Havas (eds.). 1980. Effects of Acid Precipitation on Terrestrial Ecosystems. Plenum Press, New York. 654 pp.

Jonsson, B. and R. Sundberg. 1972. Has the Acidification by Atmospheric Pollution Caused a Growth Reduction in Swedish Forests? Research Note No. 20, Department of Forest Yield Research, Royal College of Forestry, Stockholm, Sweden. 48 pp.

Junge, G.E. and R. Werby. 1958. The Concentration of Chloride, Sodium, Potassium, Calcium, and Sulfate in Rain Water Over the United States. J. Meteorol. 15:417-425.

Kovda, V.A. 1975. Biogeochemical Cycles in Nature, Their Disturbance and Study. (In Russian) Kuaka Publishing House, Moscow, USSR.

Lee, J.J., G.E. Neely, and S.C. Perregan. 1980. Sulfuric Acid Rain Effects on Crop Yield and Foliar Injury. U.S. Environmental Protection Agency, Report No. EPA-600/3-80-016. 21 pp.

Leivestad, H., G. Hendrey, I.P. Muniz, and E. Snekvik. 1976. Effects of Acid Precipitation on Fresh Water Organisms. In: F.H. Braekke, ed., "Impact of Acid Precipitation on Forest and Freshwater Ecosystems in Norway," pp. 86-111. Research Report FR 6-76. SNSF Project, Ås, Norway.

Leivestad, H. and I.P. Muniz. 1976. Fish Kill at Low pH in a Norwegian River. Nature 251:391-392.

Lewis, W.M., Jr. and M.C. Grant. 1980. Acid Precipitation in the Western U.S. Science 207:176-177.

Likens, G.E. 1976. Acid Precipitation. Chem. Eng. News 54(48):29-44.

Likens, G.E., R.F. Wright, J.N. Galloway, and T.J. Butler. 1979. Acid Rain. Sci. Amer. 241(4):43-51.

Lyons, W.A. 1980. Evidence of Transport of Hazy Air Masses from Satellite Imagery. In: T.J. Kneip and P.J. Lioy, eds., "Aerosols: Anthropogenic and Natural, Sources and Transport." Annals New York Academy of Science 338:418-433.

Malmer, D. 1973. Om effecterna på vatten, mark och vegetation av ökad svaveltillförsel från atmosfären. (On the Effects of Water, Soil, and Vegetation from an Increasing Atmospheric Supply of Sulfur.) Statens Natur Vardawerk, Lunds Universitet, Lund, Sweden. 125 pp.

Matziris, D.I. and G. Nakos. 1977. Effects of Simulated "Acid Rain" on Juvenile Characteristics of Aleppo Pine (Pinus halepensis Mill.). Forest Ecol. Mgmt. 1:267-272.

McColl, J.G. and D.S. Bush. 1978. Precipitation and Throughfall Chemistry in the San Francisco Bay Area. J. Environ. Qual. 7:352-357.

McFee, W.W. 1978. Effects of Acid Precipitation and Atmospheric Deposition on Soils. In: J.N. Galloway, E.B. Cowling, E. Gorham, and W.W. McFee, eds., "A National Program for Assessing the Problem of Atmospheric Deposition (Acid Rain)," pp. 64-73. Nat. Atm. Dep. Prog., National Research Ecol. Lab., Fort Collins, Colorado.

Munshower, F.F. 1972. Cadmium Compartmentation and Cycling in Grassland Ecosystems in the Deer Lodge Valley. Ph.D. Thesis, Univ. Montana, Bozeman, Montana.

National Academy of Sciences. 1975. Atmospheric Chemistry: Problems and Scope. Com. Atm. Sci., Nat. Res. Council, Washington, D.C. 130 pp.

Odén, S. 1968. The Acidification of Air and Precipitation and Its Consequences in the Natural Environment. Ecology Committee Bulletin No. 1, Swedish National Science Research Council, Stockholm. Translation Consultants, Ltd., Arlington, Virginia. 117 pp.

Odén, S. 1976. The Acidity Problem - An Outline of Concepts. Water Air Soil Pollut 6:137-166.

Odén, S. and T. Ahl. 1970. Forsurningen av skandenaviska vatten. (The Acidification of Scandinavian Lakes and Rivers.) Ymer, Arsbok:103-122.

Organization for Economic Cooperation and Development. 1977. The OECD Program on Long Range Transport of Air Pollutants. OECD, Paris, France.

Overrein, L.N. 1972. Sulphur Pollution Patterns Observed; Leaching of Calcium in Forest Soil Determined. Ambio 1:145-147.

Overrein, L.N., H.M. Seip, and A. Tollan. 1980. Acid Precipitation - Effects on Forests and Fish. Final report of the SNSF Project 1972-1980, Research Report FR-19. SNSF Project, Ås, Norway. 175 pp.

Pough, F.H. 1976. Acid Precipitation and Embryonic Mortality of Spotted Salamanders, Ambystoma maculatum. Science 192:68-70.

Reuss, J.O. 1975. Chemical/Biological Relationships Relevant to Ecological Effects of Acid Rainfall. EPA Report 600/3-75-032. Environmental Protection Agency, Washington, D.C.

Schofield, C.L. 1976. Effects of Acid Precpitation on Fish. Ambio 5:228-230.

Schuck, H.J., U. Blumel, I. Geier, and O. Schutt. 1979. Schadbild und atiologie des tannesterbens. Eur. J. For. Path. 10:125-135.

Shriner, D.S. 1976. Effects of Simulated Rain Acidified with Sulfur Acid on Host-Parasite Interactions. In: L.S. Dochinger and T.A. Seliga, eds., "Proc. Int. Symp. on Acid Precipitation and the Forest Ecosystem," pp.919-925. USDA For. Serv. Gen. Tech. Rep. NE-23, Upper Darby, Pennsylvania.

Shriner, D.S. 1978. Effects of Simulated Acidic Rain on Host-Parasite Interactions in Plant Disease. Phytopathology 68:213-218.

Shriner, D.S. and J.W. Johnston. 1980. The Effect of Simulated Acidified Rain on Nodulation of Leguminous Plants by Rhizobium spp. Environ. and Exp. Bot. 21:199-209.

Shriner, D.S., C.R. Richmond, and S.E. Lindberg (eds.). 1980. Atmospheric Sulfur Deposition: Environmental Impact and Health Effects. Ann Arbor Science Pub., Ann Arbor, Michigan. 568 pp.

Smith, R.A. 1872. Air and Rain: The Beginning of a Chemical Climatology. Longmans, Green, London. 600 pp.

Tamm, C.O. 1958. The Atmosphere. Handbuch fur Pflanzen-physiologie 6:233-242.

Tomlinson, G.H. 1981. Acid Rain and the Forest - The Effect of Aluminum and the German Experience. DOMTAR Research Center, Sennerville, Quebec, Canada. 22 pp.

Tyler, G. 1972. Heavy Metals Pollute Nature, May Reduce Productivity. Ambio 1:52-59.

Tyler, G. 1976. Heavy Metal Pollution, Phosphatase Activity, and Mineralization of Organic Phosphorus in Forest Soils. Soil Biol. Biochem. 8:327-332.

Tyler, G. 1980.   Metals in Sporophores of Basidiomycetes. Trans. Brit. Myc. Soc. 24:41-49.

Ulrich, B., R. Mayer, and P.K. Khana. 1980. Chemical Changes Due to Acid Precipitation in a Loess Derived Soil in Central Europe. Soil Sci. 130:193-199.

Wiklander, L.   1973.   The Acidification of Soil by Acid Precipitation. Gundforbattring 26(4):155-164.

Wittwer, S.H. and M.J. Bukovac. 1969. The Uptake of Nutrients Through Leaf Surfaces. In: K. Scharrer and H. Linser, eds., "Handbuch der Pflanzenernahrung und Dungung," pp. 235-261. Springer-Verlag, New York.

Wood, T. and F.H. Bormann. 1974. The Effects of an Artificial Acid Mist Upon the Growth of Betula alleghanien-sis Britt. Environ. Pollut. 7:259-268.

CHAPTER 2

THE CONCERN FOR ACIDIC DEPOSITION
IN THE GREAT LAKES REGION

Orie L. Loucks
The Institute of Ecology
Indianapolis, Indiana

INTRODUCTION

The Great Lakes region of the United States and Canada is an especially appropriate focus for a conference on acid rain in 1981 for one paramount reason: If the evidence is compelling that environmental alterations from acidic deposition could, indeed, be of the type and magnitude being alleged in other regions, then there are urgent decisions and actions needed now to prevent those effects from being expressed here. A principal conference goal, therefore, must be to review the evidence as it applies to the unaltered, or slightly altered, Great Lakes region, and to consider what new data or relationships still need documentation before mitigation or control strategies can be evaluated. My goal will be to outline why a lightly impacted region can be so important to the perspectives of both Canada and the United States, and to highlight the implications of certain recent insights on effects processes which may not be covered in later chapters.

IS ACIDIC DEPOSITION A PROBLEM?

One must ask, first, whether acidification of rainfall is really a problem, and, in particular, what criteria one should use to judge whether it is serious in the national or international perspective. Although questions remain about details of the atmospheric chemistry, acidic substances are known to be produced in the atmosphere when sulfur dioxide (SO_2) and nitrogen oxides (NO_x) from fossil fuel combustion combine with oxygen to produce sulfuric and nitric acids. We also know now that under appropriate weather conditions these pollutants can be carried hundreds of kilometers from their source. Some of the acid-forming emissions can be taken up by vegetation as gases, and, following sulfate formation, others drop out as particles in processes called dry deposition. Only about half of the products from acid-forming precursors are washed from the air by rain, snow, or mist, yielding acid rain.   The remainder enter into dry deposition or are carried off the continent.

But, many people ask, "What could be the source of the acidic deposition on a remote Wisconsin lake and who is culpable?" Consider the pattern of gaseous emissions from the industrial heartland and the cities of the Great Lakes region. Pollutant concentrations frequently build up and then move long distances over the rural landscape. Many questions remain as to the distance transported and ultimate effects, but techniques of source/receptor matrix analysis, and of air parcel trajectory analysis, are beginning to tell us much about the patterns of pollutant transport over periods of 12 to 24 hr before emerging as dry or wet deposition over an area such as the Great Lakes Basin.

A series of recent reviews (e.g., Last et al., 1980; Overrein, 1980) has shown that, despite some questions of rates of reactions and of control mechanisms, we know enough about the acidic deposition phenomenon now to describe the general transport and formation of acid, the areas of impact, the mechanisms of effect, and the basis for some lags in watershed response times. One can conclude now that there is a problem being expressed in the alteration of resources over time, and that it is strongly linked to acid-forming industrial and residential emissions. A principal question for the research agenda at this meeting, therefore, is whether this is a problem in the western Great Lakes states. Figure 1 shows the flow relationships among the principal ecological consequences believed to result from fossil fuel emissions and the subsequent deposition of increased hydrogen, sulfate, and nitrate on the landscape. The apparent effects from this deposition must be examined as responses to chemical alteration of a coupled system which, for simplicity, is shown here as a flow diagram.

THE REGION OF CONCERN

The resource systems of concern to a conference in the Great Lakes region are principally those of the northern tier of states from Minnesota to New York, the northern half of states bordering the Ohio River, and the Province of Ontario. The region is potentially susceptible to acid deposition (Galloway and Cowling, 1978) because its coarse-textured soils

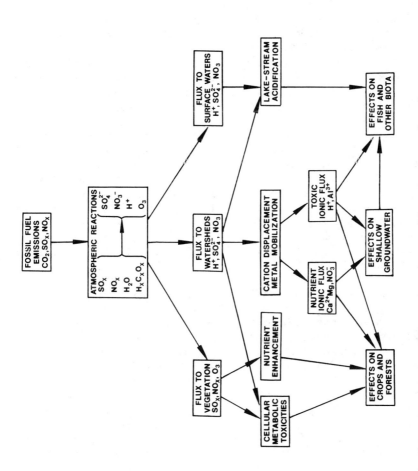

Figure 1. Flow diagram showing system linkages from acid and ozone-forming precursors through atmospheric acid formation, deposition, and, finally, effects.

are very low in calcium carbonate and are underlain by chemically inert igneous and metamorphic bedrocks and poorly buffered sandstone. The region's rich resources and industries, including recreation, forestry, farming, and mining, could be vulnerable to disruption if acid inputs reached levels presently being experienced in areas of Europe and the northern Appalachian region of the United States.

The advance of ice masses over these states during the pleistocene era scoured weathered bedrock from the higher land areas and deposited the earth and rock far to the south. What remained was only the coarsest fraction of the soil which, together with boulders and local deposits of outwash sand, makes up most of the surface deposits in the northern half of the western Great Lakes states and in much of Ontario.

Glacial activity throughout the region left thousands of depressions in the landscape, yielding an unusual abundance of lakes and waterways. These are now in high demand for recreation and fishing and are the basis for the local economy in many parts of the Great Lakes region. Here, clean water is essential, both for game fish species and human uses. Because of the outstanding quality of the aquatic resources, recreation dominates the economy locally and over relatively large areas, with large parts of the population supported to some degree by water-related industries.

Water resource inventories in this region indicate that Minnesota has a minimum of 15,000 lakes, and up to 22,000, if smaller lakes (to 2.03 ha) are counted. These lakes cover 10,360 km^2, or one-twentieth of the land area of the state. Wisconsin has some 11,000 lakes over 4.1 ha in size, mostly in the northern half of the state. Wisconsin also has 2,250 trout streams, totalling 13,982 ha of trout water. The upper peninsula of Michigan has a very large number of low alkalinity lakes likely to be sensitive to acid inputs (Schneider, 1975). Of some 8,000 lakes and ponds in the western half of the upper peninsula alone, more than 50 percent have an alkalinity in the range of 10 mg/l, indicating they are potentially susceptible to acid inputs. Ontario has estimated its lake inventory at 180,000 or more (MOE, 1980), reflecting an area some three times larger than the lake districts of the western Great Lakes states.

These states, Ontario, and the United States federal government have all tried to meet public demand for access to aquatic resources by establishing a large number of state forests, parks, and wild rivers, each of which represents a large public investment for the benefit of posterity. The state of Wisconsin estimated the use of state-owned parks and forests at more than 11 million visitors per year between 1971 and 1973 (the most recent statistic available). More than 10 million tourists each year visit the state of Michigan. The loss of even a small number of the lakes or streams in these states could adversely affect individual services, thereby influencing the industry to some degree; how much is a matter of considerable concern to each of the governments and publics involved.

The northern regions of these states and the province of Ontario also are rich in timber, a resource that also may be affected adversely by long-term nutrient losses and metal toxicities from acid deposition. The forests of Michigan cover 8.1 million hectares, or more than one-half of the state; half of the state of Wisconsin also is covered by commercial forest. Forest growth rates are relatively good throughout the region due to a combination of warmer summer temperatures and comparatively abundant moisture supplies. However, timber production depends on the nutrient status of the soil, and there is currently no general soil-maintenance or fertilizer application program for forest soils in this region.

The resources that are most likely to be affected by acid deposition, such as recreation, forests, agriculture, and mining, are largely the responsibility of state and provincial governments. Only in air quality, and to a moderate extent in interstate water management, has the federal government been given a substantial degree of authority to assure resource protection. On the other hand, the administrative agencies of state government acting alone cannot protect the renewable resources of that state or province from the prospective long-term effects of acid deposition. Thus, it may no longer be possible for state or provincial governments to meet the historic expectation that they will protect resources and related industries in their states. At the same time, decision-making institutions in adjacent states have no compelling reason to be concerned for neighboring states. The federal government has a responsibility only to set uniform standards acceptable to all parts of the country, including highly industrialized sectors such as the Ohio Valley, and more sensitive regions such as the Great Lakes states.

Thus, a further responsiblity for these proceedings, focused on the Great Lakes region, is to examine new approaches to regional air and water quality protection that may be necessary to deal with the long-term effects of acid rain. Because of the complexity of the source/receptor system, and the cause-and-effect relationships, a very substantial scientific base will be needed before region-wide government regulation or economic adjustments can be justified. Given the possible magnitude of the long-term pollutant deposition problem, however, it is urgent to develop studies that analyze the prospective impacts in some detail before more serious effects take place. It is for this reason, more than any other, that the potential impacts of moderate levels of acid deposition in the western Great Lakes must be considered so carefully.

ALTERATION OF ATMOSPHERIC CHEMISTRY

Precipitation chemistry is influenced by the patterns of chemicals released to the atmosphere, as well as by the large-scale atmospheric circulation as air masses move like large eddies over North America. All precipitation contains a wide variety of chemical constituents from sources such as sea spray, dust particles, and the natural cycling of carbon, nitrogen, and sulfur. Discharging wastes to the atmosphere increases the amounts of compounds containing elements such as nitrogen, carbon, and sulfur, and adds to the variety of compounds such as PCBs and heavy metals which are found in rainfall. The four ions of most importance to acid rain are: hydrogen (H^+), ammonium (NH_4^+), nitrate (NO_3^-), and sulfate (SO_4^{2-}).

The seasonal quantity and quality of precipitation also are important for determining the potential for acidic inputs to impact the environment. Acid pollutants accumulating in the snowpack have a higher potential for causing deleterious effects on organisms and habitats in areas with high snowfall than in areas with lower amounts of snow accumulation. This effect is due to the rapid flushing of accumulated acid during snowmelt. Thus, the distribution of precipitation during the year, the temporal behavior of rainfall, and the location of pollution sources within rainfall pathways are linked to the potential for damage to aquatic ecosystems. Here again, the gradient from moderate rainfall in the western Great Lakes region to higher rainfall in the East is an important consideration.

Distilled water in equilibrium with atmospheric carbon dioxide has a pH value of about 5.6. The presence of base substances such as dust can raise the pH of rainfall, and acidic substances can decrease it. Partial results on the present levels of chemical alteration of precipitation are presented in Figure 2, showing large areas of North America which received acidic precipitation in 1979. (Also see the paper by Barrie, Chapter 7, in this volume). The pH gradient from western Minnesota to New York is almost a classic gradient-type experimental manipulation.

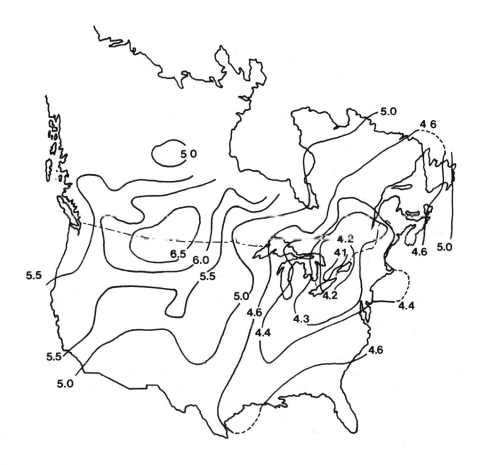

*Figure 2.   Isopleths of pH showing precipitation-weighted
averages for April, 1979, to March, 1980. NADP
and CANSAP monitoring data. (Adapted from Glass
and Brydges, 1981, and Barrie, Chapter 7 of this
volume.)*

One often asks what the historic, pre-industrialization
levels of precipitation chemistry and atmospheric deposition
might have been.   In 1979, a study conducted at the Boundary
Waters Canoe Area (BWCA) of Minnesota and the Quetico Provincial
Park in Ontario addressed this question (Glass and Loucks,
1980).   The Atikokan generating station was being planned 80
to 120 km north of the BWCA. This area, until recently, has

had especially clean air. The sulfur dioxide levels at a recording site near Fernberg, Minnesota, during 1977 never exceeded the threshold level of the instrument (10 $\mu g/m^3$) and exceeded this level only once during 1978. The suspended particulate matter, as measured by a high-volume sampler, yielded an annual arithmetic average of 15 $\mu g/m^3$, and an annual geometric average of 11 $\mu g/m^3$ (Glass and Loucks, 1980).

To evaluate historic vs. modern sources of acidity, we used measurements of bulk deposition of sulfate (wet-plus-dry fall) obtained at three locations in the extreme western end of the Great Lakes region. The longest record, at the Experimental Lakes Area (ELA) in Ontario, 180 km northwest of Atikokan, shows an average recent sulfate deposition of 10.9 kg/ha/yr. The average pH of the rainfall there during the period between 1974 and 1977 was 4.86. The pre-settlement pH for rainfall in this area appears to have been 5.7 (cf. carbonic acid equilibrium, and data for 1955 to 1956 shown by Galloway and Cowling, 1978). Data from the ELA show that 70 percent of the nitrate and sulfate in the current rainfall is balanced by cations other than hydrogen. Thirty percent is in the acid form, and the H^+ ions were assumed to divide between SO_4^{2-} and NO_3^- on a 2:1 ratio. This information yields an estimate of 7.6 kg/ha/yr of sulfate balanced by Ca^{2+}, Mg^{2+}, and NH_4^+, while 3.1 kg/ha/yr is acid sulfate, producing the net depressions in pH. Thus, depending on the historic potential of atmospheric constituents other than H^+ to balance anthropogenic acid precursors, pre-settlement loadings of sulfate must lie between 0 and 7.6 kg/ha/yr. For the purposes of the BWCA studies an intermediate value of 4 kg/ha/yr was used.

POTENTIAL ALTERATION OF THE TERRESTRIAL SYSTEM

Ozone is a pollutant often involved in long-distance transport similar to that of acid rain. Elevated levels often lead to injury, and crop yield reductions have been well documented on many crop species in the Great Lakes region. These crops include tobacco, soybeans, corn, grapes, cucumber, squash, and radish. Direct effects of acid precipitation on plants are much less than those of ozone, but effects of acid rain are known on crops for which the foliage is valued. The potential impacts include:

1. damage to protective surface structures such as cuticles;
2. interference with normal functions of guard cells;
3. poisoning of plant cells after diffusion of acidic substances through stomata or cuticle;

4. disturbance of normal metabolism or growth processes without necrosis of plant cells;
5. alteration of leaf root-exudation processes;
6. interference with reproduction processes; and
7. synergistic interaction with other environmental stress factors.

Increase in soil acidity also can be detrimental to the chemical availability of several essential nutrients. Over several decades of acidic deposition, one can expect a net loss of cations (Ca^{2+} and Mg^{2+}, both important for plant growth) from low base exchange conditions. Under conditions of low cation availability, any further loss of cations should be considered significant, however small that loss may be. Much of the northern United States and eastern Canadian forest industry is founded on these low-pH and low-nutrient soils.

The increase in soil acidity also can lead to mobilization of other elements (Al, Mn, and Fe), sometimes in quantities toxic to terrestrial plants as well as to aquatic ecosystems. One of the indirect effects of watershed acidification is the mobilization of aluminum in soils, with the potential to cause toxic effects to terrestrial organisms and the biota of streams and lakes (Ulrich et al., 1980). Aluminum solubility is pH-dependent, and increases with increasing acidity. Several reports have documented elevated aluminum concentrations in acid lakes and streams in areas known to be impacted by acid inputs, and in effluents from lysimeters treated with acid solution (Abrahamsen et al., 1976). While aluminum is typically leached from the upper soil horizon of podzol soils by carbonic acid and organic chelation, it is usually deposited in lower horizons.

Much research on aluminum toxicity has been conducted on crops and legume species (Foy et al., 1978). Excess aluminum affects cell division in roots, causing inhibition of root growth, leading to stubby and brittle roots. It generally accumulates in roots and is only found in above-ground plant parts of a few plant species. In addition, excess aluminum fixes phosphorus in less available forms in the soil and in or on plant roots. Consequently, phosphorus deficiency is often associated with toxic aluminum concentrations and is generally the form of symptom expression in above-ground plant parts.

Evaluating soil sensitivity to these processes is important for both terrestrial and aquatic ecosystems. Two systems for classifying soil sensitivity have been suggested, one by McFee (1980) and one in Canada by Coote et al. (1980). McFee's approach is based on the cation exchange capacity (CEC) of the

soil. High CEC implies strong buffering capacity, while low
CEC soils may experience rapid pH drops as a result of acid
deposition or, and if they are already acid, may release
aluminum. The fundamental criteria and assumptions for establishing the classification system are as follows:

1. acidic deposition over a 25-yr time span;
2. 100 cm precipitation per year, pH 3.7;
3. 25 cm soil layer, 1.3 gm/cm^3 bulk density;
4. acid additions equal to between 10 and 25 percent of
    CEC could be significant.

Utilizing large-scale soil maps, McFee (1980) estimated
that about half of the soils in the eastern United States are
"slightly sensitive" or "sensitive" to acid deposition and
summarized the results in a map format. The above assumptions
and criteria result in three sensitivity classes:

<u>Non-Sensitive:</u>
Soils with Free Carbonates
Soils Frequently Flooded
CEC > 15.4 meq/100 gm

<u>Slightly Sensitive:</u>
CEC = 6.2 - 15.4 meq/100 gm

<u>Sensitive:</u>
CEC < 6.2 meq/100 gm

Forest site index also can be used as a measure of terrestrial ecosystem stress from acidic deposition and an integrator
of effects from nutrient additions (or losses) and the toxicity
of aluminum, hydrogen ions, and heavy metals. A large body of
literature details the effects of pH and stress on the biotic
community; but to date, there has been no integrated study to
determine whether reductions in terrestrial biomass
productivity could be due to acid precipitation. Estimation
of the change in forest site index over 20 yr or more is one
approach to consider (Loucks et al., 1981). To determine site
index for a forest, one uses the height and the age of the
dominant and co-dominant trees (Curtis et al., 1974). Plotting
height versus age (Figure 3) of each stand results in a series
of points which can be broken down into several arrays, over
age, each array representing a separate site class. The site
index takes the form of the "height at a given time" (i.e., a
site index of 80 is a height of 80 ft at 45 yr).

On forest lands where no trees are present, soil properties
have been used in multiple regression models to predict site

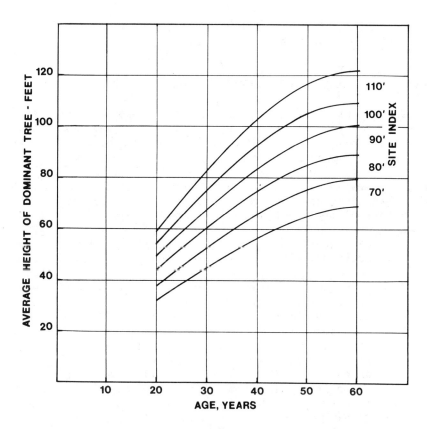

Figure 3. Site index classes determined by plotting age against the average height of dominant trees. Cation depletion by acid rain could reduce the position of some portions of the curves.

index. Fralish and Loucks (1975) developed such a model for trembling aspen in Wisconsin using the content of nutrient cations such as calcium and magnesium, which can be stripped by acid additions. The hypothesis that needs to be explored here is that depletion of this cation pool over 2 to 3 decades could alter the site index response, as estimated by the regression relationship, so that height growth, and, therefore, annual forest biomass production, was reduced significantly.

Thus, use of the site index as an integrator of acid deposition effects (via nutrient stripping and aluminum toxicity) has the advantage of being readily convertible to economic terms for cost-benefit analysis. At this time, however, we continue to lack sufficient support to test the hypothesis, even with existing data.

POTENTIAL ALTERATION OF SURFACE AND GROUNDWATER

The addition of acidic pollutants to a lake/watershed ecosystem is coming to be thought of as a large-scale quasi-titration (Henriksen, 1980), but it is a complex, uneven process, only superficially similar to a laboratory titration. Gorham and McFee (1980) and Last et al. (1980) discuss the variables involved and note that the hydrogen ions deposited with precipitation (or generated by other components of precipitation) may have four fates in the lake-watershed system:

1. where carbonates are present, hydrogen ions will be consumed, with the release of bicarbonate and calcium or magnesium;
2. hydrogen ions may be exchanged for various metal cations in soils with appreciable cation exchange capacity (McFee et al., 1976);
3. hydrogen ions may react with alumino-silicates or other minerals, releasing metal cations; and
4. any hydrogen ions not neutralized by the above processes will enter aquatic ecosystems in association with a mobile anion, usually sulfate.

The first three processes above account for the observation that surface waters are less acid than the precipitation they receive (Wright and Henriksen, 1978). However, other considerations include the mobility of anions such as sulfate, and the breakdown of ammonium ions to yield acidity. Alteration of groundwater and small-stream flows appears to represent a combination of factors.

Researchers differ and the evidence is somewhat conflicting as to whether the geochemical alteration of watersheds due to acidic inputs should be viewed as irreversible and, if so, on what scale. Irreversibility can be viewed as a failure to recover over geological time, but for natural resource systems, an incomplete recovery to a prestressed or undamaged state over a few decades, for all practical purposes, may be regarded as irreversible.

The possibility of irreversible loss of buffering capacity in lakes and watersheds is one of the most serious potential impacts of acid deposition. The ability of lake water to resist pH depression when acid is added (i.e., buffering capacity) depends on its alkalinity. The alkalinity of the water depends

on the presence of anions or other species which can take up
protons.   In most natural waters, alkalinity is largely a
function of bicarbonate, but other substances such as organic
acids, silicates, and ammonia are important in the low
alkalinity lakes most susceptible to acidification. When lake
water is titrated with acid, the change in pH per unit of added
acid (i.e., the slope) varies as successive buffers are
protonated, each acceptor taking up protons in its own
characteristic way.

When titration is considered for an entire watershed, as
may be the case from sustained long-term atmospheric acid
inputs, the same buffering processes can occur. However, the
reaction is much more heterogeneous in a watershed, and the
total quantity and variety of proton acceptors are probably
greater.   Contact between protons in water and the solid
buffering materials in the soil is not assured, depending on
hydrologic flow patterns within both the upper ("inter-flow")
and deeper soil horizons. Periods of peak runoff, when much
of the "surface" water penetrates only a few centimeters into
the soil, are often associated with acid flushing (low pH),
apparently with little buffering. This process is more likely
to occur when watershed buffering is naturally low or may have
been depleted by long-term acid input.

The hypothesis that acidification of surface waters by
atmospheric acid inputs can be viewed as a large-scale titration
has been articulated by Henriksen (1980) and supported with
data from North America and Scandinavia (Wright et al., 1980).
Nearly all lakes, except those dominated by acid bog drainage
water, are normally bicarbonate buffered with pH greater than
6.5. By stoichiometry, addition of acid from the atmosphere
will consume alkalinity, but the pH will fall only slightly as
long as some bicarbonate remains. Long-term (more than 50 yr)
records of decreasing alkalinity in a water supply reservoir
in the Adirondacks suggest that alkalinity has declined with
the continued atmospheric acid input (Schofield, personal
communication).   As bicarbonate (and any other buffers) are
depleted, and sulfate ions can no longer be retained in the
soil, Henriksen suggests lakes enter a transition phase
characterized by severe pH fluctuations (1 to 2 units) and
associated stresses on fish and other aquatic biota.   This
transition phase is a time when the lake is particularly
susceptible to seasonal pulses of acid (e.g., snowmelt) and to
the incomplete "mixing" of water and soil during flushing
events. According to this view, a lake may exist in a transition
phase until it is sulfate-saturated and the last carbonate or
other buffering above pH 5.6 in the watershed is consumed.

ARE WATERS OF THE WESTERN GREAT
LAKES ENTERING A "TRANSITION"?

The hypothesis of a relatively irreversible titration of
the geochemical components contributing buffering capacity is
a simplification of the complexities of lake/watershed
processes. The variety of potential sources of buffering, water
flow patterns, and rates of acid deposition and flushing all
tend to make each lake/watershed system unique. However, the
general pattern of observed responses is consistent with the
known sulfate saturation and low buffering capacity of the
geology and soils in the affected areas, and these conditions
do extend through Wisconsin and Minnesota.

At the present time there is no single, fully validated
methodology for projecting the alteration of aquatic systems
from acid deposition. Thus, measures of aquatic sensitivity
to acid inputs should be thought of as hypotheses constructed
with a reasonable amount of quantitative inputs, but evaluation
should not be based on any one model or measurement procedure.
The three methodologies given below are limited in scope and/or
availability of data, but taken together they present a
comprehensive view of our ability to quantify current resource
status and, therefore, make some prediction of the potential
for future change in the western Great Lakes area.

The Calcite Saturation Index (CSI)

Kramer (1976) has summarized the relationship between pH
and alkalinity in the calcite saturation index (CSI), which
expresses the relative saturation of water with calcium
carbonate:

$$CSI = p(Ca^{2+}) + p(Alk) - p(H^+) + pK$$

where $p(X) = -\log_{10}X$, $pK = 2$, (Ca^{2+}) is given as mol/l, and
(Alk) and (H^+) are given as equivalents/l. When $CSI < 1$, the
lake is nearly saturated with calcium carbonate and not
susceptible to acidification. Increasing CSI indicates
increasing susceptibility, and, at $CSI > 4$, the lake is very
likely to be affected by acid (Glass and Loucks, 1980).

The CSI has been used to classify lakes sensitive to acidification (Glass and Loucks, 1980, and others) and is a
relatively simple way to account for lake buffering capacity.
However, it is not comprehensive enough to be used as a
sensitivity index as defined earlier for watersheds because
buffering capacity is only one aspect of lake/watershed

sensitivity. Moreover, the CSI reflects only present conditions and does not relate past or future changes to loading rates. There is no indication in CSI of how a lake came to have a particular value, or of any explicit relationship between the CSI and acidic inputs.

The Dickson Relation

Another way to summarize the relationship between lake pH and atmospheric loading of acidic pollutants is to plot lake pH in relation to measured local sulfate loadings, Figure 4 (Almer et al., 1978). Various combinations of watershed characteristics are reflected in the family of curves connecting

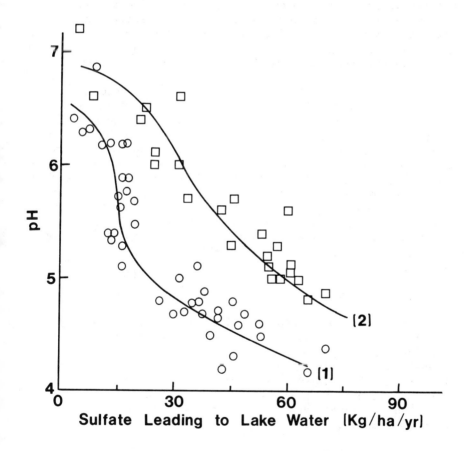

Figure 4.   Effects of various sulfate loading rates on lake pH for lakes in very sensitive (1) and somewhat less sensitive (2) surroundings in Sweden.   (From Glass and Loucks, 1980.)

the different sets of lake data. A larger number of curves may be envisioned, each representing lake/watershed systems of comparable sensitivities as they apparently respond to increases in atmospheric acid inputs.   The curves are superficially similar to titration curves, as might be expected, and each shows buffering at high pH, then a rapid drop (as this buffering apparently is depleted) to lakes with somewhat lower pH and minimal buffering. The pH in the more sensitive lakes drops sooner and more rapidly, relative to the sulfate ion loading.

A graph of sulfate deposition versus lake pH could be used to express lake/watershed sensitivity, and, therefore, acid loading   tolerance,   by   evaluating   the   position   of   the lake/watershed system on the response curves. From a knowledge of   the   response   of   very   sensitive   geochemical/biological systems in moderately impacted systems, it appears possible to anticipate   similar   responses   in   the   future   on   similarly sensitive systems if acidic inputs increase or are sustained.

Examination of Figure 4 suggests that annual sulfate loadings of less than 15 to 17 kg/ha would be unlikely to degrade lakes of the type represented in curve (1). However, if the lower envelope of the data distribution is viewed as a potential "family" of the most sensitive lakes and streams, these appear to be just barely free of potential acid loadings effects at an annual rate of 9 to 12 kg/ha/yr. Thus, two "tolerances" can be defined, one associated with a possible protection of nearly all sensitive aquatic resources, and the other with protection of something less than all sensitive lakes; e.g., only the half of the "sensitive" resources that lies above curve (1) in Figure 4.

The Henriksen Nomograph

Henriksen (1980) presents a model based on the concepts implicit in titration of a bicarbonate-buffered lake with strong acid (principally sulfuric acid) from the atmosphere. In the process, bicarbonate is depleted and lake pH can fall below 5 with consequent effects on aluminum mobilization and fish. These relations are basically the same as those in the Dickson work, and are summarized in the nomograph (Figure 5) using two key measurements:

1.   ambient concentrations of in-lake calcium (or Ca plus Mg)   as   an   estimator   of   the   pre-acidification alkalinity; and

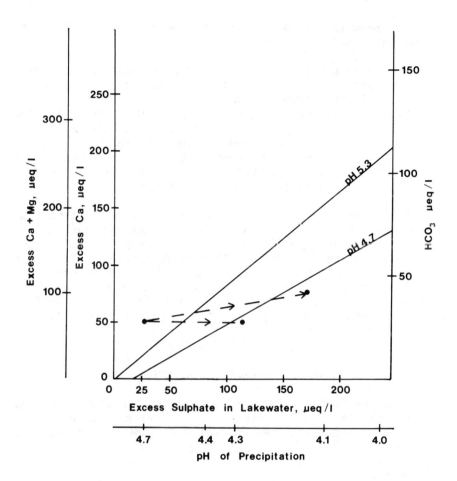

Figure 5.   *A nomograph to predict the pH of lakes given the*
sum of non-marine calcium and magnesium concen-
trations (or non-marine calcium concentration
alone) and the non-marine sulfate concentration
in lake water (or the weighted-average hydrogen
ion concentration in precipitation).   Nomograph
also can be used to predict future changes in
lake pH when precipitation acidity increases
(from Henriksen, 1980).

2.   lake sulfate concentrations (in excess of marine input) as an estimator of H^+ added to the system.

The resulting semi-predictive nomograph is divided into three sections:

1.   bicarbonate lakes, where original alkalinity was high and/or added H^+ is low, so that the lakes remain bicarbonate-buffered;
2.   acid lakes, where original alkalinity was low relative to acid inputs, and all bicarbonate appears to have been depleted by the acid addition; and
3.   transition lakes, in which bicarbonate appears to be undergoing reduction (or is almost depleted) and large pH fluctuations occur during runoff events.

The transition phase, in which the lake is shifting from a bicarbonate-buffered equilibrium at moderate pH to an aluminum-buffered equilibrium at low pH, represents the key process requiring prediction. This shift apparently is forced by H^+ and sulfate ion inputs, and Henriksen presents regression equations based on Norwegian lake and precipitation data for representing sulfate in lake water(SO_4*) in terms of sulfate in precipitation $[SO_4(p)]$ and H^+ in precipitation $[H^+(p)]$:

$$SO_4* = -19 + 1.9\ SO_4(p)$$

$$SO_4(p) = -2.7 + 1.37\ H^+(p)$$

All concentrations are in microequivalents per liter ($\mu eq/l$), and the sulfate is excess over that from marine origin. These equations have not yet been validated for North America.

CONCLUSIONS

Ultimately, even the acidification of surface water resources is of little consequence unless the productivity of resources is altered in ways that are important to human welfare. Many of the chapters which follow will address what we know and do not know about these questions. Let me turn here only to the question of the processes by which a regional experi-mental manipulation appears to be underway in the Great Lakes states: perhaps 50 yr of relatively high levels of acid additions in the eastern areas, and as little as 20 yr of moderate to low levels of acidic deposition in the western areas. Research must answer whether the duration of treatment is equivalent to intensity of treatment. How many years do we have before "significant" (yet to be defined) effects may be expressed in the lightly impacted regions?

The geography and resources of the Great Lakes region make very plain the great importance of these questions, for some states more than others. Decisions regarding industrial development in this region are being made every day, and investments in large industrial plants are relatively irreversible, at least within a generation or two. So too, apparently, are some of the effects of acid rain. The approaches that scientists and government leaders in the Great Lakes region take in evaluating these issues are likely to have considerable impact on others in the United States and Canada.

LITERATURE CITED

Abrahamsen, G., K. Bjor, R. Horntvedt, and B. Treite. 1976. Effects of Acid Precipitation on Coniferous Forest. In: F.H. Braekke, ed., "Impact of Acid Precipitation on Forest and Eco-systems in Norway," pp. 36-63. SNSF Project, Nisk, Norway.

Almer, B., W. Dickson, C. Ekstrom, and E. Hornstrom. 1978. Sulfur Pollution and the Aquatic Ecosystem. In: J.O. Nriagu, ed., "Sulfur in the Environment: Part II, Ecological Impacts," pp. 273-311. John Wiley and Sons, Inc., New York.

Coote, D.R., D. Siminovitch, S.S. Singh, and C. Wang. 1980. The Significance of the Acid Rain Problem to Agriculture in Eastern Canada. Report to the Canada-U.S. Research Consultation Group on the Long-Range Transport of Air Pollutants (LRTAP). Environment Canada, Burlington, Ontario. 85 pp.

Curtis, R.O., D.J. DeMars, and R.R. Herman. 1974. Which Dependent Variable in Site Index-Height-Age Regressions? Forest Sci. 20:74-87.

Foy, C.D., R.L. Chaney, and M.C. White. 1978. The Physiology of Metal Toxicity in Plants. Ann. Rev. Plt. Physiol. 29:511-566.

Fralish, J. and O. Loucks. 1975. Site Quality Evaluation Models for Aspen (Populus tremulodies Michx.) in Wisconsin. Can. J. Forest Res. 5(4):523-528.

Galloway, J.N. and E.B. Cowling. 1978. The Effects of Precipita-tion on Aquatic and Terrestrial Ecosystems - A Proposed Precipi-tation Chemistry Network. J. Air Pollut. Control Assoc. 28:229-235.

Glass, G.E. and O.L. Loucks (eds.). 1980. Impacts of Airbourne Pollutants on Wilderness Areas Along the Minnesota-Ontario border. EPA 600/3-80-044.

Glass, G.E. and T.G. Brydges. 1981. Problem Complexity in Predicting Impacts from Altered Precipitation Chemistry. Intl. Conf. on Acid Precip. Impacts, Cornell Univ., Ithaca, New York, August 1-4.

Gorham, E. and W.W. McFee. 1980. Effects of Acid Deposition Upon Outputs From Terrestrial to Aquatic Ecosystems. In: T.C. Hutchinson and M. Havas, eds., "Effects of Acid Precipitation on Terrestrial Ecosystems," pp. 465-480. Plenum Press, New York.

Henriksen, A. 1980. Acidification of Freshwaters - A Large Scale Titration. In: D. Drabløs and A. Tollan, eds., "Ecological Impact of Acid Precipitation," pp. 68-74. SNSF Project, Sandefjord, Norway.

Kramer, J.R. 1976. Geochemical and Lithological Factors in Acid Precipitation. U.S.D.A. Forest Service Gn. Tech. Rep. NE-23, pp. 611-618.

Last, T.F., G.E. Likens, B. Ulrich, and L. Walloe. 1980. Acid Precipitation - Progress and Problems. In: D. Drabløs and A. Tollan, eds., "Ecological Impact of Acid Precipitation," pp. 10-12. Proceedings of an international conference. SNSF Project, Sandefjord, Norway.

Loucks, O.L., R.W. Usher, R.W. Miller, W. Swanson, and D. Rapport. 1981. Assessment of Sensitivity Measures for Evaluating Resources at Risk from Atmospheric Pollutant Deposition. Final Report to the U.S. Environmental Protection Agency, ERL-Duluth. The Institute of Ecology (TIE), Indianapolis, Indiana. 87 pp.

McFee, W.W. 1980. Sensitivity of Soil Regions to Long-Term Acid Precipitation. EPA 600/3-80-013.

McFee, W.W., J.M. Kelly, and R.H. Beck. 1976. Acid Precipitation Effects on Soil pH and Base Saturation of Exchange Sites. USDA Forest Service Gn. Tech. Report NE-23, pp.725-735.

Ontario Ministry of the Environment. 1980. The Case Against the Rain. A Report on Acidic Precipitation and Ontario Programs for Remedial Action. Ontario Government Report, Toronto, Ontario, Canada.

Overrein, L.N., H.M. Seip, and A. Tollan. 1980. Acid Precipitation Effects on Forest and Fish: Final report of the SNSF Project 1972-1980. Res. Report 1980, Oslo, Norway. 175 pp.

Schneider, J.C. 1975. Typology and Fisheries Potential of Michigan Lakes. Mich. Acad. 8:59-84.

Schofield, C.L. 1981. Personal Communication. Cornell University, Ithaca, New York.

Ulrich, B., R. Mayer, and P.K. Khanna. 1980. Chemical Changes Due to Acid Precipitation in a Loess-derived Soil in Central Europe. Soil Sci. 130:193-199.

Wright, R.F., H. Dovland, C. Lysholm, and B. Wingard. 1978. Inputs and Outputs of Water and Major Ions at 9 Catchments in Southern Norway, July 1974-1975. SNSF Project TN 39/79. 60 pp.

Wright, R.F. and A. Henriksen. 1978. Chemistry of Small Norwegian Lakes, With Special Reference to Acid Precipitation. Limnol. Oceanogr. 23:487-498.

Wright, R.F., N. Conroy, W.T. Dickson, R. Harriman, A. Henriksen, and C.L. Schofield. 1980. Acidified Lake Districts of the World: A Comparison of Water Chemistry of Lakes in Southern Norway, Southern Sweden, Southwestern Scotland, the Adirondack Mountains of New York, and Southeastern Ontario. In: D. Drabløs and A. Tollan, eds., "Ecological Impact of Acid Precipitation," pp. 377-379. Proceedings of an International Conference, SNSF Project, Sandefjord, Norway.

CHAPTER 3

AN HISTORICAL RESUME OF PROGRESS IN SCIENTIFIC
AND PUBLIC UNDERSTANDING OF ACID PRECIPITATION
AND ITS BIOLOGICAL CONSEQUENCES

Ellis B. Cowling
School of Forest Resources
North Carolina State University
Raleigh, North Carolina 27650

INTRODUCTION

Some years ago the terms "acid precipitation" and "acid rain" were bits of esoteric jargon used almost exclusively by scientists in certain specialized fields of ecology and atmospheric chemistry. During the last few years, these terms have become worrisome household words in many countries around the world. How did this transition come about? Who was responsible? Why did it take so long for acid precipitation to be recognized as an important ecological and environmental problem? What factors of scientific awareness and public perception have influenced the course and development of research on acid precipitation and its biological effects?

This chapter is an attempt to illuminate some of these questions. Our approach will be to review various steps in the transformation of the concepts of acid precipitation and atmospheric deposition from the domain of scientific curiosity to the domain of public concern and debate about energy and environmental policy.

Table 1 contains an historical resume of progress in understanding the phenomenon and its biological consequences. Contributions are presented in the chronological order of their occurrence whether or not the work was recognized or accepted at the time. Names of major contributors, the country in which the research was achieved, and the principal contribution to science or public affairs are presented with appropriate references when this is possible.

Table 1

An Historical Resume of Progress in Understanding
Acid Precipitation and Its Consequences

Year(s)	Investigator and Country	Principal Contribution to Understanding
1661–1662	Evelyn, Graunt, England	Noted the influence of industrial emissions on the health of plants and people, the transboundary exchange of pollutants between England and France, and suggested remedial measures including placement of industry outside of towns and use of taller chimneys to spread the "smoke" into "distant parts" (Evelyn, 1661; Graunt, 1662; see also Gorham, 1980).
1687	Hooke, England	On the basis of experiments by Brotherton, Hooke concluded that plants have a "two-fold kind of roots, one that branches and spreads into the earth, and another that spreads and shoots into the air, both kinds of roots serve to receive and carry their proper nourishment to the body of the plant." (Hooke, 1687; see also Gorham, 1965; and Wittwer and Bukovac, 1969).
1727	Hales, England	Noted that dew and rain "contain salt, sulphur, etc. For the air is full of acid and sulphureous particles..." (Hales, 1727; see also Gorham, 1980).
1734	Linné, Sweden	Described a 500-year-old smelter at Falun, Sweden: "...we felt a strong smell of sulphur,...rising to the west of the city...a poisonous, pungent sulphur smoke, poisoning the air wide around...corrode(ing) the earth so that no herbs can grow around it." (Linné, 1734).

Table 1 (continued)

Year(s)	Investigator and Country	Principal Contribution to Understanding
1852	Smith, England	Analyzed the chemistry of rain near Manchester, England, and noted concentric zones with "three kinds of air--that with carbonate of ammonia in the fields at a distance,--that with sulphate of ammonia in the suburbs,--and that with sulphuric acid, or acid sulphate in the town." Smith also noted that sulfuric acid in town air caused fading in the color of textiles and corrosion of metals (Smith, 1852, see also Gorham, 1980).
1854–1856	Austria & Germany	Established "General Citizens Laws" prohibiting disposal of wastes by individuals on a neighbor's property. In the interest of encouraging industrialization, however, these same laws specifically excluded industries from legal liability in cases of waste disposal by pollution of water and air.
1855–1856	Way, England	Completed a very detailed series of analyses of nutrient substances in precipitation at the Rothamstead Experiment Station and showed the value of these substances in crop production (Way, 1855).
187?	Smith, England	In a remarkable publication entitled "Air and Rain: The Beginnings of A Chemical Climatology," Smith first used the term "acid rain" and enunciated many of the ideas that we now consider part of the acid precipitation problem. These ideas included regional variation in precipitation chemistry as it is influenced by such factors as combustion of coal,

Table 1 (continued)

Year(s)	Investigator and Country	Principal Contribution to Understanding
		decomposition of organic matter, wind direction, proximity to the sea, amount of rain, etc. After extensive field experiments, Smith proposed detailed procedures for the proper collection and chemical analysis of precipitation. He also noted acid rain damage to plants and materials and atmospheric deposition of arsenic, copper, and other metals in industrial regions (Smith, 1872; see Gorham, 1980).
1881	Brögger, Norway	Observed "smudsig snefeld" (dirty snowfall) in Norway and attributed it to either a large town or an industrial district in Great Britain (Brögger, 1881).
1909	Sørensen, Denmark	Developed the pH scale to describe the acidity of aqueous solutions (Sørensen, 1909).
1911	Crowther & Ruston, England	Demonstrated gradients in acidity of precipitation decreasing from the center of Leeds, England, associated the acidity with combustion of coal, and showed that both natural rain and dilute sulfuric acid inhibited plant growth, seed germination, and ammonification, nitrification, and nitrogen fixation in soil (Crowther and Ruston, 1911; see also Cohen and Ruston, 1912; and Gorham, 1980).
1919	Rusnov, Austria	Demonstrated that deposition of substances from the atmosphere accelerated the acidification of both poorly-buffered and well-buffered forest soils (Rusnov, 1919).

Table 1 (continued)

Year(s)	Investigator and Country	Principal Contribution to Understanding
1921	Dahl, Norway	Recognized the relationship between acidity of surface water and production of trout (Dahl, 1921; 1927).
1922	Atkins, England	Measured the alkalinity of surface waters and noted a relationship between alkalinity and biological productivity (Atkins, 1922).
1923	MacIntyre, USA	Made the first detailed study of precipitation chemistry in the United States (MacIntyre and Young, 1923).
1925	Shutt & Hedley, Canada	Made very early measurements of the nitrogen compounds present in rain and snow and commented on the value of these compounds for growth of crops (Shutt and Hedley, 1925).
1926	Sunde, Norway	Demonstrated the value of adding limestone to water in a fish hatchery (Sunde, 1926).
1939	Erichsen-Jones, Sweden	Demonstrated the relationship between acidity and the toxicity of aluminum to fish (Erichsen-Jones, 1939).
1939	Katz et al., Canada	Reported acidification and decrease in base saturation of soils by sulfur dioxide emissions from the lead-zinc smelter near Trail, British Columbia (Katz et al., 1939).
1939	Bottini, Italy	Detected hydrochloric acid in precipitation near the volcano on Mount Vesuvius, thus demonstrating that there are natural sources of strong acids in precipitation (Bottini, 1939).

Table 1 (continued)

Year(s)	Investigator and Country	Principal Contribution to Understanding
1942	Conway, Ireland	Completed the first modern review of precipitation chemistry (Conway, 1942).
1948	Egnér, Sweden	Initiated the first large-scale precipitation chemistry network in Europe (Egner et al., 1955; Rossby and Egner, 1955).
1950–1955	Eriksson, Sweden	Enunciated a general theory of bio-geochemical circulation of matter through the atmosphere (Eriksson, 1952; 1954; 1959; 1960).
		Expanded the regional network established by Egnér into the continent-wide European Air Chemistry Network which has provided a continuing record of precipitation chemistry for 3 decades (Emanuelsson et al., 1954).
1953	Viro, Finland	Developed a regional chemical budget by comparing analytical data for precipitation and river waters in Finland (Viro, 1953).
1953–1958	Tamm, Sweden	Demonstrated the great dependence of mosses on atmospheric sources of nutrients, especially nitrogen (Tamm, 1953) and expanded this concept to include most forest plants (Tamm, 1958).
1953–1955	Various investigators in several countries	Simultaneously investigated precipitation chemistry data for evidence of atmospheric acidity (Barret and Brodin, 1955 – Sweden; Parker, 1955, and Gorham, 1955 – England; Houghton, 1955 – USA).
1954–1961	Gorham, England	Demonstrated that acidity in precipitation markedly influenced

Table 1 (continued)

Year(s)	Investigator and Country	Principal Contribution to Understanding
		geological weathering processes and the chemistry of lake waters, bog waters, and soils (Gorham, 1955; 1958b, 1958c, 1961).
		Demonstrated that hydrochloric acid from combustion of coal rich in chlorine predominated in urban precipitation whereas sulfuric acid predominated in rural precipitation (Gorham, 1958a, 1958b).
		Established that acidity in precipitation affects the alkalinity and buffering capacity of lake and bog waters (Gorham, 1957; 1958b).
		Established that the incidence of bronchitis in man can be correlated with the acidity of precipitation (Gorham, 1959).
1957	Europe, USSR & USA	During the International Geophysical Year, a one-year study of precipitation chemistry was made in Europe, the USSR, and the USA.
1958	Junge & Werby, Jordan et al., USA	Made the first regional studies of precipitation chemistry in the United States and noted the importance of atmospheric sulfur as a source of nutrients for crops (Junge and Werby, 1958; Jordan et al., 1959).
1959	Dannevig, Norway	Recognized the relationship between acid precipitation, acidity in surface waters, and disappearance of fish (Dannevig, 1959).
1960–1963	Gordon & Gorham, Canada	Established that fumigation by sulfur dioxide and resultant acid rain contributed to the deterioration of

Table 1 (continued)

Year(s)	Investigator and Country	Principal Contribution to Understanding
		lake quality, vegetation, and soils near metal smelters (Gorham and Gordon, 1960; Gordon and Gorham, 1963).
1960-1967	McCormick, USA	The first continent-wide precipitation chemistry network in North America was operated for 6 years -- 1960-66. The data showed that precipitation is generally acidic east but alkaline west of the Mississippi River (Lodge et al., 1968).
1962	Carson, USA	Stimulated a global revolution in environmental awareness by publishing "Silent Spring" in which the term "poison rain" was used to describe concern about atmospheric transport and deposition of pollutants (Carson, 1962).
1967-1968	Odén, Sweden	Outlined the changing acidity of precipitation as a regional and temporal phenomenon in Europe (Odén, 1968; 1976). Used trajectory analysis of air masses to demonstrate that acidity in precipitation in Scandinavia was attributable largely to emissions of sulfur in England and central Europe (Odén, 1968). Demonstrated temporal trends in acidity and major cations and anions in precipitation in various parts of Europe (Odén, 1968). Demonstrated the increasing acidity of Scandinavian rivers (Odén, 1968).

Table 1 (continued)

Year(s)	Investigator and Country	Principal Contribution to Understanding
		Described biological uptake and ion-exchange processes by which natural acidification of soils would be accelerated by atmospheric deposition of ammonia and other cations (Odén, 1968).
		Postulated that acid precipitation would lead to displacement of nutrient cations, reduction in nitrogen fixation, and release of heavy metals (especially Hg) which would damage surface waters and ground waters (Odén, 1968; 1976).
		Postulated acidity in precipitation as a probable cause of decline in fish populations, impoverishment of forest soils, decreased forest growth, increased disease in plants, and accelerated corrosion and other damage to materials (Odén, 1967; 1968).
1970	Odén & Ahl, Sweden	Discovered that soluble pollutants in snow accumulate in the snow pack and are released almost totally with the first meltwater in spring (Odén and Ahl, 1970).
1970	Hultberg, Sweden	Demonstrated the effect of acidity on fish populations in two Swedish lakes (Hultberg and Stenson, 1970).
1972	Bolin, Sweden	Drafted Sweden's Case Study for the United Nations Conference on the Human Environment "Air Pollution Across National Boundaries: The Impact on the Environment of Sulfur in Air and Precipitation". Noted damage to materials as well as ecosystems by acid precipitation (Bolin et al., 1972).

Table 1 (continued)

Year(s)	Investigator and Country	Principal Contribution to Understanding
1972	Hvatum, Norway	Demonstrated increasing content of lead near the surface of Norwegian peat bogs and postulated long-distance transport as the probable cause (Hvatum, 1972).
1972	Tyler, Sweden	Reported heavy metal accumulation in forest soils and speculated on their probable effects on forest productivity (Tyler, 1972).
1972	Granat, Sweden	Described the temporal and spatial deposition of sulfate and acid over northern Europe (Granat, 1972; 1978).
1972	Jonsson & Sundberg, Sweden	Established an experimental basis for the suspicion that acid precipitation had decreased the growth of forests in Sweden (Jonsson and Sundberg, 1972).
1972	Various Agencies, Norway	Initiated the SNSF Project "Acid Precipitation: Effects on Forests and Fish (see *Ambio*, 1976 and Braekke, 1976).
1972–1980	Overrein, Norway	Demonstrated accelerated loss of calcium and other cations from soils receiving acid precipitation (Overrein, 1972). Provided leadership for the SNSF Project from its founding in 1972 until its completion in 1980 (*Ambio*, 1976; SNSF, 1980).
1972	Likens *et al.*, USA	Discussed the regional distribution of acid precipitation and its significance for aquatic and terrestrial ecosystems in North America (Likens *et al.*, 1972).

Table 1 (continued)

Year(s)	Investigator and Country	Principal Contribution to Understanding
		Indicated that nitric acid resulting from atmospheric transformation of NO_x adds to the acidity of precipitation in the eastern United States (Likens et al., 1972).
1972	Jensen & Snekvik, Norway	Showed that acidity in lakes and streams caused major decreases in salmon and trout populations in Norway (Jensen and Snekvik, 1972).
1972	Beamish & Harvey, Canada	Reported decline in fish populations due to acidification of lake waters in Canada (Beamish and Harvey, 1972).
1972–1980	Various Investigators	Showed that mushrooms, mosses, and other vegetation in forests accumulate heavy metals, especially lead and cadmium. Wildlife feeding on these plants also accumulate the metals, sometimes making both the plants and the wildlife hazardous for human consumption (Munshower, 1972; Huckabee and Blaylock, 1974; Tyler, 1980).
1973	Malmer, Sweden	Summarized research on the ecological effects of increasing sulfur deposition, especially with reference to Swedish conditions (Malmer, 1973).
1973	Dickson et al., Sweden	Reported on the pH status of 314 lakes in western Sweden (Dickson et al., 1973).
1973	Wiklander, Sweden	Proposed a general theory to account for the effects of acid precipitation on soil chemical properties (Wiklander, 1973).

Table 1 (continued)

Year(s)	Investigator and Country	Principal Contribution to Understanding
1973-1977	Ottar, Norway	Provided leadership for a continent-wide study of the long-range transport of air polllutants by the Organization for Economic Cooperation and Development in Europe (Ottar, 1976; OECD, 1977).
1974	Almer et al., Sweden	Summarized the effects of changing acidity of lake waters on fish populations in Sweden (Almer et al., 1974).
1974	Grahn et al., Sweden	Discovered that Sphagnum (peat moss) invades acidified lakes and streams, inducing a self-accelerating oligotrophication of the water body (Grahn et al., 1974).
1974	Cogbill & Likens, USA	Published maps showing changes in acidity of precipitation in the eastern United States between 1955-56 and 1972-73 (Cogbill and Likens, 1974).
1974	Hutchinson & Whitby, Canada	Established that strongly acid rain near Sudbury, Ontario, is accompanied by deposition and/or mobilization of heavy metals (especially nickel, copper, cobalt, iron, aluminium, and manganese). The toxicity of these metals was sufficient to inhibit germination and establishment of many native and agricultural species of plants (Hutchinson and Whitby, 1974; Whitby and Hutchinson,1974).
1974	Shriner, USA	Demonstrated that simulated rain acidified with sulfuric acid can accelerate erosion of protective waxes on leaves, inhibit nodulation of leguminous plants, and alter

Table 1 (continued)

Year(s)	Investigator and Country	Principal Contribution to Understanding
		host-pathogen interactions of plants (Shriner, 1974; 1976; 1978).
1975	Brosset et al., Sweden	Described the chemical transformations and trajectories that lead to "white episodes" and "black episodes" of acid deposition in western Sweden. Showed that gaseous nitric acid increases acid deposition. Established a state-of-the-art titration method for the determination of acidity in precipitation (Brosset et al.,1975).
1975-1980	Cowling, USA	Testified in Congressional hearings concerning the inadequacy of research in the U.S. on the ecological effects of acid precipitation (Cowling,1976).
		Together with many other scientists in the U.S. and Canada, began the development of a permanent network to monitor chemical changes in wet and dry deposition and to study their biological effects in various regions of the United States (Galloway and Cowling, 1978).
1975	Cragin et al., Greenland	Determined the chemistry of precipitation in Greenland from 1300 to 1975 A.D. The data show a continuing trend of increasing sulfate and lead concentration beginning about 1800 and rising even more rapidly after 1945 (Cragin et al.,1975).
1976	Schofield, USA	Reported the results of lake surveys showing decline in fish populations associated with acidification of lakes in the Adirondack Mountains of New York State (Schofield, 1976).

Table 1 (continued)

Year(s)	Investigator and Country	Principal Contribution to Understanding
1976	Kucera, Sweden	Summarized evidence that acid precipitation accelerates the corrosion of metals (Kucera, 1976).
1976	Summers & Whelpdale, Canada	Summarized earlier studies of precipitation chemistry in Canada and identified northwest Alberta, southern Ontario and Quebec, and the Atlantic Provinces as areas of present or potential impact by acid rain and snow (Summers and Whelpdale, 1976).
1976	Leivistad & Muniz, Norway	Documented a massive fish kill on the Tovdal River associated with snow melting in the spring of 1975 and established that dead and dying fish had lost control of their blood salt balance (Leivistad and Muniz, 1976; Leivistad et al., 1976).
1976	Tyler, Sweden	Demonstrated that heavy metals in the litter layer of forests inhibit microbial processes, especially decomposition of organic matter (Tyler, 1976).
1976	Galloway, USA Berry, Canada Granat, Sweden	Developed standardized protocols for precipitation collectors and collection techniques (Galloway and Likens,1976; Berry et al., 1976; Granat, 1976).
1976	Hultberg & Grahn, Sweden	Discovered a correlation between acidification of lakes and the mercury content of fish (Hultberg and Grahn, 1976).
1976	Pough, USA	Showed that reproduction of salamanders is inhibited by acidity of surface waters (Pough, 1976).

Table 1 (continued)

Year(s)	Investigator and Country	Principal Contribution to Understanding
1977	Hagström, Sweden	Showed that reproduction of frogs is inhibited by acidity of surface waters (Hagström, 1977).
1977	Rosenqvist, Norway	Enunciated a general theory that acidification of soils and surface waters is due mainly to natural processes in soils and to changes in patterns of land use (Rosenqvist, 1977).
1978	Greszta, Austria	Demonstrated accumulation of heavy metals in forest soils leading to injury to young pine and spruce seedlings (Greszta, 1978).
1978- 1980	Several Investigators, USA	Showed that acid precipitation occurs in both urban and certain rural areas in the western United States (Liljestrand and Morgan, 1978; McColl and Bush, 1978; Lewis and Grant, 1980).
1979	Altshuller, USA; McBean, Canada	Documented the transboundary exchange of sulfur and nitrogen between the United States and Canada (Altshuller and McBean, 1979c, 1980).
1979	Odén, Sweden	Showed by studies of surface waters in Sweden that acidification due to sulfur deposoition had begun during the early 1900's (Odén, 1979).
1979	Liljestrand & Morgan, USA	Completed a rigorous statistical analysis of trends in precipitation chemistry in the eastern and western United States (Liljestrand and Morgan, 1979).
1979	Cronan & Schofield, USA	Discovered that aluminum ions are leached by acid precipitation from soils into streams and lakes in

Table 1 (continued)

Year(s)	Investigator and Country	Principal Contribution to Understanding
		concentrations toxic to fish (Cronan and Schofield, 1979).
1979	Carter, USA	Established a Presidential initiative on acid rain calling for a 10 year long, $10 million per year interagency program of research on acid precipitation and its environmental consequences in the USA (Carter, 1979).
1979	Henriksen, Norway	Developed a simple descriptive model for determining the extent to which acidification decreases the alkalinity of lake waters and applied this method to predict the vulnerability of lakes to acid deposition (Henriksen, 1979).
1980	Hultberg & Wenblad, Sweden	Discovered acidification of ground water in western Sweden, postulated acid precipitation as the probable cause, established by surveys of 1300 wells the frequency of heavy metal accumulation and plumbing problems associated with these wells (Hultberg and Wenblad, 1980).
1980	Abrahamsen, Norway	Summarized many years of research on the effects of acid precipitation on forests and concluded that fertilization effects, particularly by atmospheric deposition of nitrogen, tend to offset nutrient leaching and other detrimental effects. Also emphasized that negative effects of atmospheric deposition on growth are most likely when nutrient deficiencies or imbalances are increased by acid deposition (Abrahamsen, 1980).

Table 1 (continued)

Year(s)	Investigator and Country	Principal Contribution to Understanding
1980	Wetstone, USA	Summarized the biological and materials-damage effects of acid precipitation in relation to the pollution control laws of North America (Wetstone, 1980).
1980	Schindler et al., Canada	Established an experimental system for controlled acidification of whole lakes (Schindler et al., 1980). Demonstrated that acidification eliminated organisms of several trophic levels at pH values as high as 5.8 to 6.0 (Schindler, 1980). Also demonstrated that microbial reduction of sulfate could partially protect lakes against acidification (Schindler et al., 1980).
1980	Ulrich et al., Germany	Demonstrated a significant correlation between amount of soluble aluminum in forest soils, death of feeder roots in spruce, fir, and birch forests, and widespread decline in the growth of these forests (Ulrich et al., 1980; Schuck et al., 1979).
1980	Various Investigators, USA	Established an Integrated Lake-Water Acidification Study (ILWAS), to determine detailed chemical budgets for hydrogen, sulfate, ammonium, nitrate, chloride, and other ions in three lake-watersheds of differing degrees of acidification (see Drablós and Tollan, 1980, pp. 252-267; and EPRI, 1981).

Table 1 (continued)

Year(s)	Investigator and Country	Principal Contribution to Understanding

MAJOR INTERNATIONAL CONFERENCES
DOCUMENTING FURTHER PROGRESS

Year(s)	Investigator and Country	Principal Contribution to Understanding
1950's	Eriksson, Sweden	The International Meteorological Institute sponsored a series of conferences on various aspects of atmospheric chemistry (Eriksson, 1954).
1975–1976	Dochinger & Seliga, USA	U.S. Forest Service sponsored the First International Symposium on Acid Precipitation and the Forest Ecosystem at the Ohio State University, Columbus, Ohio (Dochinger and Seliga, 1976a; 1976b).
1976	Overrein et al., Norway	SNSF Project and the Norwegian Ministry of the Environment sponsored the International Conference on Effects of Acid Precipitation, Telemark, Norway (Ambio, 1976; Braekke, 1976).
1977	Husar & Lodge, USA; Moore, England	The United Nations Environmental Program and several other organizations sponsored an International Symposium on Sulfur in the Atmosphere, Dubrovnik, Yugoslavia (Husar et al., 1978).
1978	Hutchinson, Canada	NATO sponsored an Advanced Research Institute on Ecological Effects of Acid Precipitation in Toronto, Canada (Hutchinson and Havas, 1980).
1978	Howells, England	The Electric Power Research Institute (USA) and the Central Electricity Generating Board (Great Britain) sponsored an International Symposium on Ecological Effects of Acid Precipitation (Howells, 1979).

Table 1 (continued)

Year(s)	Investigator and Country	Principal Contribution to Understanding
1978	Hendrey, USA	U.S. Environmental Protection Agency and Brookhaven National Laboratory sponsored an International Workshop on Limnological Aspects of Acid Precipitation, Sagamore Lake, New York (Hendrey, 1978).
1979	Shriner, USA	The Oak Ridge National Laboratory sponsored a symposium on the Potential Environmental and Health Effects of Sulfur Deposition, Gatlinburg, Tennessee (Shriner et al., 1980).
1979	Agle, USA & Turnbull, Canada	A group of Canadian and United States environmental protection organizations sponsored the Action Seminar on Acid Precipitation, Toronto, Canada (Reid, 1980).
1980	Overrein et al., Norway	The SNSF Project sponsored an International Conference on Ecological Impacts of Acid Precipitation, Sandefjord, Norway.
1980	Miller et al., USA	The U.S. Forest Service and other U.S. and international air pollution organizations sponsored an International Symposium on Effects of Air Pollutants on Mediterranean and Temperate Forest Ecosystems, Riverside, California.

Selecting particular contributions for inclusion in such a chronology is a hazardous business because of the likelihood that some important contributions will not be given the credit and/or priority they deserve. There is no choice but to accept the blame for whatever errors of omission and ignorance are evident in Table 1. I hope the Table will serve a useful purpose, however, and look forward to receiving comments from any person who may be interested to help describe the record more adequately.

EARLY AWARENESS OF ACID PRECIPITATION

Many features of the acid rain phenomenon were first discovered by an English chemist named Robert Angus Smith in the middle of the 19th century. In 1852, Smith published a detailed report on the chemistry of rain in and around the city of Manchester in England. In this account, Smith (1852) called attention to the changes in precipitation chemistry as one moves from the middle of a polluted city to its surrounding countryside: "We may, therefore, find three kinds of air--that with carbonate of ammonia in the fields at a distance; that with sulphate of ammonia in the suburbs; and that with sulphuric acid, or acid sulphate, in the town." Smith also pointed out that the sulfuric acid in city air caused the fading of colors in textiles and the corrosion of metals.

Twenty years later, in a remarkable book entitled "Air and Rain: The Beginnings of A Chemical Climatology," Smith (1872) first used the term "acid rain" and enunciated many of the principal ideas that are part of our present understanding of the acid rain phenomenon. On the basis of detailed studies in England, Scotland, and Germany, Smith demonstrated that precipitation chemistry is influenced by such regional factors as combustion of coal, decomposition of organic matter, wind trajectories, proximity to the sea, and amount and frequency of rain or snow. Smith proposed detailed procedures for the proper collection and chemical analysis of precipitation. He also noted acid rain damage to plants and materials in industrial regions.

Unfortunately, however, Smith's pioneering and prophetic book apparently has been overlooked by essentially every subsequent investigator of the acid precipitation phenomenon. Eville Gorham (1980) has developed the first detailed analysis of Smith's early work for a report by the National Academy of Sciences.

MODERN AWARENESS OF ACID PRECIPITATION

Modern understanding of acid precipitation and its environmental consequences had its origins in three seemingly unrelated fields of science: limnology (the study of freshwaters), agriculture, and atmospheric chemistry. Let us look at the contribution of each field of science in turn.

Progress in Limnology

The relationship between rain or snow, and the water in streams and lakes was obvious even to prehistoric man. But the relationship between changes in the chemistry of rain and snow and changes in the chemistry of lake waters remained obscure until the middle of the twentieth century.

During the first half of this century, several scientists in Europe and North America studied certain isolated aspects of atmospheric and surface water chemistry. Occasionally, interactions were demonstrated, but it was Eville Gorham, now Professor of Ecology of the University of Minnesota, who, in a long series of papers, developed the major foundations for our present understanding of the causes of acidity in precipitation and its impact on aquatic ecosystems (Gorham, 1955; 1957, 1958a-c; 1959; 1961; 1965). On the basis of research both in England and in Canada, Gorham and his colleagues demonstrated the following principles:

(1) Much of the acidity of precipitation in industrial regions can be attributed to atmospheric emissions produced during combustion of fossil fuels.
(2) Progressive loss in alkalinity of surface waters and increase in the acidity of bog waters can be traced to atmospheric deposition of acidic substances in precipitation.
(3) Much of the free acidity in soils receiving acid precipitation is due to sulfuric acid.
(4) The incidence of bronchitis in man can be correlated with the acidity of precipitation.
(5) Fumigation with sulfur dioxide and resultant acid rain contribute to the deterioration of vegetation, soils, and lake water quality around metal smelters.

Thus, by the mid 1950s and early 1960s, Gorham had established a major part of our present understanding of the sources and the limnological and ecological significance of acid precipitation. But these pioneering researchers, like those of Smith a century before, were met by a thundering silence from both the scientific community and the public at large. Lack of recognition of the environmental significance of Gorham's work resulted in a further lag in both scientific and public awareness of acid precipitation and its ecological significance.

Progress in the Agricultural Sciences

The importance of the atmosphere as a source of nutrients for the growth and development of plants was first recognized by Robert Hooke in 1687 (see also Gorham, 1965). Early experiments at the Rothamsted Experiment Station in England also demonstrated the relationship between nutrients in air and the growth of various crop plants (Way, 1855). In the mid-1940s, an imaginative Swedish soil scientist, Hans Egnér, systematized investigations on the fertilization of crops by atmospheric sources of nutrients. Working from his laboratory at the Agricultural College near Uppsala, Egnér created the first twentieth century network for the collection and chemical analysis of precipitation and air at ground level. A large number of sampling buckets were set out at experimental farms all over Sweden and the major chemical constituents in what we now call bulk deposition (rain, snow, and dust fall) were measured on a regular monthly basis. The acidity of precipitation was one of several chemical parameters that were measured. This network, initiated by Egnér in 1948, was gradually spread by other agricultural scientists, first to Norway, Denmark, and Finland and later to most of western and central Europe. This network provided the first large scale and long-term data on the changing chemistry of precipitation and its importance for agriculture and forestry (Emanuelsson et al., 1954; Egnér et al., 1955; Tamm, 1958). For the first time it became possible to discern both temporal and spatial patterns of change in precipitation chemistry. In 1956, the International Meteorological Institute in Stockholm assumed responsibility for further coordination and expansion of Egnér's network into the European Air Chemistry Network. In 1957, as part of the International Geophysical Year, the European Network spread eastward to include Poland and much of the Soviet Union. In marked contrast to monitoring efforts elsewhere in the world, most stations of the original European Network, including those in Poland and the Soviet Union, have remained in continuous operation for nearly three decades!

Progress in the Atmospheric Sciences

Under the imaginative leadership of Carl Gustav Rossby and Erik Eriksson, the science of atmospheric chemistry began in Sweden and later spread across Europe and finally to North America (NAS, 1975). Rossby and Eriksson were convinced that atmospheric processes were efficient mechanisms for the long-distance as well as the short-distance dispersal of all sorts of substances. The data from Egnér's precipitation chemistry network provided the means to test various hypotheses about the trajectories of air masses, turbulent dispersal processes, and atmospheric scavenging and deposition processes (Rossby and Egnér, 1955).

The transport and deposition of water by atmospheric processes have been well known since the earliest observations of evaporation, wind and cloud movement, and of rain and snow. But the notion that atmospheric transport and deposition processes were major means for dispersal and chemical transformation of many other substances was still only a working hypothesis in the early 1950s. Rossby and Eriksson championed these then novel ideas and initiated various experimental tests of their hypotheses. Data from the European Air Chemistry Network provided a very powerful tool for these studies. Rossby and Eriksson also sponsored a series of European conferences on atmospheric chemistry and dispersal processes. These conferences attracted the interest of scientists in many other fields including biology, forestry, agriculture, meteorology, engineering, and medicine (Eriksson, 1954).

Integration of Limnological, Agricultural, and Atmospheric Chemical Studies

The first major integration of knowledge about acid precipitation in these three fields of science was achieved by Svante Odén, a soil scientist at the Agricultural College near Uppsala in Sweden. As a young colleague of Egnér, Rossby, and Eriksson, Odén started a Scandinavian network for surface water chemistry in 1961. When data from this network were combined with those from the European Air Chemistry Network, a series of general trends and relationships began to emerge and were published by Odén in two different media: in Stockholm's prestigious newspaper Dagens Nyheter (Odén, 1967) and in an Ecology Committee Bulletin (Odén, 1968). The newspaper report outlining Odén's ideas about an insidious "chemical war" among the nations of Europe captured the attention of the press, who began the process of public education about acid precipitation in Europe. The Ecology Committee Bulletin had a similarly provocative influence on scientific interest in both the phenomenon and the ecological effects of the increasing acidity of precipitation. Suddenly, limnological, agricultural, and atmospheric scientists began talking to each other about Odén's unconventional ideas and the general theory of atmospheric influences which he enunciated. Odén's analyses of the trajectories and chemistry of air masses clearly showed that:

(1)   acid precipitation was a large scale regional
      phenomenon in much of Europe with well defined source
      and sink regions;
(2)   both precipitation and surface waters were becoming
      more acidic; and
(3)   long-distance transport (100 to 2000 km) of both
      sulfur- and nitrogen-containing air pollutants was
      taking place among the various nations of Europe.

Odén also enunciated a series of hypotheses about the
probable ecological consequences of acid precipitation with
respect to:

(1)   decline of fish populations;
(2)   decreased forest growth;
(3)   increased plant diseases; and
(4)   accelerated damage to materials.

These conclusions and hypotheses led to a veritable storm
of scientific and public concern about acid precipitation. The
Swedish government responded to growing public and scientific
concern by initiating an inquiry which culminated in Sweden's
Case Study for the United Nations Conference on the Human
Environment, "Air Pollution Across National Boundaries: The
Impact of Sulfur in Air and Precipitation" (Bolin et al., 1972).

The major ideas in both the Ecology Committee Bulletin and
the Swedish Case Study were much debated all over Europe. Two
major scientific initiatives followed in short order. The first
occurred in 1972 when three organizations in Norway joined
together in establishing the so-called SNSF Project (The
Norwegian Interdisciplinary Research Programme "Acid
Precipitation: Effects on Forest and Fish").   These three
organizations included the Norwegian Council for Scientific and
Industrial Research, the Agricultural Research Council of
Norway, and the Norwegian Ministry of Environment. The annual
budget for the SNSF Project was about 10 million Norwegian
Kroner (U.S., 2 million dollars) per year from 1972-1980. This
huge project had two comprehensive goals:

(1)   To establish as precisely as possible, the effects
      of acid precipitation on forests and freshwater fish.
(2)   To investigate the effects of air pollutants on soils,
      vegetation, and water to the extent required to
      support the primary objective.

Lars Overrein of the Norwegian Forest Research Institute
was appointed Research Director for this project and served in
this capacity from the inception to the termination of the
project in 1980.   The project produced a steady stream of

technical and scientific reports on various aspects of acid precipitation and its effects. The SNSF Project also sponsored two major international scientific conferences, one at Telemark, Norway, in June 1976 and the second at Sandefjord, Norway, in March 1980. Braekke (1976) and Ambio (1976) published major reports in connection with the Conference at Telemark. The Conference at Sandefjord was designed to provide a forum for evaluation not only of recent research within the SNSF Project, but also that being done elsewhere in the world. A final report and bibliography for the SNSF Project has been published recently (SNSF, 1980).

The second major initiative was the so-called OECD Study which provided quantitative data on the long-distance transport and deposition of atmospheric sulfur in both eastern and western Europe. Brynjulf Ottar (1976) of the Norwegian Institute for Air Research provided leadership for this continent-wide study of the long-range transport of sulfur pollutants. The results, which were published in 1977 by the Organization for Economic Cooperation and Development (OECD, 1977), confirmed many of the ideas presented in the Ecology Committee Bulletin and the Swedish Case Study regarding the long-range transport and the exchange of sulfur dioxide and sulfuric acid among the nations of Europe.

Recent analyses of air mass movements in North America have shown that similarly massive long-range transport is taking place among the various states within the United States and between the United States and Canada.

Scientific and Public Awareness in North America

Concern about acid precipitation and its ecological effects in North America developed first in Canada and then later in the United States. Initial interests were focused on the effects of sulfur dioxide fumigation and associated acid precipitation and heavy metal deposition near metal smelting and sintering operations (Katz, 1939; Gordon and Gorham, 1963), especially those near Sudbury, Ontario, the largest point source of sulfur in the world (Gorham and Gordon, 1960; Hutchinson and Whitby, 1974). During the early 1970s, interest spread to other parts of Canada as declining fish populations were discovered in more and more lakes remote from local sources of atmospheric sulfur in southern Ontario and Nova Scotia (Beamish and Harvey, 1972). Very early measurements of nitrogen compounds in rain and snow were made by Shutt and Hedley (1925); the first measurements of the pH of precipitation in Canada apparently were reported by Herman and Gorham (1957). The first national monitoring program for precipitation chemistry was initiated by the Canadian Department of the Environment in 1975. In recent years, several federal and provincial hearings on acid, sulfur, and heavy metal deposition have taken place in Canada as a result of growing public concern about acid precipitation. A summary of acid precipitation studies in Canada was published in 1976 by Summers and Whelpdale (1976).

The first detailed studies of precipitation chemistry in the United States were made by MacIntyre and Young (1923). This work was followed by the work of Junge and various collaborators during the 1950s (Junge and Werby, 1958). The first regional monitoring network for precipitation chemistry was maintained by a group of State Agricultural Experiment Station scientists during the period from 1953 to 1955 (Jordan et al., 1959). The first national monitoring program in the United States was established under the auspices of the Air Pollution Program within the Public Health Service Laboratory at Cincinnati, Ohio. The data for 1960 to 1966 were summarized by Lodge et al. (1968). As has been the case in all such studies in North America, these early programs were redirected and/or terminated so that no continuing records are available with which to evaluate long-term trends in the chemistry and acidity of precipitation.

Scientific and public interest in acid precipitation and its ecological consequences in North America were stimulated by Svante Oden in a series of 14 lectures at various institutions in the United States during the fall of 1971, and also by Torsten Ahl and Oden at the 19th International Limnological Congress in Winnipeg, Manitoba, in 1974. Further stimulus was provided by a series of publications by Gene Likens, Charles Cogbill, James Galloway, Carl Schofield, and others (Likens et al., 1972; Likens, 1976; Cogbill and Likens, 1974; Schofield, 1976; Galloway and Likens, 1976; Galloway et al., 1978; Likens et al., 1979). Experimental studies of various biological effects of acid precipitation were initiated at Cornell, North Carolina State, and other universities. David Shriner's (1976) dissertation demonstrated both direct injury to vegetation and various indirect effects through pathogens and parasites. Carl Schofield's (1976) research on extinction of fish populations in the Adirondack Mountains was especially alarming.

Growing awareness of important impacts of acid precipitation on fish populations and potential effects on forests led the U.S. Forest Service to sponsor the First International Symposium on Acid Precipitation and the Forest Ecosystem in Ohio in May 1975. The proceedings of this symposium and the associated workshop report were published by Dochinger and Seliga (1976a, 1976b). At Congressional hearings in July 1975, Cowling (1976) testified on the inadequacy of research in the United States on acid precipitation. Specifically, lack of a coordinated program of research on ecological effects and lack of a stable monitoring network were recognized as primary causes of our profound ignorance of acid precipitation. In the spring of 1976, a cadre of many scientists in various institutions and agencies throughout the United States began the process of creating a National Atmospheric Deposition Program (NADP) to meet these two critical needs (CSRS, 1977; Kennedy, 1977; Galloway and Cowling, 1978).

In the fall of 1977, the President's Council on Environmental Quality contracted with the NADP for the drafting of "A National Program for Assessing the Problem of Atmospheric Deposition (Acid Rain)." This publication (Galloway et al., 1978) provided the basis for a Presidential Initiative on Acid Precipitation which President Carter announced on August 2, 1979, in his Second Environmental Message (Carter, 1979). This initiative calls for a 10 year long, 10 million dollars per year program of research on the causes and consequences of acid precipitation in the United States.

THE PRESENT STATUS OF KNOWLEDGE CONCERNING ACID PRECIPITATION

During the summer of 1979, a cooperative agreement was established between the U.S. Environmental Protection Agency, the National Atmospheric Deposition Program, and North Carolina State University for the management and coordination of research on the Effects of Acid Precipitation on Aquatic and Terrestrial Ecosystems. One of the major responsibilities specified in this agreement is synthesis and integration of knowledge about acid precipitation and its ecological effects. Accordingly, an effort was made to summarize in a few statements of fact the present status of knowledge in this area of science. The most recent version of this status report is presented by Cowling in Chapter 1 of this book. These summary statements are providing a part of the background upon which scientists in various institutions and organizations in the United States and Canada will continue to build a scientific foundation for understanding the phenomenon of acid precipitation and its important ecological effects.

THE FUTURE OF ACID PRECIPITATION RESEARCH

We have come a long way since the earliest attempts of Smith, Gorham, and Odén to alert the scientific community and the public at large to the causes and consequences of acid rain. Much has been learned both in Europe and in North America. But much more remains to be learned about many aspects of the phenomenon and its effects. The pathway that has led to our present understanding has been illuminated by the remarkable insight of a few imaginative scientists. It has also been illuminated by many others who have filled lesser voids in our knowledge.

Today, public interest in acid rain research is at an all time high in many parts of the world. The challenge for us as scientists is to satisfy that interest by providing a still deeper understanding of the atmospheric processes, the soils transformations, the vegetational changes, the alterations in water chemistry, the materials effects, the physiological influences, and the nutrient as well as toxic effects of acid precipitation.

Research is the key to improved understanding. Improved understanding is the key to wiser public and private decisions that relate to the use of energy and to the quality of life in our society.

Let us get on with the job of learning so that the challenge of managing acid precipitation and its effects can begin as soon as possible.

ACKNOWLEDGMENTS

During the past ten years, I have developed very close friendships with many of the Scandinavian scientists who have expanded our present knowledge of acid precipitation and its various ecological effects. These contacts have inspired and encouraged my own efforts in the United States. Assistance in the preparation of this paper has been received from many of these colleagues, especially Svante Odén, Eville Gorham, William Dickson, Ivar Muniz, Hans Hültberg, Torsten Ahl, Lars Overrein, Arne Tollan, Gene Likens, Douglas Whelpdale, Tom Hutchinson, James Galloway, and Harriett Johnson.

Financial support for the preparation of this paper was provided, in part, under a cooperative agreement (CR 806912 01) between the United States Environmental Protection Agency and North Carolina State University.

LITERATURE CITED

Abrahamsen, G. 1980. Acid Precipitation, Plant Nutrients and Forest Growth. In: D. Drabløs and A. Tollan, ed., "Ecological Impact of Acid Precipitation," pp. 58-63. SNSF Project, Oslo, Norway.

Almer, B., W. Dickson, C. Ekstrom, E. Hornstrom, and U. Miller. 1974. Effects of Acidification on Swedish Lakes. Ambio 3:30-36.

Altshuler, A.P. and G.A. McBean (Co-chairman). 1979, 1980. The LRTAP Problem in North America: A Preliminary Overview Prepared by the United States-Canada Bilateral Research Consultation Group on the Long Range Transport of Air Pollutants. Atmospheric Environmental Service, Downsview, Ontario. 48 pp. Part I (1979), 48 pp.; Part II (1980), 39 pp.

Ambio. 1976. Report of the International Conference on the Effects of Acid Precipitation in Telemark, Norway. Ambio 5:200-252.

Atkins, W.R. 1922. Measurements of Acidity and Alkalinity of Natural Waters and Their Biological Relationships. Salmon and Trout Magazine:184-198.

Barrett, E. and G. Brodin. 1955. The Acidity of Scandinavian Precipitation. Tellus 7:251-257.

Beamish, R.J. and H.H. Harvey. 1972. Acidification of the LaCloche Mountain Lakes, Ontario, and Resulting Fish Mortalities. J. Fish. Res. Bd. of Can. 29:1131-1143.

Berry, R.L., D.M. Whelpdale, and H.A. Wiebe. 1976. An Evaluation of Collectors for Precipitation Chemistry Sampling. Presented at WMO Expert Meeting on Wet and Dry Deposition, Atmospheric Environmental Service, Downsview, Ontario, Canada.

Bolin, B., L. Granat, L. Ingelstom, M. Johannesson, E.Mattsson, S. Oden, H. Rodhe, and C.O. Tamm. 1972. Sweden's Case Study for the United Nations Conference on the Human Environment: Air Pollution Across National Boundaries. The Impact on the Environment of Sulfur in Air and Precipitation. Norstadt and Sons., Stockholm. 97 pp.

Bottini, O. 1939. Le pioggie caustiche nella regione vesuviana. (Acid Rain in the Region of Mt. Vesuvius.) Ann. Chim. Applic. 29:425-433.

Braekke, F.H. (ed.). 1976. Impact of Acid Precipitation on Forest and Freshwater Ecosystems in Norway. Res. Rep. 6/76, SNSF Project, Ås, Norway. 111 pp.

Brosset, C., K. Andreasson, and M. Ferm. 1975. The Nature and Possible Origin of Acid Particles Obseved at the Swedish West Coast. <u>Atmos.</u> <u>Environ.</u> <u>9</u>:631-642.

Brögger, W.C. 1881. Notes on a Contaminated Snowfall. <u>Naturen</u> <u>5</u>:47.

Carson, R. 1962. Silent Spring. Houghton Mifflin, Boston, Massachusetts. 368 pp.

Carter, J. 1979. The President's Environmental Program 1979. The President's Council on Environmental Quality, Washington, D.C. 57 pp.

Cohen, J.B. and A.G. Ruston. 1912. Smoke. Arnold, London. 88 pp.

Cogbill, C.V. and G.E. Likens. 1974. Acid Precipitation in the Northeastern United States. <u>Water</u> <u>Resources</u> <u>Res.</u> <u>10</u>:1133-1137.

Conway, E.J. 1942. Mean Geochemical Data in Relation to Oceanic Evolution. <u>Proc.</u> <u>Roy.</u> <u>Irish</u> <u>Acad.</u> <u>488</u>:119-159.

Cooperative State Research Service. 1977. Regional Research Project on Atmospheric Deposition: Chemical Changes in Atmospheric Deposition and Effects on Agricultural and Forested Land and Surface Waters in the United States. Cooperative State Research Service, U.S. Dept. Agriculture, Washington, D.C. 48 pp.

Cowling, E.G. 1976. Testimony on Research and Development Relating to Sulfates in the Atmosphere. <u>In</u>: "Research and Development Relating to Sulfate in the Atmosphere--Hearings Before the Subcommittee on the Environment and the Atmosphere of the Committee on Science and Technology," pp. 398-440. U.S. House of Representatives, 94th Congress, First Session, Washington, D.C. 1029 pp.

Cowling, E.G. 1980. Acid Precipitation and Its Effects on Terrestrial and Aquatic Ecosystems. <u>In</u>: T.J. Kneip and P.J. Lioy, eds., "Aerosols: Anthropogenic and Natural, Sources and Transport." <u>Annals</u> <u>New</u> <u>York</u> <u>Academy</u> <u>of</u> <u>Science</u> <u>338</u>:540-555.

Cragin, J.H., M. Herron, and C.C. Langway, Jr. 1975. The Chemistry of 700 Years of Precipitation at Dye-3 Greenland. U.S. Army Cold Regions Research and Engineering Laboratory, Hanover, New Hampshire. 18 pp.

Cronan, C.S. and C.L. Schofield. 1979. Aluminum Leaching Response to Acid Precipitation: Effects on High-Elevation Watersheds in the Northeast. Science 204:304-305.

Crowther, C. and H.G. Ruston. 1911. The Nature, Distribution, and Effects Upon Vegetation of Atmospheric Impurities In and Near An Industrial Town. J. Agric. Sci. 4:25-55.

Dahl, K. 1921. Undersøkelser over ørretens utdøen i del sydvestlige Norges fjeldvann. (Research on the Die-off of Brown Trout in Mountain Lakes in Southwestern Norway.) Norsk Jaeger - og Førenings. Tidsskrift 49:249-267.

Dahl, K. 1927. The Effects of Acid Water on Trout Fry. Salmon and Trout Magazine 46:35-43.

Dannevig, A. 1959. Nedborens innflytelse pa vassdragenes surhet og pa fiskebestand. (The Influence of Precipitation on the Acidity of Water Courses and On Fish Populations.) Jager og Fisker 3:116-118.

Dickson, W., C. Ekström, E. Hörnström, and U. Miller. 1973. Forsurningens inverkan på vastkustsjöar. (The Effects of Acidification of West-coast Lakes.) Swedish National Environmental Protection Board, Stockholm, Publ. No. 7. 97 pp.

Dochinger, L.S. and T.A. Seliga. 1976a. Proceedings First International Symposium on Acid Precipitation and the Forest Ecosystem. USDA Forest Service, Gen. Tech. Rept. NE-23. Northeast. For. Exp. Sta., Upper Darby, Pennsylvania. 1079 pp.

Dochinger, L.S. and T.A. Seliga. 1976b. Workshop Report on Acid Precipitation and the Forest Ecosystem. USDA Forest Service, Gen. Tech. Rept. NE-25. Northeast. For. Exp. Sta., Upper Darby, Pennsylvania. 18 pp.

Drabløs, D. and A. Tollan (eds.). 1980. Ecological Impact of Acid Precipitation. Proc. Intl. Conf., SNSF Project, Oslo, Norway. 383 pp.

Egnér, H., G. Brodin, and O. Johansson. 1955. Sampling Technique and Chemical Examination of Air and Precipitation. Kungl. Lantbrukhogskolan Ann. Uppsala, Sweden 22:369-410.

Electric Power Research Institute. 1980. The Integrated Lake-Watershed Acidification Study.   EPRI Report, EA-1825 and EA-1816. Palo Alto, California.

Emanuelsson, A., E. Ericksson, and H. Egner. 1954. Composition of Atmospheric Precipitation in Sweden. Tellus 6:261-267.

Erichsen-Jones, J.R.   1939.   The Relation Between the Electrolytic Solution Pressures of the Metals and Their Toxicity to the Stickleback. J. Exp. Biol. 16:425-437.

Eriksson, E. 1952. Composition of Atmospheric Precipitation. I.   Nitrogen Compounds. Tellus 4:215-231. II.   Sulfur, Chloride, Iodine Compounds. Bibliography. Tellus 4:280-303.

Eriksson, E. 1954. Report of an Informal Conference in Atmospheric Chemistry Held at the Meteorological Institute, University of Stockholm, May 24-26, 1954. Tellus 6:302-307.

Eriksson, E. 1959. The Yearly Circulation of Chloride and Sulfur in Nature: Meteorological, Geochemical, and Pedological Implications. Part I. Tellus 11:375-403.

Eriksson, E. 1960. The Yearly Circulation of Chloride and Sulfur in Nature: Meteorological, Geochemical, and Pedological Implications. Part II. Tellus 12:63-109.

Evelyn, J. 1661. Fumifugium. Bedel and Collins, London. 49 pp.

Galloway, J.N. and G.E. Likens.   1976.   Calibration of Collection Procedures for the Determination of Precipitation Chemistry. Water Air Soil Pollut. 6:241-258.

Galloway, J.N. and E.B. Cowling. 1978. The Effects of Precipitation of Aquatic and Terrestrial Ecosystems: A Proposed Precipitation Chemistry Network. J. Air Pollut. Control Assoc. 38:228-235.

Galloway, J.N., E.B. Cowling, E. Gorham, and W.W. McFee. 1978. A National Program for Assessing the Problem of Atmospheric Deposition (Acid Rain). Nat. Atm. Deposition Prog., Nat. Res. Ecol. Lab., Fort Collins, Colorado. 97 pp.

Galloway, J.N. and D.M. Whelpdale. 1979. An Atmospheric Sulfur Budget for Eastern North America. Atmos. Environ. 14:409-417.

Galloway, J.N., C.L. Schofield, G.R. Hendrey, N.E. Peters, and A.H. Johannes. 1980. Sources of Acidity in Three Lakes During Snowmelt. In: D. Drablфs and A. Tollan, ed., "Ecological Impact of Acid Precipitation," pp. 264-265. Proc. Intl. Conf., SNSF Project, Oslo, Norway.

Gordon, A.G. and E. Gorham. 1963. Ecological Aspects of Air Pollution From an Iron-sintering Plant at Wawa, Ontario. Can. J. Bot. 41:1063-1078.

Gorham, E. 1955. On the Acidity and Salinity of Rain. Geochim. Cosmochim. Acta 7:231-239.

Gorham, E. 1957. The Ionic Composition of Snow Lowland Lake Water From Cheshire, England. Limnol. and Oceanogr. 2:22-27.

Gorham, E. 1958a. Atmospheric Pollution by Hydrochloric Acid. Quart. J. Royal Meteorol. Soc. 84:274-276.

Gorham, E. 1958b. The Influence and Importance of Daily Weather Conditions in the Supply of Chloride, Sulphate, and Other Ions to Fresh Waters From Atmospheric Precipitation. Phil. Trans. Royal Soc. London (Series B) 247:147-178.

Gorham, E. 1958c. Free Acid in British Soils. Nature 181:106.

Gorham, E. 1958d. Bronchitis and the Acidity of Urban Precipitation. Lancet ii:691.

Gorham, E. 1961. Factors Influencing Supply of Major Ions to Inland Waters, With Special Reference to the Atmosphere. Geol. Soc. Amer. Bull. 72:795-840.

Gorham, E. 1965. Thomas Brotherton, Robert Hook, and Some Neglected Experiments in Plant Physiology During the Late Seventeenth Century. BioScience 15:412.

Gorham, E. 1980. A Historical Review of Interactions Between the Atmosphere and the Biosphere. Document submitted to the Committee on the Atmosphere and Biosphere, Commission on Natural Resources, National Academy of Sciences, Washingotn, D.C. 32 pp.

Gorham, E. and A.G. Gordon. 1960. The Influence of Smelter Fumes Upon the Chemical Composition of Lake Waters Near Sudbury, Ontario, and Upon the Surrounding Vegetation. Can. J. Bot. 38:477-487.

Grahn, O., H. Hultberg, and L. Landner. 1974. Oligotrophication—A Self-accelerating Process in Lakes Subjected to Excessive Supply of Acid Substances. Ambio 3:93-94.

Granat, L. 1972. Deposition of Sulfate and Acid With Precipitation Over Northern Europe. The University of Stockholm, Institute of Meteorology, Report AC-20, 30. 18 pp.

Granat, L. 1976. Principles in Network Design for Precipitation Chemistry Measurements. J. Great Lakes Res. 2:42-55.

Graunt, J. 1662. Natural and Political Observations Mentioned in a Following Index, and Made Upon the Bills of Mortality. Martin, Allestry and Dicas, London. 85 pp.

Greszta, J., R. Bitka, and R. Suchanek. 1978. Effect of Dusts of Nonferrous Metallurgy on the Germination of Pinus sylvestris, P. nigra, P. rigida, Picea abies, Quercus robur seeds. Sylwan 123(1):7-15.

Hagström, T. 1977. Grodornas försvinnande i en försurad sjö. (Disappearance of Frogs in an Acidified Lake.) Sveriges Natur 11:367-369.

Hales, S. 1727. Vegetable Staticks. Woodward and Peele, London. 376 pp.

Heck, W.W and C.S. Brandt. 1977. Effects on Vegetation: Native, Crops, Forests. In: A.C. Stern, ed., "Air Pollution," Vol. II, pp. 157-229. Academic Press, New York.

Hendrey, G.R. 1978. Limnological Aspects of Acid Precipitation. Proceedings International Workshop, Sagamore Lake, New York. Brookhaven National Laboratory, Upton, New York. 66 pp.

Henriksen, A. 1979. A Simple Approach for Identifying and Measuring Acidification of Freshwater. Nature 278:542-545.

Henriksen, A. 1980. Acidification of Fresh Waters - A Large Scale Titration. In: D. Drabløs and A. Tollan, eds., "Ecological Impact of Acid Precipitation," pp. 68-74. Proc. Intl. Conf., SNSF Project, Oslo, Norway.

Herman, F.A. and E. Gorham. 1957. Total Mineral Material, Acidity, Sulphur, Nitrogen in Rain and Snow at Kentville, Nova Scotia. Tellus 9:180-183.

Hooke, R. 1687. An Account of Several Curious Observations and Experiments Concerning the Growth of Trees. Phil. Trans. Royal Soc. London 16:307-313.

Houghton, H. 1955. On the Chemical Composition of Fog and Cloud Water. J. Meteorol. 12:355-357.

Howells, G. 1979. Ecological Effects of Acid Precipitation. Proceedings of a workshop held at Galloway, Scotland, Sept. 1978. General Electricity Research Laboratories, Leatherhead, Surrey, England.

Huckabee, J.W. and B.G. Blaylock. 1974. Microcosm Studies on the Transfer of Hg, Cd, and Se from Terrestrial to Aquatic Ecosystems. In: D.D. Hemphill, ed., "Proceedings Univ. of Missouri 8th Annual Conference on Trace Substances in Environmental Health," pp. 219-222. Columbia, Missouri.

Hultberg, H. and O. Grahn. 1976. Proceedings of the First Specialty Symposium on Atmospheric Contributions to the Chemistry of Lake Waters. J. Great Lakes Res. 2 (Suppl. 1):208.

Hultberg, H. and J. Stenson. 1970. Effects of Acidification on the Fish Fauna of Two Small Lakes in Bohuslän, Southwestern Sweden. Flora Fauna 1:11.

Hultberg, H. and A. Wenblad. 1980. Acid Ground Water in the Southwestern Area of Sweden. Proceedings of the International Conference on the Ecological Impact of Acid Precipitation, Sandefjord, Norway (in press).

Hutchinson, T.C. and M. Havas (eds.). 1980. Effects of Acid Precipitation on Terrestrial Ecosystems. Plenum Press, New York. 654 pp.

Hvatum, O.Ø. 1972. Fordeling av bly og en del andre tungmetaller i ombrogen torv. (Dispersion of Lead and Some Other Heavy Metals in Peat Bogs.) Jordundersøkelsens Saertrykk 175:59-70.

Interagency Task Force on Acid Precipitation. 1981. National Acid Precipitation Assessment Plan. Council on Environmental Quality, Washington, D.C. 129 pp.

Jensen, K.W. and E. Snekvik. 1972. Low pH Levels Wipe Out Salmon and Trout Populations in Southernmost Norway. Ambio 1:223-225.

Jonsson, B. and R. Sundberg. 1972. Has the Acidification by Atmospheric Pollution Caused a Growth Reduction in Swedish Forests? Research Note No. 20, Department of Forest Yield Research, Royal College of Forestry, Stockholm, Sweden. 48 pp.

Jordan, H.V., C.E. Bardsley, L.E. Ensminger, and J.A. Lutz. 1959. Sulfur Content of Rain Water and Atmosphere in Southern States as Related to Crop Needs. U.S. Dept. Agr. Tech. Bull. No. 1196. 16 pp.

Junge, G.E. and R. Werby. 1958. The Concentration of Chloride, Sodium, Potassium, Calcium, and Sulfate in Rain Water Over the United States. J. Meteorol. 15:417-425.

Katz, M. (ed.). 1939. Effect of Sulphur Dioxide on Vegetation. National Res. Council of Canada, Pub. No. 815, Ottawa. 447 pp.

Kennedy, V.C. 1977. Research and Monitoring of Precipitation Chemistry in the United States--Present Status and Future Needs. Federal Interagency Work Group of Precipitation Quality. Office of Water Data Coordination, U.S. Geological Survey, Reston, Virginia. 75 pp.

Kucera, V. 1976. Effects of Sulfur Dioxide and Acid Precipitation on Metals and Anti-Rust Painted Steel. Ambio 5:243-248.

Leivistad, H. and I.P. Muniz. 1976. Fish Kill at Low pH in a Norwegian River. Nature 251:391-392.

Lewis, W.M., Jr. and M.C. Grant. 1980. Acid Precipitation in the Western U.S. Science 207:176-177.

Likens, G.E., F.H. Bormann, and N.M. Johnson. 1972. Acid Rain. Environment 14:33-40.

Likens, G.E. 1976. Acid Precipitation. Chem. Eng. News 54(48):29-44.

Likens, G.E., R.F. Wright, J.N. Galloway, and T.J. Butler. 1979. Acid Rain. Sci. Amer. 241(4):43-51.

Liljestrand, H.M. and J.J.. Morgan. 1978. Chemical Composition of Acid Precipitation in Pasadena, California. Environ. Sci. Technol. 12:1271-1273.

Liljestrand, H.M. and J.J. Morgan. 1979. Error Analysis Applied to Indirect Methods for Precipitation Acidity. Tellus 31:421-431.

Linné, C.V. 1734. Dalaresa, Iter Dalekarlicum. Natur och Kultur, Stockholm (1964).

Lodge, J.P., J.B. Pate, W. Basbergill, G.S. Swanson, K.C. Hill, E. Lorange, and A.L. Lazrus. 1968. Chemistry of United States Precipitation. Final report on the National Precipitation Sampling Network. National Center for Atmospheric Research, Boulder, Colorado. 66 pp.

MacIntyre, W.H. and I.B. Young. 1923. Sulfur, Calcium, Magnesium, and Potassium Content and Reaction of Rainfall of Different Points in Tennessee. Soil Sci. 15:205-227.

Malmer, D. 1973. Om effecterna på vatten, mark och vegetation av ökad svaveltillförsel från atmosfären. (On the Effects of Water, Soil, and Vegetation from an Increasing Atmospheric Supply of Sulfur.) Statens Natur vardawerk, Lunds Universitet, Lund, Sweden. 125 pp.

McColl, J.G. and D.S. Bush. 1978. Precipitation and Throughfall Chemistry in the San Francisco Bay area. J. Environ. Qual. 7:352-357.

Miller, N.H.J. 1905. The Amounts of Nitrogen as Ammonia and as Nitric Acid, and of Chlorine in the Rainwater Collected at Rothamsted. J. Agric. Sci. 9:280-303.

Miller, J.M. 1979. The Acidity of Hawaiian Precipitation as Evidence of Long-Range Transport of Pollutants. In: "WMO-Symposium on the Long-Range Transport of Pollutants and Its Relation to General Circulation," pp. 231-237. WMO Report No. 538, Geneva, Switzerland.

Munshower, F.F. 1972. Cadmium Compartmentation and Cycling in Grassland Ecosystems in the Deer Lodge Valley. Ph.D. Thesis, Univ. Montana, Bozeman, Montana.

National Academy of Sciences. 1975. Atmospheric Chemistry: Problems and Scope. Com. Atm. Sci., National Research Council, Washington, D.C. 130 pp.

Odén, S. 1967. Dagens Nyheter. October 24, 1967.

Odén, S. 1968. The Acidification of Air and Precipitation and Its Consequences in the Natural Environment. Ecology Committee Bulletin No. 1. Swedish National Science Research Council, Stockholm. Translation Consultants, Ltd., Arlington, Virginia. 117 pp.

Odén, S. 1976. The Acidity Problem - An Outline of Concepts. Water Air Soil Pollut. 6:137-166.

Odén, S. 1979. The Sulphur Budget of Sweden During This Century. Nordic Hydrology:155-170.

Odén, S. and T. Ahl. 1970. Forsurningen av skandenaviska vatten. (The Acidification of Scandinavian Lakes and Rivers.) Ymer, Arsbok:103-122.

Organization for Economic Cooperation and Development. 1977. The OECD Programme on Long Range Transport of Air Pollutants. OECD, Paris, France.

Ottar, B.   1976.   Monitoring Long-Range Transport of Air Pollutants: the OECD Study. Ambio 5:203-206.

Overrein, L.N. 1972. Sulphur Pollution Patterns Observed; Leaching of Calcium in Forest Soil Determined. Ambio 1:145-147.

Parker, A. 1955. Report on the Investigation of Atmospheric Pollution. Department of Scientific and Industrial Research (U.K.), Report No. 27. 207 pp.

Pough, F.H. 1976. Acid Precipitation and Embryonic Mortality of Spotted Salamanders, Ambystoma maculatum. Science 192:68-70.

Reid, R.R. (ed.). 1980. Proceedings Action Seminar on Acid Precipitation. November 1979, Toronto, Canada. Federation of Ontario Naturalists, Don Mills, Ontario, Canada.

Rosenqvist, I.Th. 1977. Sur Jord - Surt Vann. (Acid Soil - Acid Water.) Bidrag til en analyse av geologiske materialers buffervirkning overfor sterke syrer i nedbør. Ingeniørforlaget A/S, Oslo, Norway. 123 pp.

Rossby, C.G. and H. Egner. 1955. On the Chemical Climate and Its Variation With the Atmospheric Circulation Pattern. Tellus 7:118-133.

Rusnov, P. 1919. The Removal of Lime From Soil by Smoke Gases Containing Sulfur Dioxide. Centr. Gesam. Forstw. 45:283-290.

Russell, E.J. and E.H. Richards. 1919. Amount and Composition of Rain Falling at Rothamsted. J. Agric. Sci. 9:309-337.

Schindler, D.W. 1980. Experimental Acidification of a Whole Lake: A Test of the Oligotrophication Hypothesis. In: D. Drabløs and A. Tollen, eds., "Proc. Intl. Conf. Ecological Effects of Acid Precipitation," pp. 370-374. SNSF Project, Oslo, Norway.

Schindler, D.W. et al. 1980. Special Issue on the Experimental Lakes Area. Can. J. Fish. Aquat. Sci. 37(3):313-558.

Schofield, C.L. 1976. Effects of Acid Precipitation on Fish. Ambio 5:228-230.

Schuck, H.J., U. Blumel, I. Geier, and J.T. Schutt. 1979. Schadbild und atiologie des tannesterbens. Eur. J. For. Path. 10:125-135.

Shriner, D.S. 1974. Effects of Simulated Rain Acidified With Sulfuric Acid on Host-Parasite Interactions. Ph.D. Dissertation, North Carolina State University, Raleigh. 79 pp.

Shriner, D.S. 1976. Effects of Simulated Rain Acidified With Sulfuric Acid on Host-Parasite Interactions. In: L.S. Dochinger and T.A. Seliga, eds., "Proc. First Int. Symp. on Acid Precipitation and the Forest Ecosystem," pp. 919-925. USDA For. Serv. Gen. Tech. Rep. NE-23, Upper Darby, Pennsylvania.

Shriner, D.S. 1978. Effects of Simulated Acidic Rain on Host-Parasite Interactions in Plant Diseases. Phytopathology 68:213-218.

Shriner, D.C. and J.W. Johnston. 1981. The Effects of Simulated Acidified Rain on Nodulation of Leguminous Plants by Rhizobium spp. Environ. and Exp. Botany 21:199-209.

Shriner, D.S., C.R. Richmond, and S.E. Lindberg (eds.). 1980. Atmospheric Sulfur Deposition: Environmental Impact and Health Effects. Ann Arbor Science Pub., Ann Arbor, Michigan. 568 pp.

Shutt, J.T. and B. Hedley. 1925. The Nitrogen Compounds in Rain and Snow. Trans. of Roy. Soc. of Canada 19:1-10.

Smith, R.A. 1852. On the Air and Rain of Manchester. Mem. Lit. Phil. Soc. Manchesters (Series 2) 10:207-217.

Smith, R.A. 1872. Air and Rain: The Beginnings of Chemical Climatology. Longmans, Green, London. 600 pp.

SNSF. 1980. Acid Precipitation - Effects on Forests and Fish of the SNSF Project 1972-1980. SNSF Research Report 19, SNSF Project, Oslo, Norway.

Sørensen, P. 1909. Enzymstudien. II. Uber die konzentration bei enzymatischen progressen. Biochemische Zeitschrift 21:131-200.

Summers, P.W. and D.M. Whelpdale. 1976. Acid Precipitation in Canada. Water Air Soil Pollut. 6:447-455.

Sunde, S.E. 1926. Surt vand draeper laks-og ørretyngel. (Acid Water Kills Salmon and Trout Fry.) Norsk Jaeger-og Fiskeforenings Tidsskrift 55:1-4.

Tamm, C.O. 1953. Tillvaxt, producktion och näringsekolog: i mattor av en skogsmossa, Hypocomiuim splendens. (Growth, Yield and Nutrition in Carpets of a Forest Moss.) Meddel. Statens Skogsforskningsinstitut 43(1):1-140.

Tamm, C.O. 1958. The Atmosphere. Handbuch fur Pflanzen-physiologie 6:233-242.

Tyler, G. 1972. Heavy Metals Pollute Nature, May Reduce Productivity. Ambio 1:52-59.

Tyler, G. 1976. Heavy Metal Pollution, Phosphatase Activity, and Mineralization of Organic Phosphorus in Forest Soils. Soil Biol. Biochem. 8:327-332.

Tyler, G. 1980. Metals in Sporophores of Basidiomycetes. Trans. Brit. Myc. Soc. 24:41-49.

Ulrich, B., R. Mayer, and P.K. Khanna. 1980. Chemical Changes Due to Acid Precipitation in a Loess Derived Soil in Central Europe. Soil Sci. 130:193-199.

Viro, P.J. 1953. Loss of Nutrients and the Natural Nutrient Balance of the Soil in Finland. Comm. Inst. For. Fenn. 42(1):5-51.

Way, T. 1855. The Atmosphere as a Source of Nitrogen to Plants. Roy. Agr. Soc. Jour. 16:249-267 (see also 17:123-162 and 17:618-621).

Wetstone, G. 1980. Air Pollution Control Laws in North America and the Problem of Acid Rain and Snow. Environ. Law Reporter X(2):50001-50020.

Whitby, L.M. and T.C. Hutchinson. 1974. Heavy Metal Pollution in the Sudbury Mining and Smelting Region of Canada. II. Soil Toxicity Tests. Environ. Conserv. 1(3):191.

Wiklander, L.   1973.   The Acidification of Soil by Acid
Precipitation. Grundförbättring 26(4):155-164.

Wittwer, S.H. and M.J. Bukovac. 1969. The Uptake of Nutrients
Through Leaf Surfaces. In: K. Scharrer and H. Linser, eds.,
"Handbuch der Pflanzenernahrung unds Dungung," pp.235-261.
Springer-Verlag, New York.

Wood, T. and F.H. Bormann. 1974. The Effects of an Artificial
Acid Mist Upon the Growth of Betula alleghaniensis Britt.
Environ. Pollut. 7:259-268.

PART 2

GENERAL OVERVIEW: SOURCES OF
ACID PRECIPITATION, MONITORING,
AND RELATED PROBLEMS

CHAPTER 4

THE ELECTRIC POWER INDUSTRY EMISSIONS DATA: PAST, PRESENT AND FUTURE FOR THE GREAT LAKES AND NORTH AMERICA

William L. Keepers
Wisconsin Power & Light Company
222 West Washington Avenue
Madison, Wisconsin

INTRODUCTION

Industry, government, and their scientists must join in an effort to properly assess the acid rain problem. These elements of society have a responsibility to the public to present the issues in proper perspective, provide a forum for their discussion, collectively determine policy, and develop appropriate control strategies if needed. I believe that it is absolutely imperative that the major parties involved make a special effort to communicate and seek consensus in policy making. This is particularly so when faced with a phenomenon as complex as acid deposition where the economic and environmental stakes are potentially very great. Also, since the acid rain phenomenon is not restricted to political boundaries, effective controls of acid-forming pollutants in North America will require state/federal agreements and agreements with Canada. This underscores the importance of maintaining good communications and realistic attitudes as we shape policy, research, and control programs.

One aspect of the acid rain issue that almost everyone agrees upon is that not enough is known about atmospheric chemistry, transport phenomena, and environmental effects to confidently make the hard decisions that may be required. In particular, quantification of all of the effects of acid deposition is lacking. Also, data on the amounts of acid precursors emitted to the atmosphere from both natural and man-made sources are not adequate to allow rigorous treatment of the cause-and-effect relationships.

Because of these uncertainties, dealing with the acid deposition problem cannot be handled unilaterally. Therefore, before discussing the emissions of sulfur and nitrogen compounds associated with the production of electrical energy, I will describe the cooperative efforts underway in the state of Wisconsin to investigate the question of acid deposition.

WISCONSIN'S ACID DEPOSITION PROBLEM

Background and Purpose

The state of Wisconsin cannot act entirely independently of the region or the nation in looking at the acid rain problem or formulating possible control schemes. Nevertheless, since Wisconsin is one of the areas identified as being especially vulnerable to lake acidification, state government, and those industries that are the major point sources of sulfur oxides (SO_x) and nitrogen oxides (NO_x), are keenly aware of their responsibilities in averting the economic consequences of potential severe damage to the natural resources of the state. Both the state regulatory agencies and the regulated businesses also are cognizant of the ramifications of further air pollution controls in an economy which is in need of revitalization. Wisconsin's manufacturing industry, like that of other Great Lakes states, is heavily oriented toward the automotive and building sectors, both of which are experiencing severe economic difficulties.   Any   further   burdens   or   constraints   on manufacturing activities may have serious additional adverse impacts on employment and income. On the other hand, the Wisconsin economy and that of several of our neighboring states in the Great Lakes region also are highly dependent on a healthy tourist   industry;   an   industry   that   relies   heavily   on   the continued attractiveness of our northern recreational lakes region.

The balancing of these interests and other socioeconomic factors (especially of increasing productivity and reducing inflation) against increased pollution controls and energy costs, which go hand in hand, presents a major challenge to all of us.

The possibility of an unaffordable or even an unnecessary regulatory burden is not being ignored nor taken lightly. Government and industry in Wisconsin have joined to: first, assess the potential for damage to Wisconsin recreational lakes, forests, and croplands and the industries dependent on these resources; second, to ascertain to the extent possible the actual damage to date and likely in the next 10 to 20 yr and the mechanisms/sources responsible for such adverse effects; and third, to develop policies and programs that would consider the overall economic and environmental effects on citizens of the state and the effects that activities in the state may have beyond its borders.

Organization

The parties involved in this joint effort are the Wisconsin Department of Natural Resources (DNR) and the Public Services Commission (PSC), together with all major Wisconsin electric utilities, viz., Wisconsin Power & Light Company, Wisconsin Electric Power Company, Wisconsin Public Service Company, Northern States Power-Wisconsin, Madison Gas & Electric Company, Dairyland Power Cooperative, and Lake Superior District Power Company, who have embarked on a joint research program. The Upper Peninsula Power Company in Michigan is participating in the program. The paper industry, through the Wisconsin Paper Council, also may participate.

The technical coordination of the research is provided by scientists and engineers from the DNR, PSC, and the electric utilities. This group is identified as the Joint Technical Review Committee (JTRC) and is charged with developing the conceptual framework of the research program (Figure 1). The University of Wisconsin, the U.S. Geological Survey, and the Electric Power Research Institute also are involved as investigators and/or in an advisory capacity in the watershed studies.

Subcommittees of the JTRC have been established to address specific research areas. Those areas and their main objectives are:

1. Deposition: Monitor chemistry of precipitation at several locations; integrate with other monitoring in the region.
2. Transport: Identify sources of acid precursors by precipitation event case studies, synoptic meteorology, and long-range transport models.
3. Sources and Controls: Identify point and area sources and improve emission inventories to allow predictive modeling and formulation of control strategies.
4. Watersheds: Investigation of the interrelationships between precipitation chemistry and the effects of watershed factors (soils, geology, hydrology, vegetation, and land use) on lake acidification.

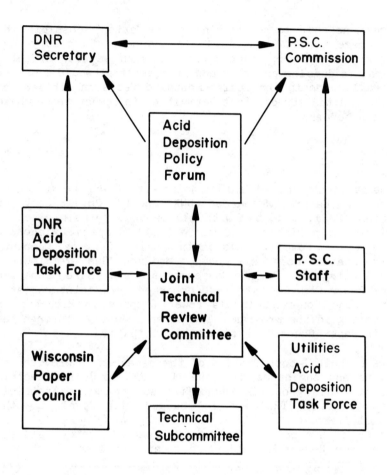

Figure 1. *Organization of the Wisconsin acid deposition
program.*

5. Susceptibility: Evaluate susceptibility of Wisconsin resources to damage from acid deposition.
6. Mitigation: Investigate methods to mitigate or reverse the effects of acid deposition.

Subcommittee responsibilities include the detailed planning of research projects and management. Each subcommittee has representation from the agencies and the utilities.

Currently, an Acid Deposition Policy Forum is being established which is comprised of people representing the major constituencies within the state that have a vital interest in the acid deposition issue. Represented are, the:

1.   Governor's Office,
2.   State Senate,
3.   State assemblies of Southern and Northern Wisconsin,
4.   Fish and game organizations,
5.   Tourist industry,
6.   Agribusiness,
7.   Environmental organizations,
8.   Paper industry,
9.   Electric utilities,
10. Public Service Commission (ex officio), and
11. Department of Natural Resources (ex officio).

The mission of the Policy Forum is to give policy advice to both the DNR and the PSC regarding the subject of acid deposition. The components of this mission are:

1.   Identification and evaluation of major policy issues of concern to Wisconsin and its resources. For example:
     a. Relationships between resource protection, economic health, and energy supply.
     b. Feasibility of alternative control strategies and mitigative measures.
     c. Determine need for controls and how they will be paid for.
     d. Role of various governmental units in implementing policies, laws, and regulations.
2.   Provide a forum or a focal point for comment by the public and special interest groups.
3.   Review the research program developed by the JTRC for completeness, for policy ramifications, and for adequacy of funding.
4.   Provide an arena for policy discussion in which participants identify areas of agreement and disagreement. It is hoped that each delegate to the forum will reflect the overall opinion or concerns of their respective constituencies. The Policy Forum will communicate its opinions to the DNR and PSC for guidance in implementing control programs that may be necessary. It also would provide guidance on interfacing with federal programs.

The role of the electric utilities in this organization and program is through the Utilities Acid Deposition Task Force (Figure 1). Representatives of the task force represent each major Wisconsin electric utility (also, the Upper Peninsula Power Company) on the Joint Technical Review Committee. These are technical people, scientists, and engineers with a broad spectrum of disciplinary backgrounds, who are working together with other technical experts from the DNR and PSC staffs.

Policy input from the electric utilities to the Policy Forum is by a person representing the collective opinions of the Utilities Acid Deposition Policy Committee, which is made up of company officers, typically vice presidents responsible for electric power production. This committee also provides guidance to the Joint Technical Review Committee.

ASSESSMENT OF ACID DEPOSITION IN WISCONSIN

Existing Information

The Wisconsin Department of Natural Resources Acid Deposition Task Force released a report in June, 1980, which summarized most of what is known about the subject as it pertains to the state of Wisconsin. In that report, the DNR concluded that Wisconsin has perhaps more aquatic resources susceptible to acidification from acid deposition than most states. Because of the geology, climate, and vegetation of northern Wisconsin, thousands of lakes apparently contain very little buffering capacity. These characteristics are similar to those of lakes in Scandinavia and the Adirondacks of the eastern United States where fish populations apparently have been lost due to acidification.

The DNR further concluded that, although data are quite limited, precipitation in northern Wisconsin is frequently more acid than that of "normal" rain. With a continued shift to greater dependence on coal fuels for electric generation and increased electrical energy needs in the state and the country, the DNR is apprehensive that significant increases in acid deposition could occur.

They further concluded that the state of Wisconsin needs answers to several questions. Among these are:

1. To what extent is Wisconsin responsible for its own acid deposition?
2. How do Wisconsin's emissions affect other states and how does the long-range transport of pollutants affect Wisconsin?

ELECTRIC UTILITY EMISSIONS

The U.S. Environmental Protection Agency (U.S. EPA) considers man-made emissions of sulfur and nitrogen air pollutants to be major sources of acid rain precursors. Although no firm, quantitative relationships have been established between pollutant emissions and the measured acidity of precipitation, contributions to acidity from man-made sources clearly occur. Contributions from natural sources are also important. In any specific locality it will be difficult (if not impossible) to verify the contributions to precipitation acidity from each of these two major source categories. That is, a precise assignment of specific contributions coming from a group of electric power plants, or from any individual point source, whether it be a power plant or some other facility, cannot be made unless the background (natural sources plus long-range transport from man-made sources) can be established. The magnitude and distribution of acid precursors from all sources must be taken into account before conclusions can be drawn regarding electric power plant emissions and control strategies on a state, regional, or national level.

Total Atmospheric Loadings of Acid Precursors

If man-made sources are to be incorporated into transport equations to determine possible contribution to specific acid deposition events and deposition loading in sensitive areas, a knowledge of natural background levels is necessary. Typically, a mass balance approach is taken where pollutants in a known air volume are examined in regard to chemical transformation and transport, scavenging and removal by washout, dry deposition, or other mechanisms.

It it not within the scope of this paper to review natural sources. A comprehensive review is provided in the Department of Energy's (1980) Acid Rain Information Book. Suffice it to say that considerable disagreement exists as to the magnitude of natural sources and how they compare with man-made sources. Generally, on a global basis, it is estimated that natural sources account for 50 to 90 percent of atmospheric sulfur, and anywhere from 40 percent to 95 percent of NO_x emissions.

The extreme disparity in these estimates is due in large part to differing assumptions on quantities of sulfur emitted by volcanoes, and the amounts of sulfur and ammonia from biogenic sources. These uncertainties have been discussed by various people, for example, Kellogg et al. (1972) and Shinn and Lynn (1979). Nor is resolution near at hand. Stephen Budiansky (1980) summarized the problem as follows: "Inconsistent data are the rule in this field of 'biogenic' emissions; attempts to resolve the inconsistencies bring into question measurement techniques, extrapolations, and our understanding of atmospheric chemistry. Much more research needs to be done, and no one is now taking bets on which way the answers will finally fall."

Regardless of the magnitude of natural emissions of major acid precursors, the tendency is for natural sources to be globally distributed, whereas man-made emissions tend to be much more local and concentrated. This is not to say, however, that man-made emissions are dominantly responsible in all locations or for any given acid precipitation event.

Emissions Inventories

At the outset, I believe it is fair to say that emission inventories are not very good, that the data available can be rather confusing and misleading in their presentation because of the very nature of the data and their derivation.

The data set that is most frequently used and is most extensive is the National Emissions Data System (NEDS) compiled by the U.S. EPA. Data from the 1977 NEDS are the most recent available for analysis (U.S. EPA, 1978). These data are estimated figures and open to question for accuracy. Some reasons for the inherent inaccuracy are that emissions from point sources such as power plants are calculated based on average fuel sulfur content values (of questionable accuracy), tons of fuel consumed, and EPA "emission factors," that contain subjective assumptions on such items as sulfur retention in the ash. Also, stack monitors and ambient air monitoring equipment routinely measure SO_2, whereas the SO_x estimates in NEDS include more substances than SO_2. This also applies to emissions of nitrogen compounds. Therefore, the monitoring that is done does not provide information which can be used directly to improve emission estimates.

Data are presented in Table 1 which provide the NEDS estimates for 1977 for all man-made sources. It is estimated that electric utilities accounted for 17.6×10^6 tonne SO_x out of a total of 27.4×10^6 tonne from man-made sources, or 64 percent.

Table 1.

Nationwide Emission Estimates, 1977 (U.S. EPA, 1978).

Source Category	SO_x	NO_x
	(10^6 tonne/yr)	
Transportation	0.8	9.2
Highway vehicles	0.4	6.7
Non-highway vehicles	0.4	2.5
Stationary fuel combustion	22.4	13.0
Electric utilities	17.6	7.1
Industrial	3.2	5.0
Residential, commercial and institutional	1.6	0.9
Industrial processes	4.2	0.7
Chemicals	0.2	0.2
Petroleum refining	0.8	0.4
Metals	2.4	0
Mineral products	0.6	0.1
Oil and gas production and marketing	0.1	0
Industrial organic solvent use	0	0
Other processes	0.1	0
Solid waste	0	0.1
Miscellaneous	0	0.1
Forest wildfires and managed burning	0	0.1
Agricultural burning	0	0
Coal refuse burning	0	0
Structural fires	0	0
Miscellaneous organic solvent use	0	0
Total	27.4	23.1

Note: A zero indicates emissions of less than 50,000 tonne/yr.

Further, electric utilities accounted for an estimated 7.1 x 10^6 tonne of NO_x out of a total of 23.1 x 10^6 tonne from man-made sources, or 31 percent.

Frequently, the news media will state that electric utilities are responsible for two-thirds of the acid precipitation! That statement only approaches the truth if one assumes that there is absolutely no contribution from natural sources to the acidity of precipitation and that emissions data for all man-made sources are accurate and complete. Such assumptions are not realistic.

In Table 2, the contribution of electric utilities' emissions is presented relative to other man-made sources and various assumed contributions from natural sources. It is to be noted that, with respect to SO_x emissions, even if one assumes a 75 percent man-made contribution to precipitation acidity, controlling electric utility emissions only addresses <u>half</u> of the problem associated with SO_x emissions and a <u>quarter</u> of the problem with NO_x emissions.

The examples above, of course, use the nationwide figures for emissions. Those who wish to blame electric utilities only usually bring up emission estimates for Ohio because of the large concentration of coal-fired power plants in that region. Ohio, in 1977, accounted for over 10 percent of total SO_x emissions in the United States. Electric utilities in Ohio accounted for 82 percent of the state emissions. If one then assumes that 90 percent of atmospheric SO_x loading in this region is man-made, then three-quarters of the sulfur loading may be allocated to power plant emissions. Even with this worst-case scenario, given the fact that acid production from sulfur oxidation to the sulfate anion is not complete or accomplished without some passage of time and transport over consid-

Table 2.

Emissions of SO_x and NO_x from Electric Utilities (1977) in Comparison to Other Man-Made Sources and to Natural Resources

	SO_x	NO_x
Percent of total inventory of man-made sources	64.0%*	31.0%*
Percent of total atmosphere loading over North America assuming 10% is from man-made sources	6.4%	3.1%
Percent of total atmosphere loading over North America assuming 30% is from man-made sources	19.2%	9.3%
Percent of total atmosphere loading over North America assuming 75% is from man-made sources	48.0%	23.0%

*Based on NEDS data in Table 1.

erable distances (and, therefore, dilution with other possible sources), one could not fairly attribute the major cause of precipitation acidity to utility emissions in downwind regions.

These statements are not made to take the "heat off" electric utilities but rather to convey the message that if controls are needed to correct the acid rain problem, then the correction cannot be accomplished by controlling only emissions from power plants.

Trends in Electric Utility Emissions

The trends of SO_x and NO_x emissions since 1970 (Table 3) reflect two counteracting factors. Implementation of air regulations along with relatively stable industrial activity produced a downward trend in total emissions of both pollutants in the 1973 to 1975 period. This trend was reversed in 1976 by growth in overall industrial activity, including new electric generation capacity and increases in highway vehicular traffic.

An unrelenting force driving emissions upward in the United States is our growing population. Nationally, population in the year 2000 is estimated to increase by 18 percent over 1979, reaching 260 million. Emissions of both SO_x and NO_x correlate fairly well (R = 0.88) with increases in population (U.S. DOE, 1980). Consequently, it may be anticipated that emissions will increase in regions where population growth is greatest, despite regulatory controls.

Table 3.

Summary of National Emission Estimates, 1970-77
(U.S. EPA, 1978)

Year	SO_x	NO_x
	----(10^6 tonne/yr)----	
1970	29.8	19.6
1971	28.3	20.2
1972	29.6	21.6
1973	30.2	22.3
1974	28.4	21.7
1975	26.1	21.0
1976	27.2	22.8
1977	27.4	23.1

Utility SO_x emissions (Figure 2) in the nation are anticipated to remain relatively constant and NO_x emissions (Figure 3) are anticipated to increase slightly over the remainder of the century (U.S. DOE, 1980).

The scenarios presented in Figures 2 and 3 are based on analyses by Hendry and Lipfert (1980) at the Brookhaven National Laboratory under sponsorship of the Department of Energy. The shaded areas in Figure 3 represent a range of scenarios for

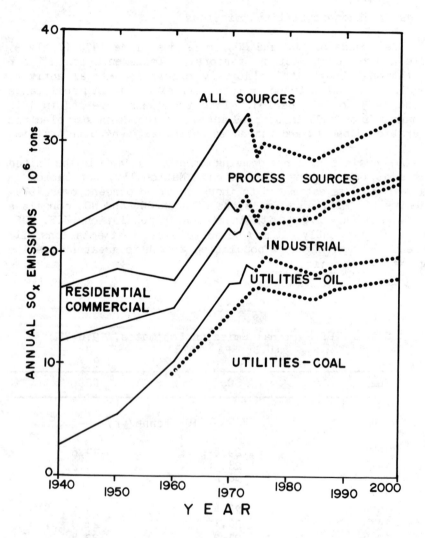

Figure 2.   *Sulfur oxides emission trends (Hendry and Lipfert, 1980).*

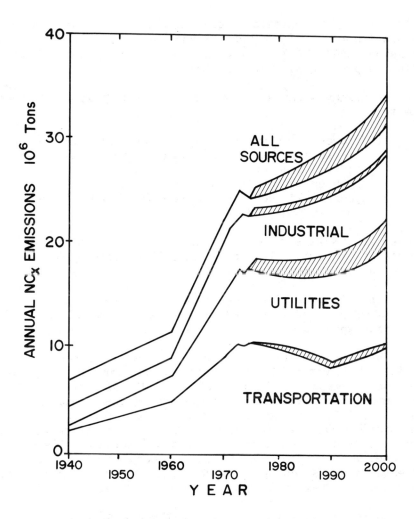

*Figure 3.   Nitrous oxides emission trends (Hendry and
            Lipfert, 1980).*

NO_x, which vary depending on assumptions made with respect to
transportation and possible emission controls on automobiles.
Other variations are associated with uncertainties of retrofit
controls on existing power plants; however, emissions shown
probably represent upper limits.

Another method of predicting total SO_x emissions is by
using the relationship between historic SO_x emissions and total
population of the United States. For example, over the 1974-77
period, this relationship was quite consistent at approximately

0.13 tonne SO_x emissions per capita. With a projected population of 260.4 million in the year 2000, the total SO_x emissions would be 33.8 million tonne. This compares closely with the projection shown in Figure 2.

In the Great Lakes region, trends may be slightly downward due to:

1. lower population growth rates relative to other regions, such as the southern and western portions of the country,
2. conservation efforts, and
3. depressed economic activity (at least in the first half of the 1980s).

These factors have already resulted in the lowering of load forecasts by many midwest utilities and the deferment or cancellation of new generating units.

Other factors affecting emissions are federally mandated fuel switching, and state/federal policies and public attitudes regarding generation of electricity by nuclear power.

The controversial "oil-backout" program pitted U.S. EPA against DOE with U.S. EPA voicing concerns about greatly increased emissions (U.S. EPA quoted 25% increase) and exacerbation of acid rain. In a study by the Edison Electric Institute (EEI), it was concluded that the increase in total emissions resulting from reconversion of generating units to coal resulted in negligible increases (less than 1% for the affected units). This change can be accomplished without major effects on emissions by using coal with lower sulfur contents than the oil fuels presently used (EEI, 1980).

In regard to nuclear units in the mix of energy sources to the year 2000, the projections made in 1979 have been substantially reduced from roughly 25 percent of the probable mix of net generating capacity down to 15 percent by the end of the century (Electrical World, 1980). The reduction is made up primarily by fossil-fueled steam electrical generating capacity. Obviously, shifting away from nuclear power has resulted in greater air pollution despite increased controls, such as scrubbers, on fossil units.

Various observers have pointed to greatly increased dependence on coal (for example, going from the present national usage rate of approximately 500 million tons per year to approximately 1250 million tons per year by the end of the century) as an argument that acid rain will be greatly intensified. However, such arguments ignore trends in coal use toward greater dependence on low-sulfur coals, and especially coals that have high alkaline ash contents which result in greater sulfur retention with the ash. For example, highly alkaline lignites may retain over 60 percent of the available sulfur in the boiler ash.

Retirement of aging high-temperature cyclone boilers and replacement with the preferred pulverized-coal boilers will gradually reduce NO_x emissions.

Thus, it is believed that this combination of factors, along with the present more stringent emission control levels applicable to new units, will result in essentially stabilized emissions levels for SO_x and NO_x over the next two decades despite increases in coal use and increased generating capacity.

CONCLUSION

In conclusion, I would like to emphasize that all projections of fuel mix, energy use, generation capacity additions, and emissions of SO_x and NO_x by the electric utility industry are extremely dependent on the vitality of the national economy.

Major federal policy changes, e.g., the emphasis on supply-side economics by the Reagan Administration and renewed emphasis on nuclear power, together with achievements in solar and wind energy technologies, may have a significant impact on the use of coal to the year 2000. The financial capability of utilities to add new capital-intensive generation capacity ultimately may be the most important factor influencing emissions of air pollutants over the next decade or so. The costs of air pollution controls are inseparable from economic and financial considerations.

The environmental controls industry anticipates a total capital expenditure of 1787 million dollars in 1981 for air pollution control equipment (electrostatic precipitators, fabric filters, mechanical collectors, and wet scrubbers) by the electric generating utilities, resulting from the establishment of standards for sulfur oxide emissions and particulates (DeLaRue, 1981). "Legislated" standards beyond existing air standards stemming from acid deposition concerns could increase capital requirements substantially. Retrofitting of older plants, if required, could be a particularly onerous burden since retrofitting is estimated to be 15 to 20 percent more costly than installation of pollution control equipment on new facilities. Also, utilities would have to consider potential retrofits on each facility on a case-by-case basis, taking into account such factors as the anticipated remaining life, the need for that capacity, the cost of replacing the old units with new units, the desired reserve margin in their systems, the availability of purchase power, etc.

With the present evidence suggesting that emission levels will not be greatly increasing (perhaps stable or even decreasing) over the next two decades, we believe that the existing air standards and the other factors discussed previously provide adequate protection against "catastrophic damage" caused by acid deposition within this period.

The electric industry is concerned about acid deposition. An extensive research program is being conducted by the Electric Power Research Institute (EPRI) in cooperation with the federal government and the scientific community.

As an industry, we are also proud of our accomplishments in providing reliable electrical service, and it is our objective to continue to do so. From a general end-use viewpoint, electricity is a cost-effective, high-quality, and clean energy. Throughout the past century, the growth rate of electricity consumption has exceeded the growth rate of total energy consumption. This trend is expected to continue. However, the historic case in rate of growth is unlikely to occur between now and the year 2000. Projected growth rates range from 3.6 to 5.2 percent per year, depending on the economic case assumed (EPRI, 1980). Some of the required capacity will be met by adding coal-fired units--a part also will be added by alternative energy sources including solar, wind, and hydroelectric.

Finally, I would like to reiterate the position on acid deposition generally held by the electric utility industry. First, we believe that a great deal of research must be conducted before:

1. the contribution of utility boiler emissions to acid rain is quantified;
2. the effects of acid rain on lake acidification and related phenomena are clearly established; and
3. geographic source-receptor relationships are unequivocally identified.

Second, we are not advocating research simply as a delaying tactic, but rather because the preponderance of evidence suggests that it would not be prudent for the United States to move headlong into a potentially very costly controls program without a more accurate definition of the problem. We believe that existing uncertainties, along with the fact that the existing air regulations applicable to new sources, coupled with the gradual replacement of older, dirtier units, are two important reasons why research now planned can shed some light on some of the more critical questions; i.e., such as whether acid rain is indeed increasing in extent and intensity.

For these reasons, viz., uncertainty of the present scientific evidence, uncertainty of future economic activity, and lack of immediate urgency to take corrective action, we believe that there is adequate time to jointly review the evidence, perform the necessary research, thoughtfully consider the costs and the benefits, and work together to develop policies that are in the best interests of the nation.

LITERATURE CITED

Anon. 1980. The 31st Annual Electrical Industry Forecast. Electrical World 194:55-70.

Budiansky, S. 1980. Biological Contributions to Air Pollution. Environ. Sci. Technol. 14:901-903.

DeLaRue, R.E. 1981. 1981--The Transition Year for the Environmental Control Markets? Pollut. Engineer. 13:25-29.

EEI. 1980. A Preliminary Analysis of Estimates of Emissions from Reconversion of Coal Capable Utility Boilers. Prepared for Edison Electric Institute by ICF Incorporated, April 23, 1980. 12 pp.

EPRI. 1980. 1981-1985 Research and Development Program Plan: Overview and Strategy. Prepared by EPRI Policy Planning Division, Electric Power Research Institute, Palo Alto, California 94304. 120 pp.

Hendrey, G.R. and F.W. Lipfert. 1980. Acid Precipitation and the Aquatic Environment. Brookhaven National Laboratory. Presented to the Committee on Energy and Natural Resources, United States Senate, May 28, 1980.

Kellogg, W.W., R.D. Cadle, E.R. Allen, A.L. Lazrus, and E.A. Martell. 1972. The Sulfur Cycle. Science 175:587-596.

Shinn, J.H. and S. Lynn.   1979.   Do Man-Made Sources Affect the Sulfur Cycle of Northeastern   States?   Environ. Sci. Technol. 13:1062-1067.

U.S. DOE.   1980.   Acid Rain Information Book.   Draft Final Report.   Prepared by GCA Corporation for the Office of Technology Impacts, Regulatory Analysis Division, U.S. Department of Energy, Washington, D.C.   Contract No. DE-AC02-79EV10273, Task Order No. 5.   221 pp.

U.S. EPA.   1978.   National Air Quality Monitoring and Emissions Trends Report, 1977.   U.S. Environmental Protection Agency, EPA-450/2-78-052, Washington, D.C.   60 pp.

WDNR.   1980.   A Review of Acid Deposition in Wisconsin; Recommendations for Studying and Solving the Problem. Prepared by DNR Acid Deposition Task Force, T.B. Sheffy, Chairman, Wisconsin Department of Natural Resources, Madison, Wisconsin. 46 pp.

CHAPTER 5

THE ECOLOGICAL CONSEQUENCES OF ACID
DISCHARGES FROM INDUSTRIAL SMELTERS

Thomas C. Hutchinson
Department of Botany
University of Toronto
Toronto, Ontario M5S 1A1
Canada

INTRODUCTION

A great deal of concern is being shown about the impacts
of acid precipitation on aquatic and terrestrial ecosystems.
It is generally recognized that the acid rain problem is the
consequence of long-distance (regional) movements of polluted
air masses to ecologically susceptible areas. A predominant
role of sulfur through sulfur dioxide emissions has been related
to the acidity of rain and snow, though nitrogen oxide emissions
have also been noted as a variable but significant component.
It is, therefore, useful to look at studies made over the past
70 yr in areas where sulfur dioxide emissions have been very
high, in order to determine the types, if not the magnitude,
of ecosystem response which might be expected. This is the
major purpose of this contribution.

SMELTER EMISSIONS

Smelters are one very obvious example of high SO_2
emissions. Since they generally represent a major point source
of sulfur dioxide, they also allow examination of the magnitude
of change along-concentration gradients. The comparisons
between long distance transport of acid precipitation and local
effects of smelters are complicated by the large role of sulfur
dioxide per se in effects on vegetation around smelters, and
the frequent complication of emissions of toxic heavy metals
and particulates from smelters. However, smelters emitting
sulfur dioxide represent a major local source of soil
acidification, and of local acid precipitation. Foliar damage
can also be caused by sulfuric acid aerosol.

Damage to local vegetation, including crops, has often been reported as a result of sulfur dioxide emissions from base metal and iron smelters, or from fluorides from aluminum smelters. Damage and contamination of crops have also been very frequently reported as a result of smelter emissions of such elements as arsenic, lead, zinc, copper, cadmium, and nickel (Haywood, 1907; Hartman, 1976; Kerin, 1975). Effects on livestock, such as horses, cattle, and sheep, have often been reported also, due to their consumption of contaminated foliage. Examples for fluoride, lead, and arsenic poisoning are the most frequently encountered (Harkins and Swaine, 1907, 1908; Schmitt et al., 1971; Taskey, 1972). The uptake of fluorides into vegetation is largely directly from the air, without a major soil route. The gaseous fluorides enter the leaf through the stomata. In contrast, a good deal of the lead, cadmium, nickel, copper, arsenic, etc., which occurs in the foliage of vegetation around smelters is absorbed from the soil, where it accumulates from airborne deposition. The particulates impacting on leaf surfaces may become trapped there, especially if the leaf surface is hairy, but much is generally washed off following heavy rainfalls. Uptake from the soil is often increased close to the smelters as a result of increased soil acidity, caused by sulfur dioxide emissions, which increases the solubility of many elements, especially heavy metals.

In North America, the major smelting developments have occurred since about 1870 (Sloane and Sloane, 1970). The effects from smelting in those early days of the industry were often quite extreme, causing devastation to the local surroundings. The impact was, and sometimes still is, visible for miles (e.g., Copper Hill in Tennessee, and Butte, Montana). Open roasting of ore was also practiced, and this caused ground level dense discharges of sulfur dioxide. The roast beds were often fueled by local cutting of timber. This added to devastation of forests. The smelters themselves often had uncontrolled or poorly controlled emissions of particulates and sulfur dioxide through low smokestacks. Damage was often channeled along valleys due to prevailing winds (e.g., at Trail, British Columbia). Hedgecock (1916) reported that around the copper smelters at Copper Hill, Tennessee, approximately 6900 ha were devastated. Beyond this zone a sequence of increasing plant species diversity occurred, radiating out from the smelter. Grasses were joined by goldenrod, then by asters and some legumes. Finally, living deciduous trees occurred. The conifers were found to be even more sensitive. The greatest damage occurred on the uplands and on slopes facing the smelters. Thomas (1965) noted that 50 yr after closure, the area around the Ducktown, Tennessee, copper smelter is still barren because of accumulated arsenic, copper, and lead in the soil resulting from aerial deposition.

As another example of combined sulfur dioxide and metal particulate effects, Harkins and Swain (1907) studied the world's largest copper smelter at Anaconda, Montana, which had a capacity of 9070 tonnes of ore per day when built in 1903. Brought in to advise because of large-scale death of livestock and crop damage in the surrounding area, they reported that, in addition to massive discharges of sulfur dioxide, over 23 tonnes of arsenic trioxide (As_2O_3) were lost each day, as well as 1700 kg of copper and somewhat larger quantities of antimony, lead, and zinc. Uptake from the soils and deposition onto the foliage led to high levels in the vegetation. Bateman and Wells (1917) reported copper concentrations up to 6200 mg/kg in plantain (<u>Plantago major</u>).

Great improvements were achieved by the 1920s at many of the large smelters in removal and retention of particulates in bag-houses with electrostatic precipitators, and in removal of sulfur dioxide through wet scrubbers. Tall smokestacks were built and a much better dispersion of the gases achieved. At Trail, British Columbia, sulfuric acid and fertilizer plants were set up in the 1930s and at many other smelters, the valuable elements were obtained from the retained particulates. However, even in the best-maintained operations, some sulfur dioxide is discharged and the finest particles escape the precipitators. At Anaconda in 1976, 635 tonnes of sulfur dioxide were discharged per day, together with 28 tonnes per day of particulates.

THE SUDBURY, ONTARIO, SMELTERS

One of the largest smelters presently operational is the nickel-copper smelter at Copper Cliff, near Sudbury, Ontario, where approximately 1.27 million tonnes of sulfur dioxide are emitted per yr into the atmosphere, through a smokestack 341 m tall. In the late 1960s, this smelter was emitting approximately 2.45 million tonnes sulfur dioxide per yr together with approximately 907 tonnes of iron, 181 tonnes of nickel, 136 tonnes of copper, 18.1 tonnes of lead, 10.7 tonnes of zinc, and 4.1 tonnes of cobalt, all per 28 days (Table 1)(Hutchinson and Whitby, 1974). Discharge was through stacks 120 m tall and much of the particulate matter was deposited within 15 km of the source. The very high concentrations of sulfur dioxide were more widely dispersed but nevertheless caused severe local acidification of the soils, as well as enormous direct sulfur dioxide damage to vegetation. The natural white pine, jack pine, spruce, and red oak forests were devastated. Local cutting of timber for the roast yards in the early part of the century added to the problem. Open roasting of nickel-copper ores emitted 272,000 tonnes of sulfur dioxide per yr in 1917 (Holloway, 1917).

Table 1

Metal Contents of Stack Emissions From Copper Cliff
Smelters, Coniston Smelter, and Iron Ore Recovery Plant
[Values Are in Tonne/28 Days for the Last Quarter of
1971 (Hutchinson and Whitby, 1977)]

	Copper Cliff Smelter	Coniston Smelter	Iron Ore Recovery Plant
	----------------tonne/28 days--------------------		
Iron	828.9	78.7	111.6
Nickel	151.4	22.0	0.73
Copper	125.0	6.4	0.054
Cobalt	3.5	0.54	0.035
Lead	15.3	1.45	0.045
Zinc	10.0	0.73	0.009

The 341 m superstack has been operational since 1972 and
has caused a much wider dispersion of the particulate matter
and sulfur dioxide.   From the mean deposition patterns of
sulfate, nickel, copper, and iron, and of gaseous sulfur
dioxide, first order estimates of the annual deposition of these
emitted pollutants falling within a 60 km radius of the Copper
Cliff, Ontario, smelter were calculated. For sulfur, only 2.3
percent of the total emitted reached the ground within this 60
km radius, while for nickel, copper, and iron the corresponding
values were 42, 40, and 52 percent, respectively (Freedman and
Hutchinson, 1980a). Thus, 97 percent of the sulfur is carried
further afield than 60 km and thus contributes to regional
problems of acid precipitation.

CONTAMINATION IN THE AREA OF THE SUDBURY SMELTER

The effects of sulfur dioxide emissions at Sudbury have
been voluminously documented (Gorham and Gordon, 1960; Linzon,
1958; Leblanc and Rao, 1966; Costescu and Hutchinson, 1972;
Hutchinson and Whitby, 1974, 1977; Leblanc et al., 1972;
Dreisinger, 1970; Whitby and Hutchinson, 1974; McIlveen and
Balsillie, 1978). Despite the improvement brought about in
pollution control over the years, damage to vegetation in the
region has been massive. This has been accompanied by increased

acidification of the already acidic incipient podzolic soils
in the inner zones, with surface soil pH values down to 3.0 in
the early 1970s; a large accumulation of toxic metals in the
soils (up to several thousand mg/kg nickel and copper); release
of soluble toxic aluminum and manganese in the soil solutions
as the result of soil acidification; and widespread soil erosion
(Hutchinson and Whitby, 1974, 1977; Freedman and Hutchinson,
1980a). Runoff, drainage, and silt-clay washing into water
bodies have also polluted local rivers with metal-contaminated
sediments, and have caused acidification of lakes, accompanied
by increases of copper and nickel concentrations to phytotoxic
levels (Whitby and Hutchinson, 1974; Fitchko and Hutchinson,
1975; Freedman and Hutchinson, 1980a, 1980b; Stokes et al.,
1973). Extinction of fish populations in the lakes has been
reported in the Sudbury area, accompanying lake acidification
(Harvey and Beamish, 1972). Algal populations have also been
adversely affected (Stokes et al., 1973; Hutchinson and Stokes,
1975; Yan and Stokes, 1978).

Dreisinger (1970) reported that over the period 1964 to
1968 the area, having an annual average atmospheric
concentration of at least 6.25 μg SO_2/m^3, covered 5600 km^2,
while 630 km^2 had average sulfur dioxide levels of 25 $\mu g/m^3$ or
more. In terms of potentially injurious fumigations occurring
during the growing season, an area of 5470 km^2 was found to be
subjected to one or more fumigations, and 1460 km^2 were subject
to 10 or more such episodes.

Hutchinson and Whitby (1974) reported surface soil levels
to contain in excess of 2000 mg/kg for both nickel and copper
within a few km of the smelters, while surface soil pH values
were as low as 2.2 within 1 km of one of the smelters, which
has since closed down. Extreme acidities with accompanying
high sulfate concentrations were reported from bulk deposition
collections during monthly sample periods in which it rained
(Table 2), while a much better dispersion of sulfur, and of
metal particulates, occurred during dry periods (i.e., very
effective scrubbing occurred in these rainfall episodes)(Tables
3, 4, and 5).

Freedman and Hutchinson (1980a) found that much of the
metal deposited was bound in the surface organic layers, this
being especially so for copper. Concentration of nickel up to
4000 mg/kg occurred in the humus horizon of soils 3 km from
Copper Cliff, Ontario, smelters, while copper levels reached
6500 mg/kg in samples of this same soil collected in 1976.

Table 2

pH, Conductivity (μmhos), and Sulfate Content (mg/m^2/28 days)
of Dustfall-Rainfall Collected in the Coniston-Sudbury Area
in 1971, May 12 to June 10 (Hutchinson and Whitby, 1977)

Distance and Direction From Coniston (km)	Rain Volume (ml)	pH	Conductivity	Sulfate
0.8 S	1250	2.93	1850	5456
1.6 N	1200	2.76	2170	1401
10.4 E	1080	3.53	320	620
13.4 S	960	3.70	335	620
56.0 S	1225	4.25	170	533
104.0 S	1100	4.10	94	131

Table 3

pH, Conductivity (μmhos), and Sulfate Content (mg/m^2/14 days)
in Dry Precipitation Occurring in Transects From the Coniston
Smelter During the Period July 21 to August 4, 1970 — No
Rainfall Fell in This Period — Volumes Were Made to 45 ml
With Deionized Water for pH and Conductivity — Dilute Nitric
Acid (5 ml) was Added for Sulfate and Metal Analyses
(Hutchinson and Whitby, 1977)

Distance and Direction (km)	pH	Conductivity	Sulfate
1.6 S	1.60	11,100	94
1.6 S	2.79	930	62
1.6 N	2.36	2,725	198
1.7 S	2.09	3,650	136
1.9 S	2.74	950	75
7.4 S	3.80	129	87
13.4 S	4.64	42	74
19.2 E	6.16	65	93

Table 4

Heavy Metal Concentration of Rainfall-Dustfall Collections
Along Transect From Coniston Smelter, Collected From May 13
to June 10, 1971 - Concentrations Are Given in
mg/m^2/28 Days (Hutchinson and Whitby,1977)

Distance and Direction (km)		Rain Volume (ml)	Ni	Cu	Co	Fe	Zn	Pb
0.8	S	1250	288	205	5.5	53	6.9	8.9
1.6	N	1200	127	67	2.6	1515	3.6	5.7
10.4	E	1080	13	5	0.9	75	3.6	1.2
13.4	S	960	11	15	0.1	129	3.5	5.1
56.0	S	1225	3	3	0.3	69	2.5	3.5
1040.0	S	1100	1	6	nd[a]	74	5.1	2.9

[a]nd = not detected

Table 5

Metal Analyses of Dry Precipitation-Dustfall Collected At
Various Distances From the Coniston Smelter During the
Period July 21 to August 4, 1970 - Data Are Given as
mg/m^2/14 Days (Hutchinson and Whitby, 1977)

Distance and Direction (km)		Ni	Cu	Co	Fe	Zn	Pb
1.6	S	2.6	2.5	0.2	10	0.7	nd[a]
1.6	S	4.1	2.8	0.3	10	1.1	nd
1.6	N	23.3	10.2	0.6	28	1.0	nd
1.7	S	4.1	5.1	nd	9	0.7	nd
1.9	S	2.3	3.5	0.1	8	0.6	nd
7.4	S	1.7	3.9	0.4	7	0.4	nd
13.4	S	0.4	0.4	nd	0.4	nd	nd
19.2	E	0.9	0.7	0.1	3.0	0.5	nd

[a]nd = not detected

PHYTOTOXIC EFFECTS

Not only has metal accumulation occurred, but the levels of soluble (available) metals in the soils were found to be very high.   Water extractable concentrations of nickel and copper, originating as deposition of smelter emissions, were over 100 mg/l and 50 mg/l in soil solution.   It is of great importance to know if such levels are toxic.   In tests using the ability of the roots of a wide range of test species to elongate in water extractions of these soils, and in "artificial" solutions made up specifically in distilled water to contain the range of copper, nickel, and cobalt found in the soil solutions, but with each element present singly, it was found that concentrations of 2 mg/l of all 3 metals singly reduced root growth by up to 75 percent. Concentrations greater than between 5 and 10 mg/l (depending on the species) killed the seedlings (Whitby and Hutchinson, 1974).   Interestingly, increased acidity induced in the soils by smelter fumes caused aluminum, normally bound to the mineral fraction of the soil, to increase in solubility.   Consequently, concen-trations of soluble aluminum in excess of 50 mg/l occurred in soils within 2 km of the smelter, and up to 15 mg/l occurred even at 10.4 km from the smelter. Root elongation tests of cabbage, radish, lettuce, and tomato showed that even 2 mg/l aluminum in solution caused 80 percent reduction in root growth (Whitby and Hutchinson, 1974).

In the Sudbury area then, we have very widespread acidic deposition, combined with that of metals smelted locally. The metals are sufficiently water-soluble to create widespread phytotoxic conditions for the growth of roots, and for the establishment of seedlings. While the decline in the forests has largely been due to sulfur dioxide damage, the continued inability of the forests to regenerate seems to be largely a function of soil metal toxicity. This toxicity is compounded by the increased solubility of aluminum (and to a lesser extent Mn) as a result of increased soil acidity. Any one of the ions of copper, nickel, cobalt, or aluminum, depending on concentration, could cause inhibition of growth. Synergistic interactions, especially between nickel and copper, further compound the problem.

EFFECTS ON FLORA AND SOIL MICROBES

In the studies of Gorham and Gordon (1960) and Freedman and Hutchinson (1980b), species diversity was found to increase with distance from the smelters both in terrestrial and in aquatic systems. The recovery in diversity at Sudbury was more rapid in the herbaceous layers than in the tree species. This is partly due to the particular sensitivity of coniferous species to sulfur dioxide, but also to the trees acting as major above-ground targets for sulfur damage. Seed germination and seedling establishment were affected adversely (Whitby and Hutchinson, 1976).

In a number of studies, the enzymatic activity in soils and smelters has been shown to be inhibited. This has been found in Europe, the United States, and Canada. It has been found for a variety of crucial enzyme systems involved in nutrient cycling, especially of phosphorus cycling. Inhibitory effects have been attributed to zinc, lead, nickel, copper, and cadmium accumulations, though in studies where multiple element sources exist copper appears to have had a predominant influence. The relative role of acidity per se as opposed to metal accumulations is often not clear; but since microbial activity is influenced by soluble ions, the effect of increased acidity is very probable. Examples of such studies include the work of Tyler (1974, 1975, 1976), Watson (1976), Hartman (1976), Jordan and Lechevalier (1975), and Freedman and Hutchinson (1980c). This lack of activity for soil acid phosphatases, etc., is often accompanied by an accumulation of litter at the sites. It is as though, for a period, litterfall continues as before, but this litter fails to break down as rapidly as before. Such accumulations of litter have the clear implication that reduced forest growth will occur as a consequence of inhibited nutrient cycling over the longer term (Strojan, 1978; Watson et al., 1976).

OTHER EFFECTS

While the focus in this report has been on directly demonstrable, harmful effects, two other categories of impact are worthy of much greater attention. These are:

(1) effects on the structure and properties of soil organic matter, and
(2) the effects of dry deposition and especially particulates on the growth and survivorship of plants around smelters (and coal-fired power plants).

Soils

In the studies of Whitby (1974) and Hutchinson and Whitby (1977), the chemical and physical properties of organic matter extracted from soils collected 1.6, 7.4, and 52.0 km from a Sudbury smelter have been studied. This was done in the laboratories of M. Schnitzer of Agriculture Canada. Attention focused on the properties of fulvic acid extracts obtained from the Sudbury soils. Humic substances facilitate the transport and availability of nutrients, especially trace metals, and improve aeration and drainage. Factors which affect the functioning of humic substances in a deleterious way require study because they influence soil fertility.

The tests on the Sudbury fulvic acid from sites at 1.6 and 7.4 km from Coniston smelter showed that profound changes had taken place. The buffering capacity of the organic matter had changed. In purified organic matter, up to 118,000 mg/kg nickel and 23,900 mg/kg copper occurred in the ash. The titratable acidity of purified podzolic fulvic acid was greatly reduced. This suggests significant differences in the acidic functional groups, such as carboxylic acid (-COOH) and phenolic-OH. Indeed, the sulfur content of the organic matter from 1.6 km was 30.6 percent compared with 1.3 percent for the control. This, together with gas chromatographic data, suggests that the sulfur has replaced carbon and it is likely that the central benzene rings are now surrounded by sulfonic groups (SO_3H) rather than carboxylic groups. The organic matter now tends to act as a modified sulfonic acid resin, with very different properties to that of normal soil organic matter. These changes may be simply the result of the very severe conditions imposed on the local environment by the Sudbury smelters. This is likely but no one knows because no other similar studies have been made.

Dry Deposition

Dry deposition of sulfur dioxide, particulates, and aerosols, is a subject in which a great deal of interest is now being expressed. Metals in dust particulates are available for uptake by leaves (Krause and Kaiser, 1977). Foliar uptake was found for cadmium, lead, and copper from a mixture of their oxides when dusted onto the leaf surfaces of 3 plant species. Apparent foliar injury was increased by simultaneous fumigation of 0.21 mg SO_2/m^3. In the Sudbury area, substantial foliar deposition occurs. My analyses of stack dust fines from the Copper Cliff smelter, courtesy of INCO Canada Ltd., showed the dust, when mixed 1:1 with water on a volume basis, to have a pH of 1.0. It also contained 18.7 percent sulfur, 9,900 mg/kg copper, and 12,400 mg/kg nickel, of which 95 and 77 percent, respectively, were water-soluble. A few particles of this dust placed on geranium leaves overnight produced severe foliar burns. The damage was expanded in area when the particles were moistened. In the field, this strongly acidic particulate deposition may have an effect. Cox and Hutchinson (1981)

studied the factors controlling the spread of the grass
Deschampsia cespitosa in the inner Sudbury area since 1972.
Since sulfur dioxide levels were reduced, it has appeared in
the area and now covers hundreds of hectares. Apart from the
remarkable tolerances of this population of D. cespitosa to
elevated levels of copper, nickel, aluminum, and cobalt in water
culture, and elevated tolerances also to zinc, lead, and silver
(Cox and Hutchinson, 1980; Hutchinson and Cox, 1981) it was
found that in the field the rate of spread was most closely
correlated with amount of deposition of particulate matter from
the smelter, rather than with soil conditions or even
atmospheric sulfur dioxide levels (i.e., in areas of high

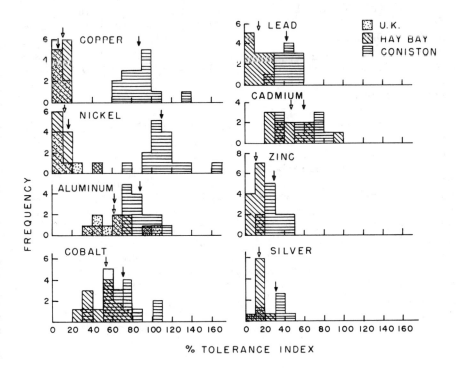

Figure 1.   The distribution of (a) copper, (b) nickel, (c)
            aluminum, (d) cobalt, (e) lead, (f) cadmium, (g)
            zinc, and (h) silver tolerances for clones of
            Deschampsia cespitosa sampled from around a copper-
            nickel smelter at Coniston, Ontario, and from an
            uncontaminated site at Hay Bay, Ontario - a lime-
            stone location. Arrows are mean values for the
            populations. Only a few UK clones were tested,
            from uncontaminated sites.

particulate deposition, the rate of spread was always slower)(Figure 1). This may be due to the accumulation of nickel and copper or of strong acidity around the basal meristem (growing point) of the plant as the particles are washed from the leaves. No visible damage is apparent.

LOW-LEVEL SULFUR DIOXIDE EFFECTS

While acute sulfur dioxide toxicity has received most of the attention in the past in smelter areas, a lot of attention is now focused on chronic or low-level sulfur dioxide effects. Studies on the grass Lolium perenne have shown that decreased yields in this species are due to long-term chronic exposure to low sulfur dioxide levels in the range of 20 to 86 µg SO_2/m^3. Effects on leaf growth and on tiller production are involved (Bell and Clough, 1973; Crittenden and Read, 1978; Bell et al., 1979). In our D. cespitosa studies, we suggest that in grasses the soluble acidic smelter particulates are washed to the base of the leaves, where they accumulate and inhibit intercalary meristem activity.

Despite the normally harmful effects of the combination of metal accumulation and acidity, some interesting evolutionary responses have been found. Metal tolerances occur in higher plants, especially grasses and sedges which invade mining and smelter sites. Numerous studies have shown this, following the study by Bradshaw (1952). Multiple-metal tolerances occur (Cox and Hutchinson, 1980). Adaptations have been described for algae, fungi, mosses, ferns, and bacteria, as well as for various soil fauna. In addition, frequently it has been reported that certain organisms have developed a strongly positive growth response to levels of an element which would normally be toxic. Such an effect is shown in Figure 2 for the D. cespitosa from Sudbury to copper.

SUMMARY

Overall, the impacts that have occurred from smelter discharges have been somewhat local but severe. The phyto-toxicity of sulfur dioxide has been a key factor. Accumulations of metals in the soils create long-term problems and cause ecosystem breakdown. Runoff and drainage create aquatic problems and increased long-distance dispersal of pollutants. Dry deposition and particulate impaction have the potential for substantial harmful effects but are rather poorly researched. Wet deposition creates severe local problems while dry deposition spreads pollutants regionally. The effect of severe acid deposition into forest soils has been sufficient to change

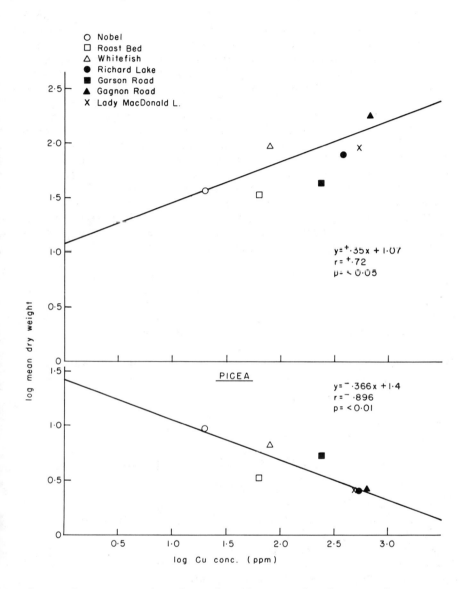

Figure 2. Regression data for the growth of a metal-
tolerant *Deschampsia cespitosa* population and a
Sudbury area *Picea mariana* population when grown
on 7 different soils differing in copper content,
collected from the Sudbury area. Data are of log
mean dry weight on log soil copper concentration.

soil organic matter structure at Sudbury, while changes in soil microbial activity have detrimentally influenced litter decomposition, nutrient cycling, and soil enzyme activity. Residence times for many metals in soils are believed to be hundreds to thousands of years; therefore, there is a substantial need to be concerned that forest growth could be reduced.

Plants have responded by the evolution of metal- and acid-tolerant ecotypes in certain instances. Chronic effects from low-level sulfur dioxide exposures ranging between 2.5 and 62.5 µg/m^3 may occur and also need exploration. Lichens and mosses, lacking a protective cuticle, are especially vulnerable to aerial pollutants. Their absence around many smelters is a consequence. Soil erosion can be a severe effect when vegetation is killed. This is one of the greatest damage effects from deforestation by smelter fumes. Food chain effects through metal accumulations in foliage have been reported for livestock.

Thus, there is a need to have an enhanced awareness of the profound long-term effects of pollutant accumulation in the organic (humus) layers of the soil on ecosystem functioning, when attempting to assess impacts of acid precipitation. This is brought home by studies of ecosystems near smelters. The influence of soil acidification on loss of bases and on solubilization of potentially toxic elements such as aluminum, lead, and copper are also factors which require careful monitoring.

LITERATURE CITED

Bateman, W.G. and L.S. Wells. 1917. Copper in the Flora of a Copper Tailing Region. J. Amer. Chem. Soc. 39(1):811-819.

Bell, J.N.B. and W.S. Clough. 1973. Depression of Yield in Ryegrass Exposed to Sulphur Dioxide. Nature 241:47-49.

Bell, J.N.B., A.J. Rutter, and J. Relton. 1979. Studies on the Effects of Low Levels of Sulphur Dioxide on the Growth of Lolium perenne L. New. Phytol. 83:627-743.

Bradshaw, A.D. 1952. Populations of Agrostis tenuis Resistant to Lead and Zinc Poisoning. Nature 169:1098-1099.

Costescu, L.M. and T.C. Hutchinson. 1972. The Ecological Consequences of Soil Pollution by Metallic Dust From the Sudbury Smelters. In: "Proc. Inst. Environ. Sci. 18th Annual Technical Meeting," pp. 540-545. New York, May 1972.

Cox, R.M. and T.C. Hutchinson. 1980. Multiple Metal Tolerances in the Grass Deschampsia cespitosa (L) Beauv. From the Sudbury Smelting Area. New Phytol. 84:631-647.

Cox, R.M. and T.C. Hutchinson. 1981. Environmental Factors Influencing the Rate of Spread of the Grass Deschampsia cespitosa Invading Areas Around the Sudbury Nickel-copper Smelters. Water Air Soil Pollut. 16:83-106.

Crittenden, P.D.and D.J. Read. 1978. The Effects of Air Pollution on Plant Growth With Special Reference to Sulphur Dioxide. New Phytol. 80:49-62.

Dreisinger, B.R. 1970. SO_2 Levels and Vegetation Injury in the Sudbury Area During the 1969 Season. Dept. of Energy and Resources Management, Province Ontario. 45 pp.

Fitchko, J. and T.C. Hutchinson. 1975. A Comparative Study of Heavy Metal Concentrations in River Mouth Sediments Around the Great Lakes. J. Great Lakes Res. 1(1):46-78.

Freedman, B. and T.C. Hutchinson. 1980a. Pollutant Inputs From the Atmosphere and Accumulation in Soils and Vegetation Near a Nickel-copper Smelter at Sudbury, Ontario, Canada. Can. J. Bot. 58:108-132.

Freedman, B. and T.C. Hutchinson. 1980b. Effects of Smelter Pollution Near a Nickel-copper Smelter on Surrounding Forest Communities. Can. J. Bot. 58:2123-2140.

Freedman, B. and T.C. Hutchinson. 1980c. Smelter Pollution Near Sudbury, Ontario, and Effects on Litter Composition. Can. J. Bot. 58:1722-1736.

Gorham, E. and A. Gordon. 1960. Some Effects of Smelter Pollution Upon the Aquatic Vegetation Near Sudbury, Ontario. Can. J. Bot. 41:371-378.

Harkins, W.D. and R.E. Swain. 1907. Papers on Smelter Smoke. I. The Determination of Arsenic and Other Solid Constituents of Smelter Smoke, With Study of the Effects of High Stacks and Large Condensing Flues. J. Amer. Chem. Soc. 29:970-998.

Harkins, W.D. and R.E. Swain. 1908. The Chronic Arsenical Poisoning of Herbivorous Animals. J. Amer. Chem. Soc. 30:928-946.

Hartman, L. 1976. Fungal Flora of the Soil as Conditioned by Varying Concentrations of Heavy Metals. Ph.D. Thesis, Department of Botany, University of Montana, Missoula, Montana.

Harvey, H.H. and R.J. Beamish. 1972. Acidification of the La Cloche Mountain Lakes Ontario and Resulting Fish Mortality. J. Fish. Res. Bd. Can. 29:1131-1143.

Haywood, J.K. 1907. Injury to Vegetation and Animal Life by Smelter Fumes. J. Amer. Chem. Soc. 29:998-1009.

Hedgecock, G.G. 1914. Injury by Smelter Smoke in Southeastern Tennessee. Wash. Acad. Sci. J. 4:70-71.

Holloway, G.T. 1917. Report of the Ontario Nickel Royal Commission, With Appendix. Legislative Assembly of Ontario.

Hutchinson, T.C. and R.M. Cox. 1981. Tolerances of the Native Grass Deschampsia cespitosa for Possible Mine Tailings Revegetation. ALUR Report, Dept. Indian Northern Affairs, Ottawa. 77 pp. (in press).

Hutchinson, T.C. and L.M. Whitby. 1974. Heavy-metal Pollution in the Sudbury Mining and Smelting Region of Canada. I. Soil and Vegetation Contamination by Nickel, Copper and Other Metals. Environ. Conserv. 1(2):123-132.

Hutchinson, T.C. and L.M. Whitby. 1977. The Effects of Acid Rainfall and Heavy Metal Particulates on a Boreal Forest Ecosystem Near the Sudbury Smelting Region of Canada. Water Air Soil Pollut. 7:421-438.

Hutchinson, T.C. and P.M. Stokes. 1975. Heavy Metal Toxicity and Algal Bioassays. In: "Water Quality Parameters," pp. 320-343. Special Tech. Pub. No. 573, ASTM.

Jordan, M.J. and M.P. Lechevalier. 1975. Effects of Zinc Smelter Emissions on Forest Soil Microflora. Can. J. Microbiol. 21:1855-1865.

Kerin, Z. 1975. Relationship Between Lead Content in the Soil and in the Plants Contaminated by Industrial Emissions of Lead Aerosol. In: T.C. Hutchinson, ed., "Intern. Conf. Heavy Metals in the Environment," pp. 487-502. Institute of Environmental Studies, University of Toronto, October, 1975, Vol. 2.

Krause, G.M. and H. Kaiser. 1977. Plant Response to Heavy Metals and Sulphur Dioxide. Environ. Pollut. 12(1):63-71.

Leblanc, F. and D. Rao. 1966. Reaction du quelques lichens et mousses epiphytique a l'anhydride sulfureux dans la region de Sudbury, Ontario. Bryologist 69:338-346.

Leblanc, F., G. Robataille, and D.N. Rao. 1974. Biological Response of Lichens and Bryophytes to Environmental Pollution in the Murdochville Copper Mine Area, Quebec. J. Hattori Bot. Lab. 38:405-433.

Linzon, S.N. 1958. The Influence of Smelter Fumes on the Growth of White Pine in the Sudbury Region. Ontario Lands and Forest Department, and Ontario Department of Mines. 45 pp. Illustr.

McIlveen, W.D. and D. Balsillie. 1978. Air Quality Assessment Studies in the Sudbury Area. Vol. 2. Effects of Sulphur Dioxide and Heavy Metals on Vegetation and Soils - 1970-77. Ontario Ministry of the Environment, Northeast Region, Sudbury, Ontario.

Schmitt, J., G. Brown, E. Devlin, A. Larsen, E. McCausland, and J. Seville. 1971. Lead Poisoning in Horses. Arch. Environ. Health 23:185-192.

Sloane, H.N. and L.L. Sloane. 1970. Pictorial History of American Mining. Crown Publishers, Inc., New York. 341 pp.

Stokes, P.M., T.C. Hutchinson, and K. Krauter. 1973. Heavy-metal Tolerance in Algae Isolated From Contaminated Lakes Near Sudbury, Ontario. Can. J. Bot. 51:2155-2168.

Strojan, C.L. 1978. Forest Leaf Litter Decomposition in the Vicinity of a Zinc Smelter. Oecologia. 32:203-212.

Taskey, R.D. 1972. Soil Contamination at Anaconda, Montana: History and Influence on Plant Growth. M.Sc. Thesis, Department of Botany, University of Montana, Missoula, Montana.

Thomas, M.D. 1965. The Effects of Air Pollution on Plants and Animals. In: G.T. Goodman, R.W. Edwards, and J.M. Lambert, eds., "Ecology and the Industrial Society Special Symp. Brit. Ecol. Soc. 5," pp. 11-34. J. Wiley and Sons, Inc., New York.

Tyler, G. 1974. Heavy Metal Pollution and Soil Enzymatic Activity. Plant and Soil 41:303-311.

Tyler, G. 1975. Heavy Metal Pollution and Mineralization of Nitrogen in Forest Soils. Nature 255:701.

Tyler, G. 1976. Heavy Metal Pollution, Phosphatase Activity and the Mineralization of Organic Phosphorus in Forest Soils. Soil Biol. Biochem. 8:327-332.

Watson, A.P. 1976. Trace Element Impact on Forest Floor Litter in the New Lead Belt Region of South-Eastern Missouri. In: D.D. Hemphill, ed., "Trace Substance in Environmental Health," pp. 9, 227-237. University of Missouri, Columbia, Missouri.

Watson, A.P., R.I. Van Hook, D.R.Jackson, and D. Reichle. 1976. Impact of a Lead-Mining Complex on the Forest Floor Arthropod Fauna in the New Lead Belt Region of South Eastern Missouri. Publ. 881. Oak Ridge National Laboratory, Environmental Science Division, Oak Ridge, Tennessee. 161 pp.

Whitby, L.M. 1974. The Ecological Consequences of Airborne Metallic Contaminants From the Sudbury Smelters. Ph.D. Thesis, Department of Botany, Univeristy of Toronto, Toronto, Ontario.

Whitby, L.M. and T.C. Hutchinson. 1974. Heavy-Metal Pollution in the Sudbury Mining and Smelting Area of Canada. II. Soil Toxicity Tests. Environ. Conserv. 1:191-200.

Yan, N. and P.M. Stokes. 1978. Phytoplankton of an Acid Lake and Responses to Experimental Alterations of pH. Environ. Conserv. 5:1-8.

CHAPTER 6

RELATIONSHIP BETWEEN MESOSCALE ACID
PRECIPITATION AND METEOROLOGICAL FACTORS

Kenneth W. Ragland and Kenneth E. Wilkening
Department of Mechanical Engineering
University of Wisconsin
Madison, Wisconsin 53706

INTRODUCTION

This chapter deals with sulfur dioxide and sulfate deposition from the atmosphere to the ground. Sulfur dioxide may be transformed to sulfate compounds in the atmosphere and in vegetation, soil, and water. The extent to which sulfur deposition changes the pH of the soil and water is not addressed here. Rather, sulfur loading is taken as an indicator of potential for acidification. Nitrogen compounds may also produce acidification; however, less is known about these processes and they are not considered here.

Acid precipitation can occur when particles and gases are removed from the atmosphere (1) by rain and snow (wet deposition), and (2) by impaction on water, soil, and vegetation surfaces (dry deposition). The transformation of gases to particles in the atmosphere and the deposition rates of gases and particles are a function of meteorological conditions over a period of days as well as the distribution of emission sources.

To justify controls on emissions of acid precursors, the relationship between a particular source and the deposition pattern that it produces needs to be understood. This is difficult to accomplish by field monitoring because monitoring data register many sources, some of which may be hundreds or thousands of kilometers away. Mathematical models are the only way to estimate the relative contribution of different sources to the total deposition at a particular receptor.

Atmospheric deposition models may be divided into (1) long-range transport models, where the region of interest extends over thousands of kilometers, say the eastern half of the U.S. and Canada, and (2) mesoscale models where the analysis region extends over hundreds of kilometers, say the state of Michigan. The focus here is on mesoscale models. Because mesoscale models can provide higher resolution and more detailed simulation of meteorological conditions, they can predict the influence of a specific emission source on deposition at a nearby receptor point. However, mesoscale models show only incremental changes within the region rather than absolute deposition rates.

The purpose of this chapter is to briefly describe our mesoscale deposition model (UWATM-SOX) and present the results of the application of this model to the Rainy Lake watershed, which is located northwest of Lake Superior. Annual, seasonal, and weekly calculations of sulfate deposition are presented and compared with field monitoring data.

DESCRIPTION OF THE MODEL

The UWATM-SOX model is a time-dependent, cell-type model which numerically solves coupled sulfur dioxide and sulfate ion diffusion in an Eulerian (fixed) frame of reference in the atmospheric boundary layer. Included are the processes of advection, turbulent diffusion, chemical reactions, source input, and dry and wet deposition (Figure 1). The numerical solution is a first-order fully implicit finite difference scheme.

The model is developed for a mesoscale (10s to 100s of kilometers) analysis. The region of study is divided into a uniform horizontal two-dimensional grid network. The rectangular grid size and number depends on the extent of the region, the desired cost of the computer runs, and the desired accuracy of analysis. Above the grids are six vertical cells which extend from ground level to an upper limit called the mixing height. The model uses three lower fixed-height (50 m) and three upper variable-height cells. Therefore, the space enclosing the region of study is completely partitioned by a three-dimensional cell structure. The single cell is the basic unit within which the model calculations are performed.

The model inputs emissions once at the beginning of the simulation and inputs meteorological data every hour. Calculations are performed every twenty minutes within an hour for time periods up to one year. Each emission source is assigned to one of the grids. A plume rise is calculated and the resulting "effective stack height" is placed within the appropriate vertical cell. The meteorological condition, known

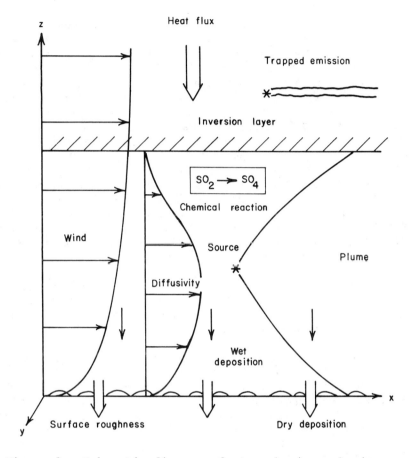

Figure 1.   Schematic diagram of atmospheric mechanisms.

as lofting (effective stack height above mixing height) is accounted for.

The meteorological data are obtained on tapes from the National Climatic Center (NCC) in Asheville, North Carolina. The tapes required are: hourly surface observations, twice daily mixing heights, and hourly precipitation. These data are consolidated and pre-processed by three tape-handling programs and a program (METDATA) which converts the data to the hourly format required by the model.

The atmospheric boundary layer is characterized by a set of wind speed and eddy diffusivity profile equations which are calculated from ground level to the top of the mixing layer (Ragland et al., 1977). The profiles describe the transport and

diffusion   behavior of the pollutants and are assumed to be
vertically variable but horizontally uniform across the region
of study over a 1-hr period. The profiles are dependent on the
wind speed and stability of the atmosphere. The boundary layer
is subdivided into a surface layer and a second layer extending
up to the mixing height. The surface layer equations are based
on the Monin-Obukov similarity theory. In the upper layer, the
wind profile is a form of the power-law and the diffusivity is
assumed   constant.   The   profiles   are   calculated   from   the
geostrophic wind speed, net heat flux to the ground and surface
roughness. The geostrophic wind is calculated from the surface
wind speed using a modified Regula Falsi iteration scheme. The
net heat flux is calculated by first using the NCC surface
observations data and the Pasquill-Turner method of stability
classification   to   determine   the   stability   class,   and   then
assigning a heat flux value to each class. A single surface
roughness value is chosen to characterize the region.

The chemical conversion from sulfur dioxide to sulfate ion
is assumed to be first-order. At present, separate day and
night reaction rates are used although hourly and seasonal
changes are possible if data are available.

The dry deposition is handled by means of a deposition
velocity which is calculated as a function of both terrain and
meteorological conditions. Deposition velocity changes hourly,
but is uniform across the region. It is determined from an
aerodynamic   resistance   which   depends   on   meteorological
conditions close to the ground, and a surface resistance which
depends on pollutant species and the nature of the surface
(snow, water, vegetation, etc.). Surface resistances are input
by the user and aerodynamic resistance is internally calculated
from wind profile information.

The wet deposition scheme assumes all precipitation forms
according to the Bergeron or cold cloud process. Ice crystals
form in the upper portion of the cloud, and as they sweep
downward through the cloud, they capture cloud droplets containing sulfate and dissolved sulfur dioxide. The sulfur eventually
incorporated into the precipitation is drawn in through the
cloud base.   Three equations for the wet deposition removal
rate are used; one for rain or snow removal of sulfate, and one
each for rain and snow removal of sulfur dioxide (Glass and
Loucks, 1980). Precipitation is assumed to occur uniformly
across the region.   At present, there is no analysis for
fractional area covered by a precipitation event.

There are several computer techniques worth noting. First,
to handle the numerical calculations for arbitrary wind

directions, a dual grid system is used, one fixed in position and one which overlaps the fixed grid and rotates with the wind angle. Transfers of concentrations are performed back and forth between the two systems as the wind direction shifts. This procedure minimizes so-called numerical diffusion (Lio, 1980). Second, the main program has a "stop-start" capability such that progress can be stopped anywhere in the stream of calculation and started exactly where it left off with no outside intervention other than restarting the run. Third, extensive use has been made of common blocks and the Fortran procedure (PROC) statement to eliminate unnecessary interprogram storage locating calls and to facilitate internal changes which affect many programs at once. Fourth, a very complete record of the values of all significant variables is stored on tape hourly for future analysis. The stored information includes the date, ground level concentration and deposition arrays, and certain meteorological, emissions, and regional mass balance data.

The model results include 3-hr, 24-hr, seasonal and annual average ground level concentrations, and dry, wet, and total depositions of sulfur dioxide and sulfate. The results are printed grid by grid in an array format and isopleths are computer-plotted. The results also include a frequency distribution of concentration/deposition values for any selected grid point. Finally, a regional atmospheric sulfur budget is computed which analyzes, by total mass and percentage of sulfur input, the flow of sulfur (SO_2 and SO_4^{2-}) into, out of, and within the region.

The model is written in ASCII Fortran and has been run on the Madison Academic Computing Center's SPERRY UNIVAC 1100/82 series computer. In an application of the model in which a cell structure of 11 x 13 x 6 was used with hourly changes in meteorological data and a 20-min time step, the main program required 50K core space, and a 1-yr simulation took 6 hr of CPU time.

An overview of the model is shown schematically in Figure 2. A summary of the user-supplied model input parameters and the values which we have used is given in Table 1.

CASE STUDY - THE RAINY LAKE WATERSHED

The Rainy Lake watershed is located in northeastern Minnesota and western Ontario, and includes the Quetico Provincial Park, the Boundary Waters Canoe Area (BWCA), and the Voyagers National Park. This area lies within 2 days travel of 50 million people and yet contains extensive prime wilderness

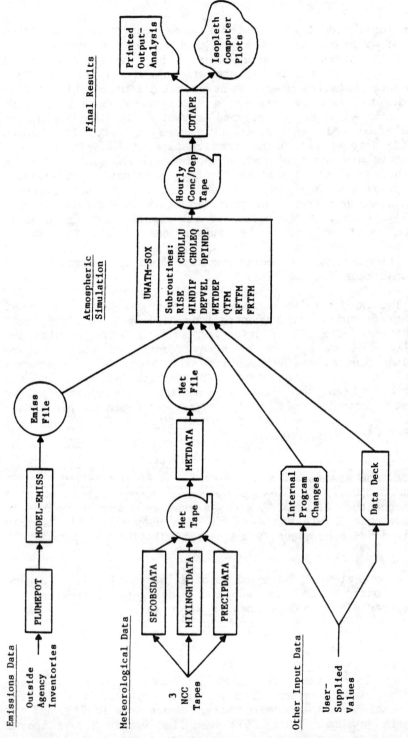

Figure 2. Schematic of the modeling operation.

Table 1

Summary of Model Parameters Used

Parameter	Value
SO_2 chemical transformation rate (%/hr):	Night 0.5, day 2.0
SO_2 dry deposition surface resistance (sec/cm):	Summer 0.5, winter 3.0
SO_4^{2-} dry deposition surface resistance (sec/cm):	Summer 0.5, winter 5.0
SO_2 rain deposition rate (%/hr): where P = rain rate (mm/hr)	$33.0\ P^{0.90}$
SO_2 snow deposition rate (%/hr):	$16.5\ P^{0.90}$
SO_4^{2-} rain and snow deposition rate (%/hr):	$33.0\ P^{0.63}$
Mixing height (m):	Hourly values calculated from twice daily soundings.
Wind data (m/sec):	Six levels calculated hourly from wind speed, direction, temperature, and cloud cover surface observations.
Diffusivity data (m^2/sec):	Six levels calculated hourly from wind speed, temperature, and cloud cover surface observations for vertical and laterial diffusivity.
Cell size (km):	30 x 30
Number of cells:	11 x 13

areas. The BWCA and adjoining areas are believed to be sensitive to acid precipitation (Glass and Loucks, 1980).

The modeling area we have chosen is the 330 x 390 km region shown in Figure 3. The Rainy Lake watershed is shown in dotted lines and major emission sources in Duluth, the Missabe Iron Range, International Falls, and Atikokan are indicated. The sulfur dioxide emissions inventory for the region is given in Table 2 (Price, 1980). The total emissions of sulfur dioxide due to point sources in the region in 1976 were 3232 gm/sec or 102,000 tonne/yr. In 1978, the sulfur dioxide emissions were 3139 gm/sec or 99,000 tonne/yr. The area source emissions of

Table 2

Emission Inventory for the Study Region

Source	Location	Stack Height (m)	1976 Sulfur Dioxide Emissions (gm/sec)	1978 Sulfur Dioxide Emissions (gm/sec)
Portlach NW	Cloquet	91	34	0
Portlach NW	Cloquet	11	22	0
Continental Oil	Cloquet	27	20	27
Continental Oil	Cloquet	41	21	22
Continental Oil	Cloquet	22	8	0
Conwed	Cloquet	107	3	0
Erie Mining	Taconite Harbor	67	542	400
Minn. P&L	Clay Boswell	213	939	965
Boise Cascade	Intl. Falls	61	26	11
Two Harbors W&L	Two Harbors	46	101	0
Reserve Mining	Silver Bay	66	101	74
Erie Mining	Hoyt Lakes	40	35	24
Minn. P&L	Aurora	91	192	215
Minn. P&L	Duluth	101	20	67
Minn. P&L	Duluth	66	29	0
Duluth Steam	Duluth	72	23	14
Hibbing-Public UT	Hibbing	12	10	42
Virginia Public UT	Virginia	40	54	43
Superwood	Duluth	24	7	6
U.S. Air Force	Duluth	9	4	0
Reserve Mining	Babbit	17	3	0

U.S. Steel Coke	Duluth	69	109	123
U. of Minn.	Duluth	18	3	0
U.S. Steel Shipping	Duluth	8	9	9
Ontario-Minn. P&L	Fort Francis	76	26	27
Steep Rk.-Caland Ore	Atikokan	85	861	863
Lake Superior P	Ashland	43	45	80
Murphy Oil	Superior	63	58	41
Superior W, L&P	Superior	46	13	14
U. of Wisconsin	Superior	69	9	0
CLM Corp.	Superior	20	9	24
Hanna Minn.	Hibbing	37	0	3
U.S. Steel	Mt. Iron	51	0	5
Eveleth Tac.	Eveleth	47	0	6
Hibbing Tac.	Hibbing	39	0	20
Inland Steel	Virginia	41	0	10
American Can	Ashland	22	0	4
TOTAL			3232	3139

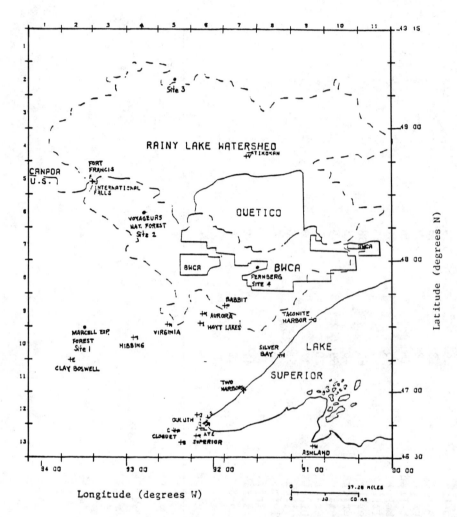

Figure 3.   Map of the study region.

sulfur dioxide due to space heating and small industrial and commercial sources were not available, but are believed to be small in this region.   Direct sulfate emissions were assumed to be 3 percent of the sulfur dioxide emissions by weight, which is a generally observed result from stack sampling.

The regional sulfur dioxide emissions may be compared to the Minnesota state-wide total of 521,000 tonne/yr, Wisconsin 937,000 tonne/yr, and Illinois 2,344,000 tonne/yr. The question to be answered is how much of the regional sulfur emissions are deposited within the region and how does this compare to the monitoring data on an annual, seasonal, and weekly basis.

Before discussing the deposition results, the ambient air concentrations will be briefly mentioned. The annual average sulfur dioxide concentrations away from any sources were calculated to be about 0.5 $\mu g/m^3$, while ambient sulfate concentrations were about 10 times smaller. The 24-hr worst-case sulfur dioxide concentrations away from any sources were about 5 $\mu g/m^3$ and the 24-hr worst-case sulfate concentrations were also about 5 $\mu g/m^3$

The annual (1976) wet deposition isopleths for the region are shown in Figure 4 and the annual wet plus dry deposition isopleths are shown in Figure 5. To express deposition as kilograms of sulfate per hectare, the sulfur dioxide deposition was multiplied by 1.5 (the ratio of molecular weights) and combined with the sulfate values. The regional mass balance, presented in Table 3 for 1976, shows that 17 percent of the sulfur emissions was deposited in the region due to dry deposition and 1.3 percent was retained due to wet deposition, whereas 82 percent of the sulfur emissions was transported out of the region annually. Of the sulfur deposited in the region, 8.5 x 10^9 gm out of 9.4 x 10^9 gm (77%) were deposited by dry deposition of sulfur dioxide.

In this region, rain occurs from May through October, while snow occurs from November through April. During 1976, there was 27.2 cm of rain compared with a normal of 40 cm, and there

Table 3.

Regional Mass Balance for 1976

Origin	Sulfur Dioxide	Sulfate	Sulfur
	-------------10^9 gm/yr-------------		
Emissions	102	3.1	52.3
Chemical reaction	6	9.8	-
Dry deposition	17	0.6	8.7
Wet deposition	1	0.5	0.7
Transport out of region	78	11.8	42.9

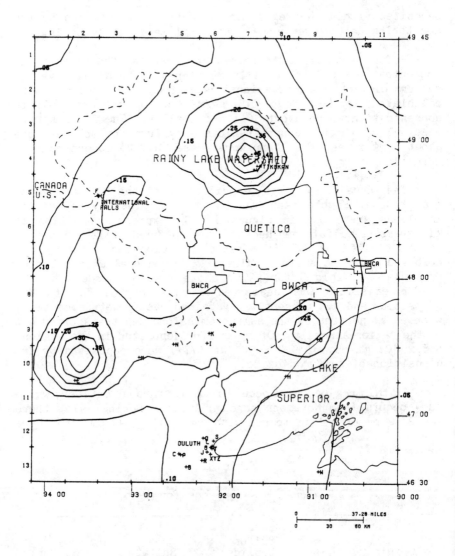

*Figure 4.   Computed annual wet deposition of sulfate and
sulfur dioxide (kg SO_4^{2-}/ha).*

was 9.0 cm of snow compared with a normal of 15 cm. Hence,
there was 34 percent less precipitation than normal during 1976.

It is of interest to compare the deposition during the
snow season (November to April) with the rain season (May to
October). The calculated total deposition (wet plus dry) of
sulfur dioxide and sulfate plotted as sulfate is shown in Figure

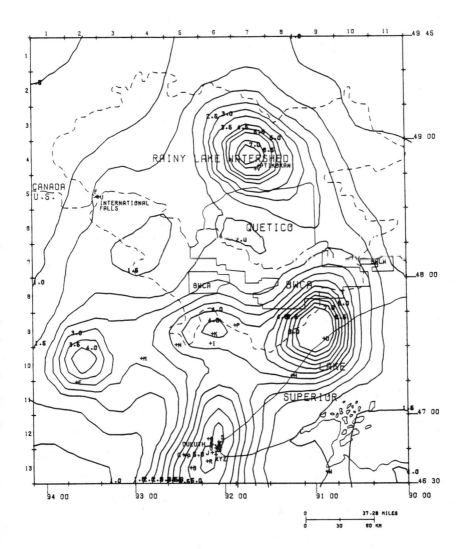

Figure 5. Computed annual wet plus dry deposition of sulfate and sulfur dioxide (kg SO$_4^{2-}$/ha).

5. The calculations are summarized for grid (2,9) the Marcell, Minnesota, site in Table 4. The greatest percentage of sulfur deposition occurs by dry deposition of sulfur dioxide during the summer. Wet deposition of sulfate was comparable in winter and summer, although total wet deposition of sulfate and sulfur dioxide was greater in the summer. It is believed that the reason wet deposition of sulfate is similar in winter and summer

Table 4

Calculated Regional Deposition at the Marcell,
Minnesota, Site and Comparison with Measurements

	Dry	Wet	Wet & Dry
	-------------kg/ha-------------		
UWATM-SOX MODEL, 1976			
Winter SO_2 deposition	0.24	0.01	0.25
Summer SO_2 deposition	1.23	0.10	1.33
Winter SO_4 deposition	0.02	0.03	0.05
Summer SO_4 deposition	0.02	0.03	0.05
Winter deposition as SO_4	0.36	0.05	0.41
Summer deposition as SO_4	1.86	0.18	2.04
Annual deposition as SO_4	2.22	0.23	245
MEASUREMENTS			
Annual deposition as SO_4	-	15.6[a] (11.5[b])	-
Snow deposition as SO_4	-	1.6[a] (2.2[b])	1-2[c]
EASTERN NORTH AMERICA MODELS[d]			
Annual deposition as SO_4	1.8-7.2	1.9-4.5	3.7-11.7

[a]National Atmos. Deposition Program data July 1979 through June 1980.
[b]NADP data April 1978 through March 1978.
[c]Snow cores (65) in BWCA, 1978-1978 (Glass, 1979).
[d]Preliminary report (Ferguson and Machta, 1981).

is that there are more hours of precipitation in winter than in summer although the total amount of precipitation is greater in the summer.

The computer calculations of sulfate deposition are compared with measured deposition for the Marcell, Minnesota, site (Table 4). Unfortunately, the computer calculations for the time period of the measurements (1978 to 1980) have not been completed so that calculations for 1976 will have to be used. The precipitation for 1978 was 96 percent of normal.

The National Atmospheric Deposition Program (NADP) has a site in the region at Marcell, Minnesota, starting in July 1978. The NADP site provides wet sulfate deposition on a weekly basis. The total wet sulfate deposition at the Marcell site was calculated to be 0.23 kg/ha, whereas the monitoring value was 15.6 kg/ha for 1979 to 1980 and 11.5 kg/ha for 1978 to 1979. The main difference is due to transport of sulfate from outside the region. During the snow season, the calculated wet sulfate deposition was 0.05 kg/ha, compared to measured values of 1.6 and 2.2 kg/ha. Again, the measured value represented inter-regional and intraregional transport.

Snow core samples taken at the end of the snow season provide valuable data on wet plus dry deposition of sulfate. Sixty-five snow cores taken in April, 1979, in the Boundary Waters Canoe Area were mostly in the range of 1 to 2 kg/ha sulfate, which may be compared to the calculated value of 0.5 kg/ha (Glass, 1979).

Finally, Table 4 presents deposition calculations from long-distance transport models of eastern North America which show a range of 1.9 to 4.5 kg/ha for wet deposition at the Marcell site and 3.7 to 11.7 kg/ha for wet plus dry deposition (Ferguson and Machta, 1980). Our regional model gave 2.4 kg/ha for the total annual sulfate deposition.

The wet and dry deposition fluxes may also be compared on a weekly basis. In Table 5, the regional mass balance is summarized for the extreme or worst-case conditions of a clear summer week, a wet summer week, a clear winter, and a winter week with snowfall during 1976. The rainy week had 7 hr of rainfall, producing 4.3 cm of water. The snowing week had 32 hr of snowfall, producing 1.7 cm of water. The dry deposition is three times greater in summer than in winter, whereas the surface resistance for sulfur dioxide was six times less in summer than in winter (Table 1). The wet deposition is four and one-half times greater in summer than in winter. During the wet summer week, 65 percent of the sulfur emissions is transported out of the region, while during a winter week, more than 90 percent is transported out of the region.

During the wet summer week (7/19/76 to 7/25/76), the calculated wet deposition flux of sulfur dioxide and sulfate as sulfate was 0.06 kg/ha at the Marcell site. By way of comparison, a wet deposition flux of 2.2 kg SO_4^{2-}/ha was recorded during a week in July of 1979 at the Marcell site. Since the total measured for the year was 11.5 kg SO_4^{2-}/ha at Marcell, it is clear that a few big events coming from extraregional buildup of sulfate produce much of the wet deposition loadings.

Table 5

Mass Balance for Worst-Case Weeks

	Dry Deposition	Wet Deposition	Transport Out of Region
	--------------10^8 gm S--------------		
Clear summer week	3.0	0	7.1
Wet summer week	2.6	0.9	6.6
Clear winter week	1.0	0	9.1
Snowing winter week	0.7	0.2	9.2

ACKNOWLEDGMENT

This work was supported by a grant from the U.S. Environmental Protection Agency, Environmental Research Laboratory, Duluth, Minnesota, under the direction of Dr. Gary E. Glass.

LITERATURE CITED

Ferguson, H.L. and L. Machta. 1980. Preliminary Report of Atmospheric Modeling Work Group 2. U.S. Environmental Protection Agency, Washington, D.C. 20460.

Glass, G.E. 1979. Unpublished Data. U.S. Environmental Protection Agency, Environmental Research Laboratory, Duluth, Minnesota 55804.

Glass, G.E. and O.L. Loucks (eds). 1980. Impacts of Airborne Pollutants on Wilderness Areas Along the Minnesota-Ontario Border.   EPA-600/3-80-044, U.S. Environmental Protection Agency, Environmental Research Laboratory, Duluth, Minnesota 55804.

Lio, P.I. 1980. A Two-Dimensional Sulfur Dioxide/Sulfate Air Quality Model and Comparison With a Three-Dimensional Model. M.S. Thesis, University of Wisconsin, Madison, Wisconsin 53706. 104 pp.

Price, B. 1980. Personal Communication, Minnesota Pollution Control Authority Emission Inventory Data. Minnesota Pollution Control Authority, Minneapolis, Minnesota.

Ragland, K.W., R.L. Dennis, and K.E. Wilkening. 1977. Boundary Layer Model for Transport of Urban Air Pollutants. AIChE Symposium Series 165, 73:1-10.

CHAPTER 7

ENVIRONMENT CANADA'S LONG RANGE TRANSPORT OF
ATMOSPHERIC POLLUTANTS PROGRAM: ATMOSPHERIC STUDIES

Leonard A. Barrie
Atmospheric Environment Service
Downsview, Ontario

INTRODUCTION

The acidification of terrestrial and aquatic ecosystems is occurring around the globe as a consequence of oxides of sulfur and nitrogen being transported hundreds, even thousands of kilometers from industrial areas and deposited on the earth's surface. Barren lakes in recreational, wooded areas of Canada, the United States, and Scandinavia are ugly reminders that a truly serious regional pollution problem exists.

In August, 1976, the Canadian Department of Environment established the Long Range Transport of Air Pollutants program (LRTAP) to investigate the nature, extent, and magnitude of "acid rain." In June, 1980, the program was expanded to include research projects in five additional departments. The Atmospheric Environment Service in Toronto maintained its role as lead agency as well as its responsibilities for atmospheric studies.

The structure of the LRTAP program is dictated by the nature of the "acid rain" problem. It is best viewed within the framework of the atmospheric cycle of acid-related pollutants (Figure 1). While enroute through the atmosphere between source and receptor, substances are dispersed and chemically or physically transformed before being removed by precipitation scavenging and dry deposition processes. As indicated by the level of funding from a 5.5 million dollar budget in 1981/82, the intensity of LRTAP research on emissions, atmospheric processes, and terrestrial-aquatic effects, is 14.4, 29.7, and 42 percent, respectively. This paper will focus on results of atmospheric research projects in LRTAP. Aquatic

141

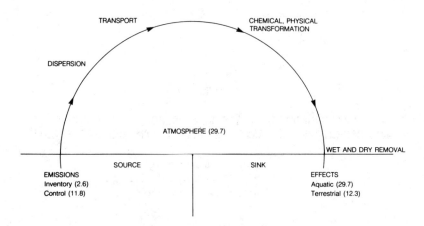

*Figure 1.   A Schematic diagram of the atmospheric cycle of
             pollutants.   In brackets is the fraction (%) of
             Canada's 1981-82 "acid rain" funding allocated
             to each aspect of the problem.*

effects studies are discussed later in these proceedings (Kelso,
1981.

     Activities in the air program are concentrated in three
areas: monitoring,   modeling,   and   supporting processes
research. The main objective is to provide information on the
rate of deposition of acidic pollutants to those assessing the
effects of acidity.   A second objective is to establish a
capability to mathematically simulate the atmospheric cycle of
acid substances. Models being developed are used in either a
diagnostic mode to gain an understanding of the interaction of
various components of the atmospheric cycle or in a prognostic
mode to assess the impact of energy scenarios in North America.

     Air and precipitation monitoring on a regional basis is
essential in achieving the above objectives.   It provides a
data base with which:

     (1)   deposition rates and patterns can be determined,
     (2)   models can be improved, and
     (3)   areas requiring   more   intensive study   can   be
           identified.

     Supporting   processes   research   helps   to   refine   both
modeling and monitoring.

MONITORING NETWORKS

In April, 1977, the first nationwide precipitation moni- toring network was initiated. The Canadian Network for Sampling Precipitation (CANSAP) collects monthly samples at about 50 sites across Canada using wet-only collectors. Samples are analyzed for pH and major ions using standard wet-chemical techniques. The analysis laboratory participates routinely in international quality-control intercomparisons. A summary of results for the first two years of operation is available (Berry, 1979). From CANSAP precipitation data, it quickly became evident that wet deposition of acidity is greatest in eastern Canada (east of the Ontario-Manitoba border) and, to a lesser extent, occurs on the west coast, and in Alberta and northern Saskatchewan (Figure 2). It also became apparent from a sulfur budget in eastern North America (Galloway and Whelpdale, 1980) that dry deposition processes may contribute as much sulfur, and presumably acidity, to Canadian ecosystems as precipitation does. Since estimates of dry deposition rates require atmospheric pollutant concentrations, an air monitoring effort was deemed necessary. At the same time, the need for information on the chemical composition of precipitation from individual events, rather than from monthly samples provided by CANSAP, was emphasized by modeling and effects groups. In answer to these needs, the Canadian Air and Precipitation Monitoring Network (APN) was established in November, 1978. It presently consists of six stations (Figure 3) collecting daily precipitation, suspended particulate, and sulfur dioxide samples at sites carefully selected to be regionally representative. Precipitation samples are analyzed for pH and major ions using the same techniques as those used for CANSAP samples. Details of network operations can be found elsewhere (Barrie et al., 1980). Results from four sites operated in 1979 will be discussed: Long Point, Chalk River, ELA-Kenora, and Kejimkujik National Park (Figure 3).

Atmospheric Concentrations of Oxides and Sulfur

Monthly mean concentrations of sulfur dioxide and particu- late sulfate at each site in eastern Canada for 1979 are given in Figure 4. On an annual basis, the lowest concentrations occurred at ELA-Kenora in northwestern Ontario (SO_2 - 1.5; SO_4 - 1.6 ug/m^3) while the highest concentrations prevailed at Long Point on the north shore of Lake Erie (SO_2 - 16; SO_4 - 8.1 ug/m^3). In the mid-latitudinal westerlies ELA is upwind of major North American sulfur sources. Long Point, on the other hand is located in an industrialized section of Ontario and due north of major emissions in Ohio and Pennsylvania. At all sites in eastern Canada, atmospheric sulfur dioxide levels undergo a

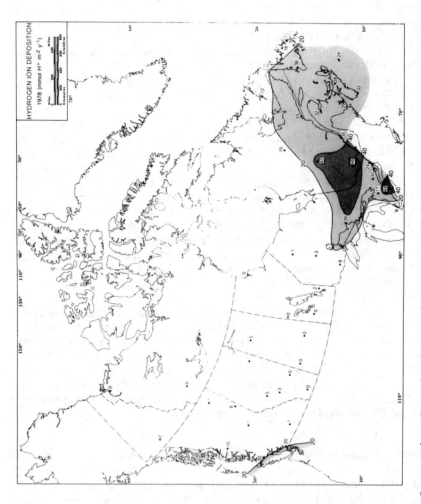

Figure 2.   Total hydrogen ion deposited by rain and snow in Canada during 1978 as measured by the CANSAP monthly precipitation network.

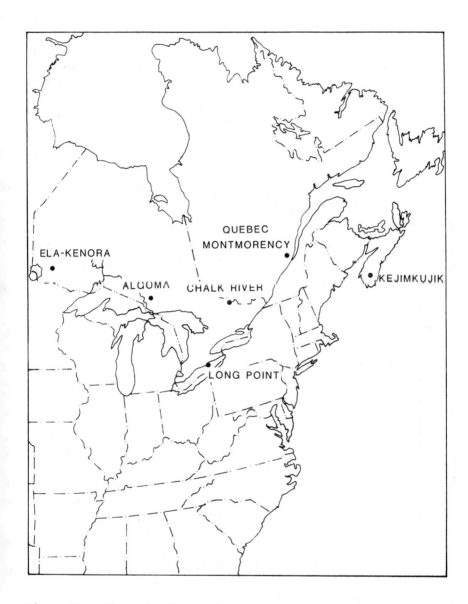

*Figure 3.   Sites in the Canadian Air and Precipitation Moni-
toring Network (APN) at which daily air and precipitation samples are collected.*

seasonal variation. They are maximum in winter, minimum in
summer. On a percentage basis, the amount of variation about
the annual mean concentration depends on location. It is lowest
close to source regions and highest remote from source regions.
The percent standard deviation above the annual average

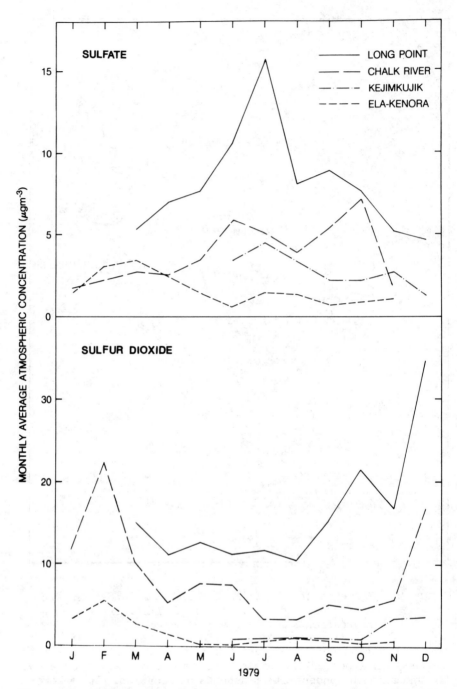

Figure 4.   Temporal variations of the monthly average concentration of atmospheric sulfate and sulfur dioxide at APN sites during 1979.

concentration is 43 at Long Point on Lake Erie; 63 at Chalk River, 500 km away; 75 at Kejimkujik National Park; and 108 at ELA.

Particulate sulfate concentrations are highest in summer at all APN sites except ELA-Kenora. A winter maximum in both sulfur species at Kenora owes its existence largely to meteorological factors. Maximum transport westward from eastern North America in winter coincides with a late winter peak of background sulfur concentrations in arctic air masses (Barrie et al., 1981) that prevail at this mid-continental location during winter.

At APN sites located in the continental pollution plume, the sulfate seasonal cycle is one hundred eighty degrees out-of-phase with the sulfur dioxide cycle. Indications are that the sulfate summer maximum owes its existence to a summer maximum in the conversion rate of sulfur dioxide to sulfate (Barrie et al., 1982). One manifestation of higher sulfur dioxide conversion to sulfate in summer is that, at all stations, the fraction of total airborne sulfur (SO_2 + SO_4) that consists of sulfur dioxide is lowest in summer (e.g., Figure 5a for Long Point).

On a daily basis, the concentrations of sulfur oxides are highly episodic regardless of location in eastern Canada. Polluted and non-polluted episodes of 3 to 6 days duration alternate regularly (Figure 5b). The dry deposition rates of these acidic substances, which is roughly proportional to their atmospheric concentrations, is equally episodic. A comparison of monthly wet and dry deposition rates of sulfur compounds is made later in this section.

Spatial/Temporal Variations of Major Ions in Precipitation

The seasonal variation of monthly precipitation-weighted mean concentrations of the four major ions--hydrogen, ammonium, sulfate, and nitrate--at each APN station during 1979, is shown in Figure 6. At sites central to the major pollutant source region (i.e., Long Point, Chalk River), sulfate concentrations were highest in the summer-half of the year and lowest in the winter-half, while hydrogen and nitrate ion concentrations showed no significant trend. A similar seasonal variation of these ions in precipitation has been observed in northeastern United States (MAP3S/RAINE, 1981). At ELA-Kenora, a site more remote from the influence of continental sources, sulfate did not peak in any particular season, nitrate concentrations varied little, and precipitation acidity was highest in winter. The latter was paralleled by a minimum in ammonium ion

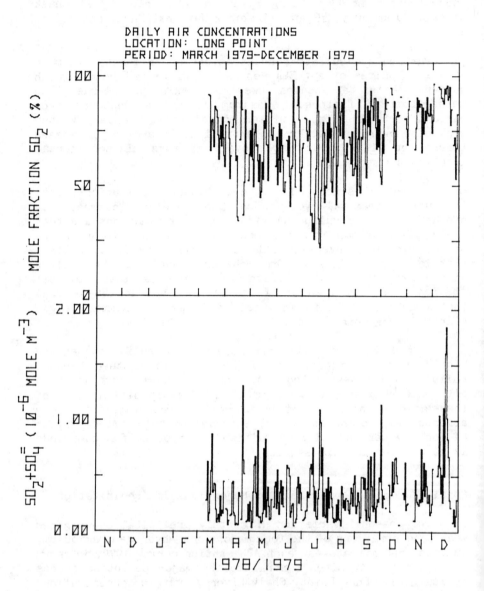

Figure 5.   (a)   The fraction of total airborne sulfur exist-
ing as sulfur dioxide at Long Point on the north
shore of Lake Erie on a daily basis.

(b)   The concentration of total airborne sulfur
at Long Point on a daily basis.   Note the
prevalence of episodes of elevated sulfur levels
of 3 to 6 days duration.

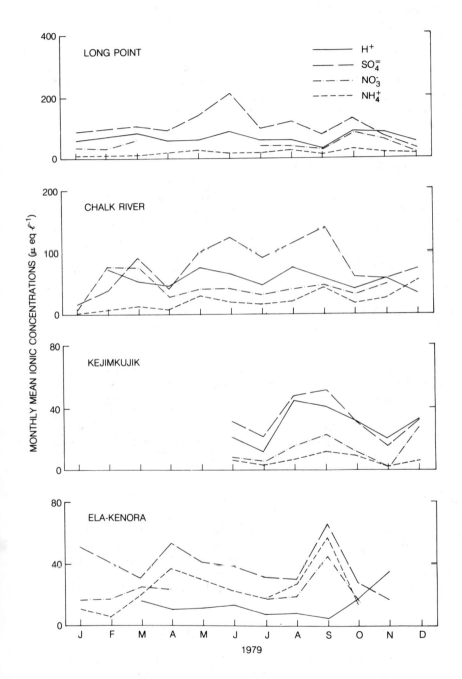

Figure 6.   The temporal variation of monthly precipitation - amount weighted mean concentration of four major ions at APN sites during 1979 (Figure 3).

concentration. On an ion-equivalents basis, sulfate was more abundant than nitrate throughout the year at all stations except during winter at Chalk River when they were equal in concentration.

The spatial distribution of precipitation-amount weighted-mean hydrogen ion concentration in precipitation in eastern North America for 1979 is shown in Figure 7. The distribution was reconstructed using data from several networks: APN, MAP3S, and NADP.

Wet Versus Dry Deposition

Monthly wet and dry deposition of sulfur oxides were derived from APN air and precipitation data, and are shown in Figure 8. Dry deposition was calculated from air concentrations measured by the APN network (Figure 4) and from the deposition velocities listed in Tables 1 and 2. Deposition velocities were estimated using monthly frequency distributions of Pasquill-Gifford stability classes and the technique of Sheih et al. (1979) modified to include higher particulate sulfate surface-resistances. Wet deposition was calculated from the product of precipitation amount and ionic concentrations (Figure 7).

In general, dry deposition of sulfur is not negligible compared to wet. By all rights, the "acid rain" problem would be more properly called an "acidic deposition" problem. The ratio of dry to wet deposition is highest close to source regions. It is also higher in winter than in summer since sulfur dioxide concentrations are highest, and precipitation amount is usually lowest during winter.

Summary

The above results on air and precipitation quality in eastern Canada are examples of the uses of network data. They provide a better understanding of atmospheric and biospheric interactions and a valuable data base, against which atmospheric models must be verified if they are to attain credibility.

MODELING LONG-RANGE TRANSPORT OF AIR POLLUTANTS

The long-range transport model used in LRTAP is described by Voldner et al. (1981) in an intercomparison of model predictions and observations. It is a trajectory box model employing gridded sulfur emissions (127 x 127 km grid),

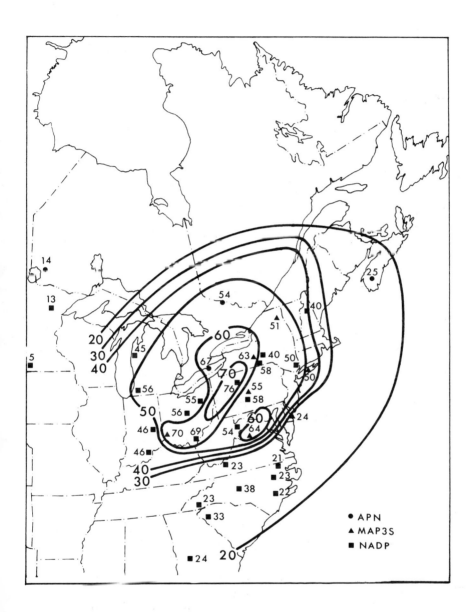

*Figure 7.   The spatial distribution of precipitation-amount
weighted-mean hydrogen ion concentration μmol/l)
in precipitation in eastern North America during
1979 as determined using data from Canadian (APN)
and American (MAP3S, NADP) networks.*

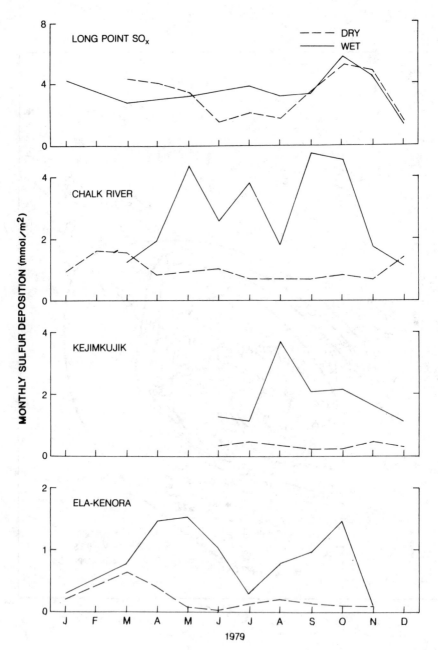

*Figure 8.   The temporal variation of monthly wet and dry
deposition of oxides of sulfur at APN sites (Figure 3).   Dry deposition is calculated from air
concentrations (Figure 4) and deposition velocities (Tables 1 and 2).   Wet deposition is measured.*

Table 1

Estimated Monthly Average Dry Deposition Velocities (cm/sec) of Sulfur Dioxide at APN Sites

Site	Month											
	J	F	M	A	M	J	J	A	S	O	N	D
ELA	0.15	0.15	0.45	0.43	0.43	0.43	0.42	0.42	0.41	0.43	0.43	0.15
Long Point	0.15	0.15	0.71	0.68	0.64	0.41	0.37	0.37	0.55	0.58	0.65	0.15
Chalk River	0.18	0.18	0.29	0.27	0.22	0.22	0.18	0.27	0.22	0.22	0.23	0.18
Kejimkujik	0.18	0.18	0.44	0.37	0.37	0.36	0.30	0.30	0.25	0.28	0.29	0.17

Table 2

Estimated Monthly Average Dry Deposition Velocities (cm/sec) of Particulate Sulfate at APN Sites

Site	Month											
	J	F	M	A	M	J	J	A	S	O	N	D
ELA	0.08	0.08	0.19	0.19	0.20	0.15	0.15	0.15	0.14	0.15	0.14	0.08
Long Point	0.08	0.08	0.10	0.10	0.09	0.08	0.08	0.03	0.10	0.10	0.11	0.08
Chalk River	0.10	0.10	0.32	0.29	0.27	0.30	0.30	0.29	0.18	0.18	0.19	0.10
Kejimkujik	0.09	0.09	0.33	0.33	0.34	0.24	0.25	0.23	0.17	0.17	0.17	0.09

meteorological wind fields (on a 381 x 381 km grid),
precipitation-amount fields, and linear parameterizations of
pollutant transformation and removal processes. It is used to
predict the concentration of oxides of sulfur in air as well
as wet and dry deposition to a particular receptor. A prediction
consists of the reconstruction of the pathway followed by an
air parcel over a 4-day period prior to the date of prediction.
Then, assuming an initially clean air mass, the air parcel is
moved forward in 6-hr jumps. At each interval, the following
processes are mathematically simulated:

(1)   The air parcel receives an injection of sulfur dioxide
       whose magnitude is determined by its location in the
       emission grid.
(2)   A fraction of airborne sulfur dioxide is converted
       to particulate sulfate using a constant hourly
       conversion rate (%/hr).
(3)   A fraction of airborne sulfur dioxide and sulfate is
       removed by dry deposition. Deposition per unit area
       and time is determined by the product of deposition
       velocity and air concentrations. Deposition
       velocities are input parameters.
(4)   A fraction of airborne sulfur dioxide and sulfate is
       removed by precipitation scavenging. The amount
       removed is directly proportional to the product of a
       washout ratio (an input parameter), air concentration
       (a model-predicted parameter), and precipitation
       amount (from observations).

The air parcel is assumed to be well mixed and capped by
an inpenetrable barrier whose height (the height of the mixing
layer) is varied seasonally (average annual depth 1181 m). The
air trajectory used in each calculation is an average of those
at three levels: ground level, 750 m, and 1500 m.

The box trajectory model can be used either in a diagnostic
mode or in a prognostic mode. As an example of its practical
application, the results of a diagnostic exercise conducted by
Olson (1980) to reconstruct the sulfur budget of the eastern
Canadian provinces are presented. The model was run using
emissions for 1978 and the input parameters shown in Table 3.
The results of the budget (Figure 9) are by no means conclusive
but are a first estimate of the impact of one region on another.
For instance, it predicts that the net influx of sulfur to
eastern Canada was 0.65 Tg (million tonnes) in 1978, or 36
percent of total eastern Canadian emissions.

At present, the model is being refined to predict seasonal
variations and is being compared with other models in a joint
Canadian-American intercomparison. In the future, the model

Table 3

Model Input Parameters Used by Olson (1980)
to Construct a Sulfur Budget for Eastern Canada (Figure 9)

INPUT PARAMETERS

SO_2 transformation rate	1%/hr
SO_2 dry deposition velocity	1 cm/sec
SO_4^{2-} dry deposition velocity	0.1 cm/sec
SO_2 washout ratio (volumetric)	5×10^3
SO_4^{2-} washout ratio (volumetric)	8.5×10^5

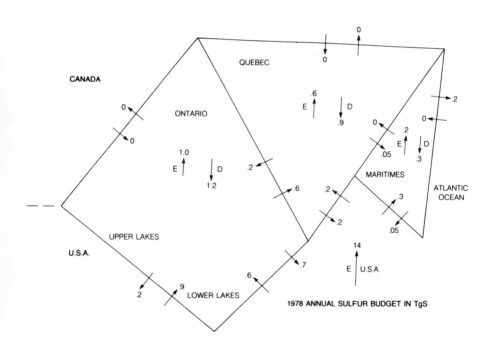

Figure 9.   A sulfur budget for eastern Canada in 1978 computed using a box trajectory model (Olson, 1978). E-emissions; D-deposition.

will be checked and improved using a data base of air and
precipitation concentrations from the APN and CANSAP networks.

SUPPORTING PROCESSES RESEARCH

Several atmospheric studies in LRTAP that are of a
fundamental nature and of relevance to the "acid rain" problem
cannot be classified under modeling or routine monitoring. A
brief description of several studies follows.

Measurements of Particle Deposition to
Snow and Artificial Collectors

Particulate dry deposition is an important pathway of
pollutants between the atmosphere and ecosystems. Methods of
measuring dry deposition are varied, fraught with difficulties
and in some cases are so highly technical that their widespread
use is presently impossible (Hicks et al., 1980). Little
information exists on deposition rates to natural surfaces
(forests, crops, lakes) as a function of particle size. A
common misconception among scientists in the environmental
field is that particle deposition to an exposed bucket is
representative of deposition to the surrounding natural
surface. A tracer study of particle deposition to snow and to
bucket-type receptors (Ibrahim et al., 1980) has shown that
this is not the case. Using ammonium sulfate particles of a
diameter characteristic of that of windblown dust (5 to 10 μm),
it was shown that a bucket collects two to three times more
particles than a flat snow surface (Table 4). Furthermore,
deposition of submicron acidic particles to a bucket is so small

Table 4

Measured Dry Deposition Velocities of 5 to 10 μm
Diameter Ammonium Sulfate Particles to Natural and
Artificial Collectors for Two Tracer Experiments

Snow	Harwell Collector*	Bucket
0.16	0.27	0.78
0.096	0.18	0.30

*A horizontally exposed filter paper (see Barrie,
1980).

relative to soil particle deposition that it is masked by the contribution of supermicron soil sulfate even if the acidic sulfate to soil-sulfate mass ratio is 10:1. This leaves the investigator with an unrepresentative dry deposition catch and leads to the conclusion that buckets are unrepresentative collectors of acidic particulates: Investigation of submicron particle deposition to snow and forests is presently underway.

A Climatological Study of Acidic Snowmelt Shock Potential

A potentially serious environmental problem arises in snow-covered areas during a melt period. Runoff is acidified by pollutants that have accumulated in the snowpack during the cold spell preceding the thaw. Terrestrial and aquatic life is subjected to acidic runoff. The problem is compounded further as most of the snowpack acids are leached by the first 30 to 40 percent of snowpack water to melt, resulting in a pulse of acid at the beginning of a discharge. This acid-pulse is several times more acidic than a melted snowcore and can have severe environmental effects (Leivestad and Muniz, 1976; Jeffries et al., 1979).

A climatological study of acid snowmelt shock has been undertaken to delineate its magnitude and probability in Canada. Coupled with estimated wet and dry deposition rates of acidity, a hydrometeorological model of a snowpack modified to include pollutants is being used to generate snowmelt chemistry statistics (frequency and duration of acid melts). At least 30 yr of historical precipitation and temperature data are input to the model. Details of the results are reported elsewhere (Wilson and Barrie, 1981). The model will also prove useful as a tool for investigating documented cases of fish kill.

Monitoring Nitric Acid and Ammonia

Since nitric acid and ammonia play an important role in the atmospheric acidity cycle, there is an urgent need for information on their ambient concentrations. As part of the joint Canadian-American study of Prolonged Elevated Pollution Episodes (PEPE), ambient concentrations of these substances were monitored routinely on a six hour basis at three rural locations in Ontario during July and August, 1980. Results are forthcoming (Wiebe, 1981).

A Snowpack Chemistry Survey in the Canadian Shield Region

In late January, 1981, a snowpack chemistry survey of the Canadian Shield was undertaken. Using helicopters, snowcores were collected at approximately twenty sites between Dorset, Ontario, on the west, Sept Iles, Quebec, on the east, the St. Lawrence River on the south, and Val D'Or/Chibougamau, Quebec, on the north (Figure 10). The purpose of the survey was twofold:

(1) To determine the concentration and spatial distribution of major ions and trace elements (insoluble and soluble) in a snowpack layer that had been unleached by snowmelt.

(2) To obtain an estimate of the relative magnitude of the wet and dry deposition rate of major ions by comparing total deposition (wet plus dry) measured in the snowpack with wet deposition measured by precipitation networks.

The weather during early winter 1980/81 favored the survey. By late January, temperatures had been persistently sub-zero for a 6- to 12-week period, depending on location. Snowcores were then taken, melted, and analyzed for major ions and trace elements including metals. Preliminary results for snowmelt acidity are shown in Figure 10.

Future Research Efforts

The long-range transport of acid-related substances will continue to be the focus of environmental research for many years to come. In addition, the closely related problem of atmospheric oxidants (e.g., ozone, peroxyacetylnitrate) associated with nitrogen oxide/hydrocarbon emissions and of contaminants such as mercury and organic compounds will receive more attention.

LITERATURE CITED

Barrie, L.A. 1980. The Fate of Particulate Emissions from an Isolated Power Plant in the Oil Sands Area of Western Canada. Ann. New York Acad. Sci. 338:434-452.

Barrie, L.A., H.A. Wiebe, P. Fellin, and K. Anlauf. 1980. APN, The Canadian Air and Precipitation Monitoring Network: A Description and Results for November 1978 to June 1979. Int. Rep. #AQRB-80-002-T, Atmospheric Environment Service, Downsview, Ontario.

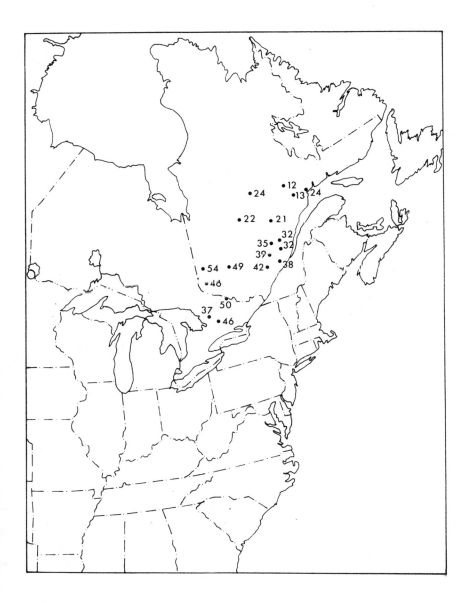

*Figure 10. The spatial variation of snowpack acidity
(μmole H⁺/l) in the eastern Canadian Shield
in late January, 1981, after a 6 to 12 weeks
melt-free period of pollutant accumulation.*

Barrie, L.A., R.M. Hoff, and S.M. Daggupaty. 1981. The Influence of Mid-Latitudinal Pollution Sources on Haze in the Canadian Artic. Atmos. Environ. 15 (in press).

Barrie, L.A., H.A. Wiebe, K. Anlauf, and P. Fellin. 1982. Acidic Pollutants In Air and Precipitation at Selected Rural Locations in Canada. Proc. Amer. Chem. Soc. Symposium on Acid Precipitation, Las Vegas, Nevada, March 28-April 2, 1982.

Berry, R.L. 1979. An Assessment of the CANSAP Project After Two Years of Operation. Int. Rep. #LRTAP-79-9, Atmospheric Environment Service, Downsview, Ontario.

Hicks, B.B., M.L. Wesely, and J.L. Durham. 1980. Critique of Methods to Measure Dry Deposition. Workshop Summary, EPA-600/9-80-050, Research Triangle Park, North Carolina. 271 pp.

Ibrahim, M., L.A. Barrie, and F. Fanaki. 1980. An Experimental and Theoretical Investigation of Particle Deposition to Snow and Artificial Collectors. Int. Rep. #AORB-80-013-T, Atmospheric Environment Service, Downsview, Ontario.

Jeffries, D.S., C.M. Cox, and P.J. Dillon. 1979. Depression of pH in Lakes and Streams in Central Ontario During Snowmelt. J. Fish. Res. Bd. Canada 36:640-646.

Kelso, J.R.M. 1981. Whole Lake Acidification (Chapter 8 of this book).

Leivestad, H. and I.P. Muniz. 1976. Fish Kill at Low pH in a Norwegian River. Nature 259:391-392.

MAP3S/RAINE. 1981. The MAP3S/RAINE Precipitation Chemistry Network: Statistical Overview for the Period 1976-1980. Submitted to Atmospheric Environment by the MAP3S/RAINE research community, coordinator J.M. Hales, Battelle P.N.W. Labs.

Olson, M. 1980. A Computed Sulphur Budget for Eastern Canada. Presented at AGU/CMOS Conference, Toronto, Ontario.

Sheih, C.M., M.L. Wesely, and B.B. Hicks. 1979. Estimated Dry Deposition Velocities of Sulphur Over the Eastern United States and Surrounding Regions. Atmos. Environ. 13:1361-1368.

Voldner, E.C., M.P. Olson, K. Oikawa, and M. Loiselle. 1981. Comparison Between Measured and Computed Concentrations of Sulphur Compounds in Eastern North America. *J. Geophy. Res.* (in press).

Wiebe, H.A.   1981.   Personal communication, Atmospheric Environment Service, Downsview, Ontario.

Wilson, E.E. and L.A. Barrie. 1981. A Climatological Study of Acidic Snowmelt Shock Potential. Proc. Conf. on Long Range Transport of Airborne Pollutants, Albany, New York, May 1. Sponsor, Environment Canada - American Met. Soc.

PART 3

THE EFFECTS OF ACID PRECIPITATION
ON AQUATIC ECOSYSTEMS

CHAPTER 8

CHEMICAL AND BIOLOGICAL STATUS OF HEADWATER LAKES
IN THE SAULT STE. MARIE DISTRICT, ONTARIO

John R.M. Kelso, Robert J. Love, and James H. Lipsit
Great Lakes Biolimnology Laboratory
Sault Ste. Marie, Ontario P6A 2B3

Ronald Dermott
Great Lakes Biolimnology Laboratory
Burlington, Ontario L7R 4A6

INTRODUCTION

In Canada, the perturbative effect of atmospheric deposition on freshwater biotic systems was first recognized from local conditions at Sudbury, Ontario (Gorham and Gordon, 1960; Gorham and Gordon, 1963; Beamish and Harvey, 1972; Conroy et al., 1974). Dillon et al. (1978) indicated that acidic precipitation either already has, or soon will, cause a severe environmental problem in, at least, south central Ontario, which is considerably removed from the sources in Sudbury.

Unlike Scandinavia (Eriksson, 1955; Haapala et al., 1975; Wright et al., 1976; Wright and Snekvik, 1978), Canadian regional effects of long-range transport of atmospheric pollutants are poorly understood. In north central Ontario, 300 km from Sudbury and 500 km from south central Ontario, atmospheric precipitation is only slightly less acidic than in either south central Ontario (Dillon et al., 1978), the eastern United States (Likens et al., 1979), or Scandinavia.

Since vast areas of northern Ontario are in geologically sensitive terrain, and deposition is known to be considerably above expected background levels, we attempted to relate the well being of extant biological communities to chemical status in sensitive headwater lakes within the Ontario Ministry of Natural Resources administrative district of Sault Ste. Marie.

GEOLOGY OF THE SAULT STE. MARIE DISTRICT

The Sault Ste. Marie district lies entirely within the Canadian Shield. The oldest rocks are the metavolcanic and metasedimentary rocks of Early Precambrian (Archean) greenstone belts (Figure 1). Two major greenstone belts exist, one extending in a northeast direction along the Batchawana River and a discontinuous arcuate belt extending north and westward centrally through the district to the south shore of Batchawana Bay. These greenstone belts consist mainly of basic metavolcanic (metamorphosed basalt) and acid metavolcanic rocks (metamorphosed dacite and rhyolite). Metamorphosed sedimentary rocks (graywacke) are locally prominent. Iron formations (mainly magnetite-chert) and intrusive equivalents of the Early Precambrian metavolcanics are present but not significant.

The greenstone belts are surrounded by large areas of Early Precambrian granitic rocks which were emplaced about 2500 million years ago.

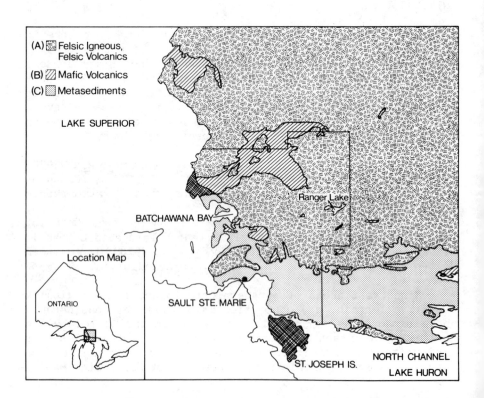

Figure 1.   Bedrock geology of the Sault Ste. Marie district.

Many diabase dikes intrude both the granitic rocks and the older greenstone belts. A limestone-bearing formation extends eastward beyond the city of Sault Ste. Marie. Several lesser geologic formations exist (e.g., mafic volcanics, Keweenawan volcanic rocks, and sandstones of the Jacobsville Formation) but bear few lakes.

Southeastward from Batchawana Bay, the Early Precambrian greenstone and granitic rocks are unconformably overlain by Middle Precambrian rocks of the Huronian Supergroup. The Huronian rocks consist mainly of sedimentary rocks including sandstone, siltstone, mudstone, and conglomerate.

The thin widespread deposits of glacial ground moraine closely reflect bedrock composition (Boissonneau, 1968) and, therefore, should exert only limited influence on the aquatic systems.

METHODS

Between mid-August and mid-November, 1979, 85 lakes were visited either by land (27) or by Bell Jetranger 206B helicopter (58). Most lakes were sampled by helicopter between mid-October and mid-November. We sampled lakes that were not perturbed by logging, mining, or extensive cottage development, and that were not dystrophic bog systems (Welch, 1952).

Monthly "wet only" precipitation samples were collected with automatic moisture-activated samplers operated as part of the Great Lakes Precipitation Chemistry Network. Chemical analyses of samples were performed at the Canada Centre for Inland Waters according to Environment Canada (1974).

All lakes were sampled identically for water chemistry and biota. First, soundings were taken perpendicular to the long axis of the lake and, at the deepest location, temperature was measured using a battery-operated telethermometer. An integrated tube sample of water was taken (MOE, 1979) and at least 1.5 l of water retained. Of the 1.5 l sample, 500 ml was preserved with Lugol's solution for later phytoplankton identification and enumeration; 500 ml was stabilized with 1 ml concentrated nitric acid for trace metals analyses; 250 ml was retained unaltered for major ion analyses; and 250 ml remained for pH, conductivity, and alkalinity analyses.

Conductivity and pH were determined potentiometrically. Alkalinity was determined by a Gran titration (Stumm and Morgan, 1970) using a digital pH meter coupled to an analog strip chart recorder. Additions of standard acid (0.02 N H_2SO_4) were made manually using a 100 μl Eppendorf pipet. Analyses of all other chemical constituents in water followed Environment Canada (1974).

Phytoplankton were identified and counted using the Utermohl inverted microscope technique (Vollenweider, 1969). A diversity index for each lake was calculated using Shannon's formula (Pielou, 1969). The reciprocal averaging ordination method of Hill (1973) was used to determine the similarity of lakes in relation to their species composition.

Fifteen lakes were sampled for benthos during late winter or upon disappearance of ice cover to lessen temporal variation and insure collection of insect larvae prior to their emergence. Eleven of the lakes were sampled the first week of May, 1980, by helicopter from the approximate center of each lake. Comparative samples from the three uppermost lakes of the Turkey Lake watershed (latitude 47°03', longitude 84°23') were collected through the ice during late March. The Upper Headwater Lake of this group is composed of two basins, referred to as Upper Headwater basin 1 (UH1) and Upper Headwater basin 2 (UH2), separated by a beaver dam on a narrow sill.

Four replicate benthic samples were collected from each lake using an Ekman dredge. For each lake, the in situ pH was measured in the top 2 cm of undisturbed sediment using a Metrohm pH meter and narrow probe. The probe was inserted into the sediment through the upper lids of the Ekman dredge. A 100 ml subsample of surface sediment was retained for analyses of sand, water, and organic content (from loss on ignition at 450 °C over 2 hr). Sediment samples, because of their consistency, were sieved through a 500 μ-mesh net. Residues and organisms retained were preserved in 10 percent formalin buffered with $NaHCO_3$. Organisms were stained with Rose Bengal to enhance contrast in the large volume of dark residue and hand picked over a white background. Taxonomic references include Pennak (1978), Clark (1973), and Brinkhurst and Jamieson (1971), and Oliver et al. (1978).

In 31 of the lakes, a bottom set gill net was installed perpendicular to shore and lifted 24 hr later. Each net was 64 m long and consisted of 7 panels, 1.5 m deep by 9.2 m long, ranging in mesh size from 3.8 to 12.7 cm in 1.3 cm intervals. Two net sets were made in each lake.

Captured fish were identified, measured, weighed, a scale sample taken, and frozen individually. Each fish was homogenized by being passed thrice through a grinder following removal of gonads and stomach contents.   Subsamples were analyzed for:

(1)   total lipid by fat extraction;

(2)   PCB's, DDT, and metabolites, dieldrin, hepta-chlorepoxide, chlordane, and HCB by the Ontario Ministry of Agriculture and Food, Pesticide Residue Testing Laboratory, Guelph, Ontario;

(3)   mercury using a modification of the method described for sediments (Environment Canada, 1974); and

(4)   calcium, magnesium, potassium, sodium, copper, nickel, zinc, and lead by digestion of ashed tissue in a nitric acid matrix and detection by atomic absorption or emission spectrophotometry.

RESULTS AND DISCUSSION

In all, 150 lakes were visited and 85 sampled. Chemical results are presented for only 75 of the sampled lakes as 10 were later found either not to be headwaters or to receive significant bog drainage. A charge balance constructed from our limited chemical analyses showed that the cations and anions which we measured are the dominant ions in the system. The discrepancy in the charge balance was less than 200 µeq/l in all but 12 percent of the lakes, indicating, essentially, that analytical techniques were sound.

Precipitation Chemistry

Monthly mean pH's were within a narrow range (Figure 2) and annual mean values ranged from a high of 4.52 in 1976 to a low of 4.25 in 1979 (Table 1). The pH in precipitation was generally lowest during late winter and early spring (Figure 2) and coincided generally with the temporal deposition pattern observed in Scandinavia (Barrett and Broden, 1955).

Sulfate concentrations ranged between 1.4 and 7.9 mg/l (Table 1) with mean annual values between 2.75 (1977) and 3.94 (1978). High concentrations (5 mg/l or greater) occurred in each year, usually in April, May, or June. Mean annual values were lower than the 8 mg/l reported by Sanderson and LaValle (1979) for southern Ontario, but similar to those reported by Dillon et al. (1978) for south-central Ontario and Fisher et al. (1968) for New Hampshire. It is likely that results of Sanderson and LaValle (1979) were influenced by the lithology and well-buffered soil of southern Ontario.

Table 1

Concentrations of Ions and Metals in "Wet Only" Precipitation Samples from the Sault Ste. Marie District - Concentrations of Ions are in mg/l and Metals are in µg/l - Metal Data are Available Only from 1976 and 1977

Chemical Parameter		Year			
		1976	1977	1978	1979
		------mg/l------			
pH	average	4.52	4.42	4.26	4.25
	range	3.90-6.20	4.0-4.7	3.9-4.5	4.02-4.45
	S.D.	0.79	0.20	0.17	0.19
Sodium	average	1.04	0.21	0.15	0.14
	range	0.10-5.4	0.04-0.70	0.10-0.30	0.05-0.30
	S.D.	1.95	0.20	0.08	0.10
Calcium	average	0.50	0.31	0.46	0.67
	range	0.20-1.3	0.02-0.6	0.10-1.7	0.12-1.8
	S.D.	0.41	0.17	0.51	0.69
Magnesium	average	0.17	0.04	0.07	0.17
	range	0.05-0.75	0.01-0.06	0.04-0.13	0.02-0.65
	S.D.	0.26	0.02	0.03	0.27
Potassium	average	0.18	0.10	0.10	0.21
	range	0.05-0.47	0.05-0.22	0.01-0.22	0.10-0.47
	S.D.	0.15	0.05	0.07	0.17
Chloride	average	1.92	0.22	0.60	0.37
	range	0.23-8.9	0.08-0.45	0.13-2.4	0.22-0.50
	S.D.	0.31	0.12	0.76	0.10

Sulfate

average	3.51	2.75	3.94	2.88
range	1.4-6.9	1.5-5.0	2.2-7.9	1.4-4.8
S.D.	1.90	1.25	1.90	1.47

Metals Average

	Cadmium	Copper	Iron	Lead	Nickel
			----μg/1----		
Average	1.6	7.3	14.5	8.2	2.1
S.D.	3.5	4.8	7.9	3.9	2.3
Range	0.2-12.0	1.0-16.0	3.0-26.0	1.0-14.0	0.5-9.0
Number	11	13	13	13	13

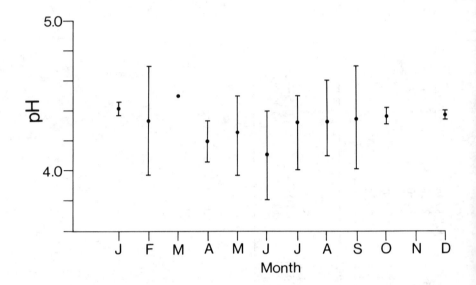

*Figure 2.   Monthly pH values in precipitation at one site in
the Sault Ste. Marie district.   Data are for 1976-
1979 inclusive and bars represent ranges.*

Sodium, potassium, calcium, chloride are all (Table 1) similar to the inland "polluted" regions of Scandinavia and eastern United States summarized by Likens et al. (1979).

The trace metals cadmium, copper, iron, lead, and nickel were always present above detection limits during 1976 and 1977 (Table 1).   Concentrations of iron were highest followed by lead and copper. Cadmium was least abundant. Loadings of these metals were greater for copper and cadmium (85 and 294%, respectively), lower for iron and lead (91 and 45%, respectively), and similar for nickel to the rural areas of Copenhagen (Andersen et al., 1978).

In general, although deposition of heavy metals tends to reflect hemispheric demography, it is apparent that a large area of Ontario extending in at least an arcuate belt from Muskoka-Haliburton area through Sudbury to the Sault Ste. Marie district is under a high acidic loading similar to the eastern United States and Scandinavia.

Geological Influence

The acid-neutralizing capacities of natural waters are expected to depend upon the carbonate-bicarbonate buffer system. The Canadian Shield is expected to foster oligotrophy in lakes; however, the lithology in the Canadian Shield is broadly variable and well-buffered lakes exist (Conroy and Keller, 1976). The geological influence within the Canadian Shield was assessed through categorization of lakes into those situated on:

(1) high silicate: sodium- and potassium-enriched rocks; e.g., granites, felsic volcanics;

(2) low silicate: calcium, magnesium, and iron-enriched rocks; e.g., gabbro, diabase, mafic volcanics; and

(3) metasediments; e.g., graywacke, arkose, chert, siltstone.

Most lakes sampled were on granitic substrates (approximately 60% of lakes sampled) and without exception were poorly buffered (Figure 3). Probably because of the diversity in lithology within stratigraphic groupings (Giblin and Leahy, 1967), remaining lakes showed a much broader range in buffering capacity. Invariably, however, better buffered systems were on non-granitic or conglomerate rock types.

General Lake Characteristics

Lakes in the sampled district lie in two major drainage basins flowing into southern Lake Superior and northwestern Lake Huron. The combined area of these two watersheds is 862,000 ha. Lakes sampled ranged in size (Figure 4) from 1.6 to 110 ha (mean 17 ha $\pm$ SD 20 ha). Lakes in this size range constitute 98.7 percent of all lakes within the district (Cox, 1978). The Ontario Ministry of Natural Resources (OMNR) sampled lakes in a sport fish lake inventory program and lake sizes in their program ranged from 4.3 to 1149 ha (mean 98 ha $\pm$ SD 40 ha). Neither data set is fully representative of lake size in the district, according to Cox (1978). Lakes in the two major watersheds in the study area fall into four major size categories (from Cox, 1978):

Lake Size (ha)	Number of Lakes
>1000	3
100 - 999	60
10 - 99	784
1 - 9	3124
<1	960

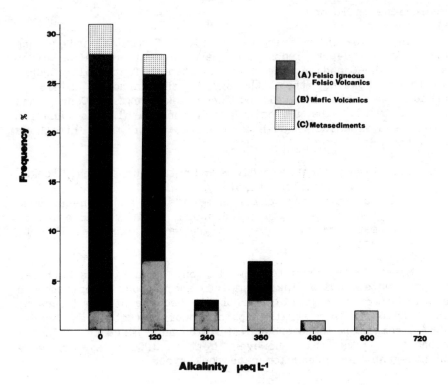

*Figure 3.   Alkalinity, ueq/l, in relation to bedrock type
(see text and Figure 1 for description type A,
B, C) for the Sault Ste. Marie district.*

OMNR sampled lakes are larger, accessible, and usually support
viable sport fish populations. We selected headwater lakes,
many of which were inaccessible, by land with unknown fisheries.

However, if conductivity (the most reliable chemical
parameter in the OMNR data set) is generally indicative of
sensitivity to acid addition, the two data sets are
representative of lake chemistry within the district (Figure
4). From both data sets, it is apparent that a large proportion
of lakes (69% of our lakes, 46% of OMNR sampled lakes) have
conductivites below 30 umhos/cm and must be considered highly
sensitive (Environment Canada, 1979) to acid deposition.

In the Sault Ste. Marie district, lakes lie between elevations of 290 and 549 m (mean 406 m ± SD 62 m) above mean sea
level. Lake depth ranged from 2 to 28 m (mean 8.7 m ± SD 5.1
m). Since we expected depth to severely restrict biotic

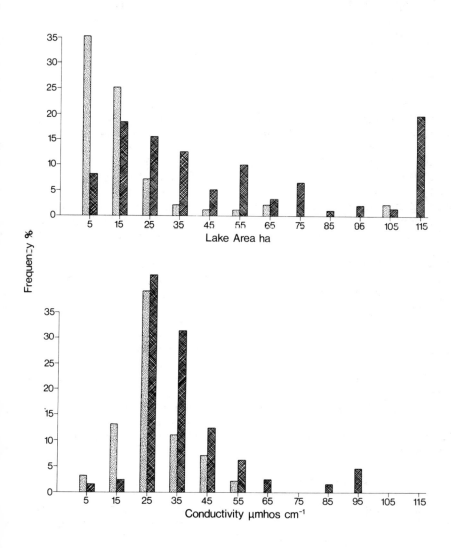

Figure 4. *Frequency of occurrence for area (ha) and con-*
ductivity (μmhos/cm) in OMNR fisheries (cross-
hatched) and our (stippled) data.

communities, we selected lakes with depths of at least 3 m;
only one lake had a depth less than this minimum.

Although all lakes in our sample were headwaters and the
ratio of drainage basin area to lake area was expected to be

small, the ratio ranged from 1.5 to 130.1 (mean 7.6 m $\pm$ SD 15.7). Those lakes with large drainage basins (ratio> 20) were few (4% of lakes) and were with low relief yet possessed no characteristics forming a basis for rejection.

Lake Chemistry

Principal component analysis was applied to the 19 variables which reflected morphometry (lake area, drainage basin area, ratio of drainage basin area to lake area, "maximum" detected depth, elevation), major ion chemistry (pH, conductivity, alkalinity, Ca, Mg, Na, K, SO_4, Cl), and metal concentrations (Al, Pb, Zn, Cu, Ni) for 75 lakes. This analytical approach served to eliminate variables (e.g., pH) when that variable loaded heavily on more than one factorial component or when variance was unique to that variable (e.g., potassium and elevation). Just as important, principal component analysis served to group variables that loaded heavily on the same factor in the terminal solution (e.g., lead and aluminum). Following elimination of variables, the groupings then led to an iterative examination of lake chemistry and morphometry.

Calcium, Magnesium, Carbonate-Bicarbonate, Sulfate

In most relatively dilute freshwaters, it is expected that calcium and magnesium are the dominant cations and that the bicarbonate exceeds the sulfate ions except in acid lake waters (Hutchinson, 1957). In pristine poorly buffered systems in Canada and Europe, this appears to be the case (Henriksen,1979; Schindler et al., 1980).

In our poorly buffered systems (Figure 5), two cations (Ca^{2+} and Mg^{2+}) and two anions (HCO_3^- and SO_4^{2-}) are the overall dominant ions in the aquatic system. The balance among these four ions persisted over the range of alkalinities although the majority of data was from dilute systems.

Henriksen (1979) contends that since alkalinity of water is derived essentially from weathering of carbonate materials, the calcium and magnesium content should balance the bicarbonate component in systems unaffected by acidic precipitation. Henriksen (1979) determined that this was the case in Norway, and Dillon (unpublished) has determined that this also applies to unaffected lakes in Canada. This expected balance exists in only a few of the lakes (5 to 10%) in our study area (Figure 6) as most lakes, particularly below alkalinities of 400 μeq/l,

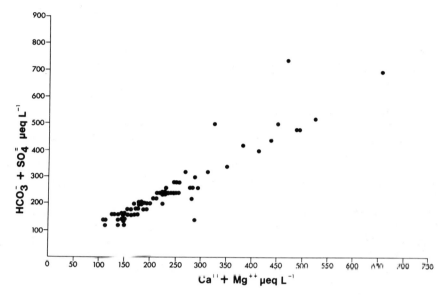

Figure 5. The relation between calcium plus magnesium and
bicarbonate plus sulfate for 75 lakes in the
Sault Ste. Marie district.

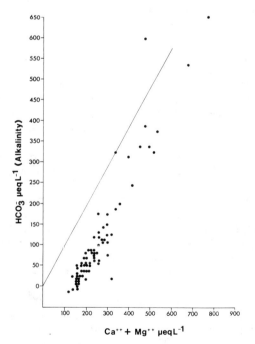

Figure 6. The relation between calcium plus magnesium and
bicarbonate for 75 lakes in the Sault Ste. Marie
district.

have an obvious bicarbonate deficiency. Since the methodologies used have been corroborated and the charges balance, we believe that the deviation from expected is real in these sensitive headwater systems and likely results from atmospheric deposition. If, as suggested by Dillon (unpublished data), bicarbonate depletion reflects extent of acidification, the alterations to the Sault Ste. Marie lakes roughly parallel those within the direct influence of the Sudbury stack.

In our lakes, calcium consistently exceeded magnesium by a factor of 2 to 4 (Figure 7b) until alkalinity was greater than approximately 100 to 120 μeq/l. At this point, calcium concentrations increased up to 12 times greater than magnesium. Further, at alkalinities less than 100 to 120 μeq/l, bicarbonate was no longer dominant (Figure 7a), and pH values tended to decrease (Figure 7c). It appears that although weathering may play an important role, particularly in aquatic systems above 200 μeq/l, sulfate contributed by the atmosphere above a base availability (Dillon, unpublished data) is a balancing force in poorly buffered waters.

Our analysis of buffer status is notable for its exclusion of pH. Because of its apparent plasticity and its unique variance in principal component analysis, pH was not extensively used to describe system status.

Metals

Concentrations of metals in freshwater systems may be increased directly through deposition (Beamish and VanLoon, 1977) or indirectly through mobilization by the acidification process within the watershed (Wright and Gjessing, 1976).

In our lakes, both aluminum and lead were significantly correlated ($P < 0.05$) with the bicarbonate ion concentration and increased with decreasing alkalinity (Figure 8). Aluminum increased from relatively low levels of 10 to a maximum of 240 μg/l detected at 0 μeq/l bicarbonate. Concentrations appeared to increase when bicarbonate fell below 100 μeq/l. Mean total aluminum, 53 μg/l SD $\pm$ 40, was only slightly higher than that detected by Scheider et al. (1979) and, as they discussed, reflects a position intermediate between severely affected and slightly affected systems in Canada and Norway. Lead followed a similar pattern and increased from a low of 1 to a maximum of 67 μg/l, again observed when bicarbonate was non-existent. Since both these metals showed a high negative correlation ($P < 0.05$) with alkalinity (Figure 8), it appears the behavior of both metals is influenced by the buffer status within the system, since deposition may be assumed constant within the district.

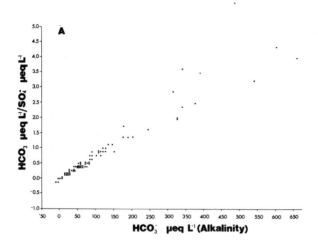

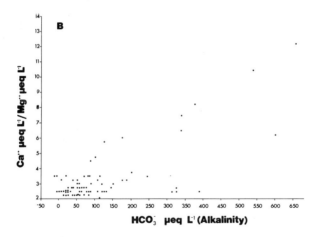

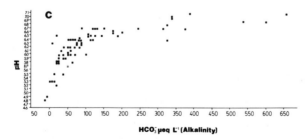

Figure 7. The relation between (a) the ratio of bicarbonate
          to sulfate ions and bicarbonate ions, (b) the
          ratio of calcium to magnesium ions and bicarbonate
          ions, and (c) pH and bicarbonate ions for 75 lakes
          in the Sault Ste. Marie district.

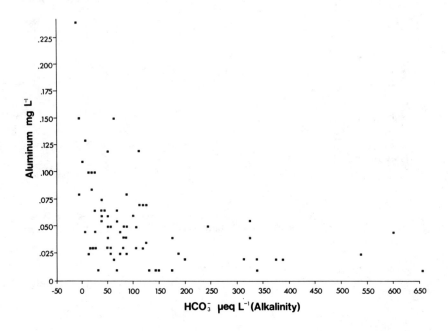

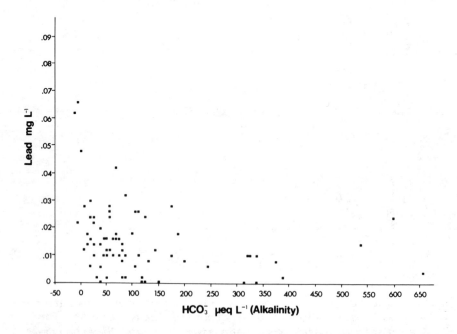

*Figure 8.   The relation of aluminum and lead to alkalinity in
75 lakes from the Sault Ste. Marie district.*

Concentrations of copper, nickel, and zinc were low and minimally influenced by difference in pH and buffering:

Concentration (µg/l)

Metal	Mean	Range	SD
Copper	2	1 - 8	2
Nickel	2	1 - 5	1
Zinc	5	1 - 13	3

These three metals were also lower than concentrations detected in lakes of southern Ontario (5.7, 3.6, 12.6 µg/l mean concentrations for Cu, Ni, and Zn, respectively) by Dillon et al. (1977). We have no explanation for these low levels even though our study area is only 350 km from Sudbury and levels in precipitation were higher (Cu) or similar (Ni) to Scandinavia.

Phytoplankton

Only the phytoplankton species that occurred in the 56 lakes sampled in October and November will be discussed in detail.

Chrysophyta. The Chrysophyceae dominated the algal composition in almost every lake encountered. The ubiquity of the genus Chromulina was striking. Other important genera were Ochromonas, Pseudokephyrion, Chrysochromulina, and Dinobryon. The Dinobryon species were those identified in other acidic locations: D. bavaricum, D. divergens, D. sertularia, and D. suecicum var. longispinum (Johnson et al., 1970; Almer et al., 1974). The Bacillariophyceae were not an important group. The most prominent species in this group were Cyclotella comta, C. kutzingiana, C. stelligera, Synedra radians, and S. vaucheriae.

Chlorophyta. The green algae were the second most important group in the lakes. Oocystis pusilla and O. lacustris were by far the most important species. Other important species were Selenastrum minutum, Sphaerocystis schroteri, Crucigenia tetrapedia, Tetraedron minimum, Chlorella sp., Ankistrodesmus falcatus, and Quadrigula closterioides.

Pyrrhophyta. The yellow-brown algae were unimportant in all lakes, a stark contrast to lakes in Sweden (Almer, 1974) and Sudbury district of Ontario (Yan and Stokes, 1978). Of the Dinophyceae, the genus Gymnodium was only occasionally

present.   Of the Cryptophyceae, Rhodomonas minuta and
Cryptomonas erosa were present in most lakes.

Cyanophyta. Blue-green algae were absent from four lakes
while Agmenellum quadruplicatum and Coccochloris stagnina were
dominant in two.   The percent composition of blue-green algae
in terms of numbers of individuals was generally low. The most
common species besides those above were Anacystis cyanea, A.
thermalis, and Oscillatoria planktonica.

The additional 14 lakes sampled in August resulted in an
increase in the total number of species from 150 to 185.
However, the increase was mainly the result of single
occurrences of Bacillariophyceae and Chlorophyta. The overall
species composition remained unchanged from the fall samples
despite an increase in the number of organisms per ml.

The lakes in this study correspond to those classified as
types B and C by Schindler and Holmgren (1971) in which the
Chrysophyta were dominant (Kling and Holmgren, 1972; Schindler
et al., 1973; Findlay and Kling, 1975).   Comparable species
composition has been found in other shield lakes (Ostrofsky and
Duthie, 1975).

Studies of acid-stressed lakes in Scandinavia and Ontario
have shown Chrysophyceae to be less important than other groups
in their contribution to the species composition (Johnson et
al., 1970; Almer et al., 1974; Scheider et al., 1975; Scheider
and Dillon, 1976; Hendrey and Wright, 1976; Wright et al., 1976;
Kwiatkowski and Roff, 1976; Conroy et al., 1976; Yan and Stokes,
1978).   Figure 9a demonstrates the dominance of the
Chrysophyceae at every pH interval. The contribution by Chrysophyceae to the composition, over 60 percent at every pH
interval, is an apparent contradiction of the results of other
researchers who have found a shift in species composition
concomitant with a change in pH.   Johnson et al. (1970) and
Hendrey and Wright (1976) screened their samples through a 24
μ mesh net prior to analysis. This treatment would not retain
the small Chrysophyceae which were important in our lakes.

Kwiatkowski and Roff (1976) found phytoplankton diversity
indices ranging from 1.32 to 3.80 in acid lakes in the LaCloche
Mountains of Ontario. This coincides well with the range of
1.45-3.96 found in our study.   However, they found a direct
relationship between diversity index and pH at pH < 5.6 and a
somewhat constant relationship at pH ≥ 5.6. No such relationship
existed in our data.   The regression of diversity upon pH
produced a weak correlation of 0.44 (P < 0.05).

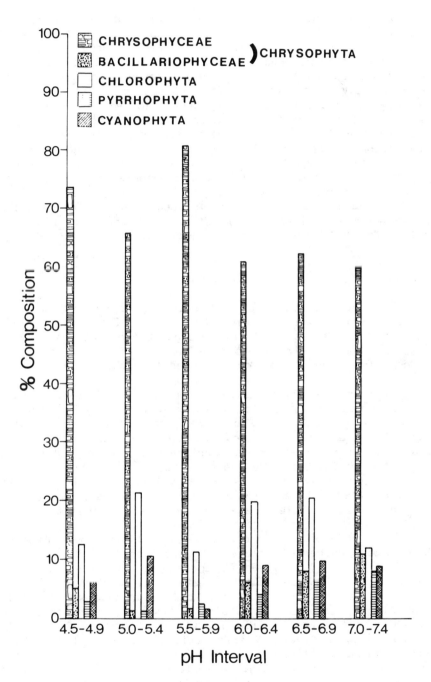

Figure 9.   Effect of pH on phytoplankton (a) percent composition (from numbers of individuals), (b) number of species.

The technique of reciprocal averaging ordination was employed to orient lakes on a continuum based on species presence or absence. Culp and Davies (1980) considered reciprocal averaging ordination an important tool in delimiting site grouping (in this case lakes) through an interpretation of community composition. The ordination of all 70 lakes produced a separation of those lakes sampled in August from those sampled in October-November. This separation would be due to the additional 35 species present in the August samples. Ordination of the 56 fall samples does not produce a major separation of those lakes. The cluster of 43 of the 56 lakes indicates the homogeneity of the fall samples. Phytoplankton appear to be great integrators of the physical-chemical lake features discussed in previous sections. The species composition varied only slightly despite the wide range of values found for the parameters measured.

Benthos

The benthic fauna was not diverse, comprising only 42 genera, which were dominated by the Chironomidae and Chaoboridae. The relevant physical and biological features of the lakes have been summarized in Table 2. One lake, in which no organisms were found, was excluded from the analysis.

Sediments in all lakes were composed of soft dark refractory material having a high water content (92 to 96% wet wt) and organic content (35 to 56% dry wt) characteristic of Canadian Shield lakes. These lakes fall into Hamilton's (1971) type C and D categories in terms of surface area, depth, and a fauna dominated by Diptera. Mean depth sampled was 9.8 m, yet the total abundance averaged $273/m^2$ (SE = 137.8) and the standing stock averaged 96.5 mg/m^2 (SE = 37.5). This is low in comparison to other Canadian Shield lakes (Rawson, 1960; Hamilton, 1971).

Parsons (1968) and Hendrey and Wright (1976) found a reduction in the number of species in waters of low pH. Mossberg and Nyberg (1979) observed no correlation between number of individuals and pH in 7 Swedish lakes with a pH range between 4 and 5. In the present study, total abundance was slightly reduced in lakes of low pH (Table 3); however, the relationship was not satistically significant. Likewise, a poor relationship existed between diversity index (Pielou, 1969) and pH. Roff and Kwiatkowski (1977) found only a slight decrease in diversity over the pH range examined, in spite of a significant decline in zooplankton diversity below pH 5.3. Alkalinity and total benthic abundance in our lakes were related (r = 0.6) while the diversity index was negatively correlated with sample depth (r = -0.55).

Table 2

Some Physical and Biological Aspects of the Sault Ste. Marie Study Lakes — Means for Each Category are Given Including the Percent Composition

Category pH	Lake	Sample Depth	Z_r %	pH Water	pH Sed.	Total n/m²	Chir.	Chaob.	Olig.	Total Wt. mg/m² dry
>6.0	Little Turkey	11	–	6.5	6.1	276	180	89	0	275.3
	JR-01	16	0.4	6.5	6.0	11	–	–	–	1.6
	JR-28	10	3.2	6.0	5.7	283	234	42	0	119
	Evans	18	–	7.0	6.6	42	0	0	42	54
	Maple	14	4.4	6.9	6.2	10	0	0	10	0.2
	Hult-7	12	2.1	6.7	6.3	0	–	–	–	0
	$\bar{x}_5$	13	2.6	6.6	–	124	66%	21%	8%	–
5.9–5.4	Upper Head 2	9	2.6	5.5	6.0	107	25	33	0	11
	Lower Head	5	1.1	5.6	5.9	898	378	283	67	159
	JR-35	5	1.2	5.9	5.8	2025	1049	430	104	543
	Roussain	14	2.9	5.8	5.7	63	0	41	22	10
	$\bar{x}_4$	8	1.9	5.7	–	773	47%	26%	5%	–
<5.3	Upper Head 1	9	–	4.8	5.6	107	29	3	74	38
	JR-02	14	3.6	4.9	6.0	95	0	95	0	53
	JR-04	4	1.9	5.2	5.9	94	62	22	0	57
	JR-05	3	0.8	4.6	5.9	56	14	0	28	121
	JR-36	4	1.4	5.3	5.7	42	10	32	0	6
	$\bar{x}_5$	7	1.9	4.9	–	78	30%	38%	26%	–

Table 3

Early Spring Abundance of Benthic Fauna (Excluding Diptera) in Some Sault Ste. Marie District Lakes are Arranged in Order of Increasing Minimum Surface pH – An Asterisk Indicates Occurrence of Organisms in Littoral Samples from Lakes in the Turkey Lake Watershed

Benthic Fauna	Lake													
	JR5	UH1	JR2	JR4	JR36	UH2	LH	R	JR35	JR28	JR1	LT	M	E
	$number/m^2$													
Nematoda	--	--	--	--	--	--	34	--	52	--	--	--	--	--
Lumbriculus	37	--	--	--	--	--	--	--	--	--	--	--	--	--
Enchytraeidae	--	65	--	--	--	--	--	--	2	--	--	--	10	--
Imm. tubificids	--	--	--	--	--	--	21	22	48	--	--	*	--	--
Tubifex tubifex	--	4	--	--	--	--	--	--	--	--	--	--	--	--
Limnodrilus hoffmeisteri	--	3	--	--	--	--	--	--	11	--	--	--	--	32
Uncinais uncinata	--	--	--	--	--	--	34	--	--	--	--	--	--	--
Nais spp	--	--	--	--	--	--	11	--	27	--	--	--	--	--
Slavina appendiculata	--	--	--	--	--	--	3	--	17	--	--	--	--	--
Dero spp	--	--	--	--	--	--	3	--	--	--	--	--	--	--
Sperchonopsis	--	4	--	--	--	--	*	--	--	--	--	*	--	--
Unionicola	--	--	--	--	--	28	--	--	31	--	--	--	--	--
Pisidium spp	--	--	--	--	--	9	*	--	167	--	--	5	--	--
Hyalella	--	*	--	--	--	*	10	--	--	--	--	*	--	--

Taxon													
Caenis	—	*	—	—	—	42	—	9	*	—	—	—	*	—
Leptoceridae	—	*	—	—	—	42	—	*	—	—	—	—	—	—
Hydroptilidae	—	—	11	—	—	95	—	*	*	—	—	—	—	—
Dytiscidae	—	*	—	—	—	—	—	—	—	—	—	—	—	—
Elmidae	—	5	—	—	—	—	—	—	—	—	—	—	—	—
Sialis	—	*	—	—	—	—	—	11	*	—	—	—	—	—
Diversity H Number	0	0	1.4	0	1.0	2.7	0.5	2.3	1.7	0.6	1.0	0	1.0	0.6

Variance of the total abundance, numbers of Chaoboridae, and total standing stock (mg/m^2 dry wt) was greater within the lakes of a pH category than between the pH categories, indicating that factors other than pH had a significant effect (P < 0.01).

Hierarchical cluster analysis of the percent similarity of community using Ward's minimum variance (Figure 10) delimited the lakes with respect to the diversity index.   The Upper Headwater basin 2, Lower Headwater, and JR-35 all had a diversity above 1.5, while the group including JR-4 and JR-28 had diversity above 0.9.   In the discrete group of lakes JR-1, JR-5, Maple, and Evans, the diversity index was zero except JR-5, where the two species present were not found in any of the other lakes.

Although oligochaetes were not abundant, the tubificids, which are considered tolerant of acidic conditions (Parsons, 1968; Roff and Kwiatkowski, 1977), were present over a wide range in lake pH (Table 3). The Naididae were collected only in lakes of pH > 5.6, and being typically littoral, at depths < 5m.

Pisidium <u>casertanum</u>, <u>P.</u> <u>variable</u>, and <u>P.</u> <u>nitidum</u> were collected in lakes of pH > 5.5.   In two oligohumic lakes in Sweden, Mossberg (1979) found a great reduction in the abundance in <u>Pisidium</u> since the 1940s, during which time the pH had decreased from 6.3 to 5.2

The amphipods <u>Hyalella</u> <u>azteca</u> and <u>Crangonyx</u> <u>richmondensis</u> were abundant in the littoral zone of both basins of Upper Headwater lake, although they were not collected in Ekman samples.   According to Bousfield (1958) this latter, large gammarid is often abundant in small, typically acidic, lakes and bogs of the Laurentian Shield.   Okland (1969), Sutcliffe and Carrick (1973), Roff and Kwiatkowski (1977) observed <u>Gammarus</u> absent in water having a pH below 5.9 and the presence of <u>Crangonyx</u> in water with a nominal pH below 5.5.   This indicates a wider range of pH tolerance in the family Gammaridae.

Diptera and Chironomidae, in particular, are considered tolerant to a wide range of acidic conditions (Parsons, 1968; Wiederholm and Eriksson, 1977).   Chaoborids were collected from 8 of the 15 lakes with <u>G.</u> <u>americanus</u> and <u>C.</u> <u>brunskilli</u> restricted to the Upper Headwater lake.   The distributon of the genus <u>Chaoborus</u> showed a slight increase in percent abundance with lower pH.   The distribution of the four species (Table 4) is consistent with their requirements (Hamilton, 1971; Pope <u>et</u> <u>al</u>., 1973), the major discriminating factors being depth and presence of fish.

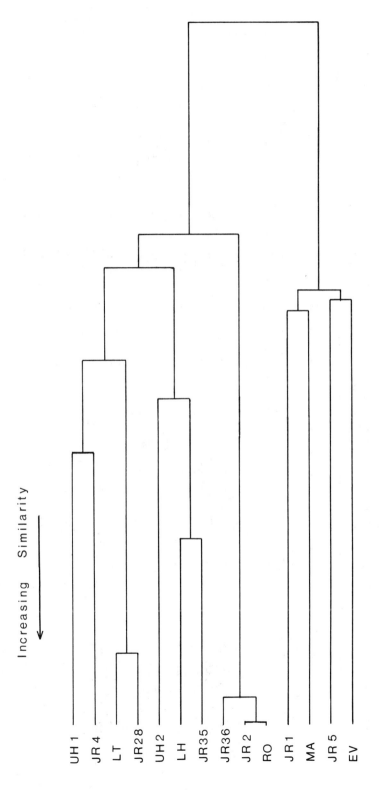

Figure 10. Percent similarity of community cluster dendogram of the benthic communities clustered by Ward's minimum variance.

Table 4

Early Spring Abundance of Diptera in Some Sault Ste. Marie District Lakes - Arrangement of the Lake is in Order of Increasing pH - An Asterisk Indicates the Occurrence of Organisms in Littoral Samples From Lakes in the Turkey Lake Watershed

Diptera	Lake										
	JR5	UH1	JR2	JR4	JR36	UH2	Lh	R	JR35	JR28	Lt
	--number/m^2--										
Ceratopogonid	--	--	--	--	--	--	45	--	--	--	--
Chaoborus americanus	--	3	--	--	--	--	--	--	--	--	--
C. brunskilli	--	--	--	--	--	33	--	--	--	--	--
C. flavicans	--	--	95	--	36	--	--	--	42	--	89
C. punctipennis	--	--	--	21	--	--	283	42	431	--	--
Procladius spp	18	--	--	--	--	--	--	--	95	--	--
Ablabesmyia	--	--	--	--	--	*	*	--	--	--	*
Psectrocladius	--	--	--	--	--	*	11	--	--	--	*
Eukiefferiella	--	--	--	--	--	--	10	--	--	--	36
Zalutschia	--	--	--	21	--	--	--	--	11	--	--
Heterotanytarsus	--	--	--	--	--	*	--	--	59	--	--
Cladotanytarsus	--	--	--	--	--	*	*	--	109	--	--
Tanytarsus spp	--	--	--	--	--	*	82	--	194	--	*

	1	2	3	4	5	6	7	8	9	10	11
Cryptochironomus	—	—	—	—	—	17	21	—	40	—	*
Paracladopelma	—	—	—	—	—	*	17	—	—	—	*
Cladopelma	—	29	—	54	10	—	21	—	—	—	—
Chironomus anthracinus	—	—	—	—	—	8	—	—	73	—	23
C. salinarius gr.	—	—	—	—	—	—	—	—	160	—	118
Dicrotendipes	—	—	—	—	—	*	10	—	207	—	*
Glyptotendipes	—	—	—	—	—	*	11	—	69	—	—
Nilothauma	—	—	—	—	—	—	92	—	—	—	—
Microtendipes	—	—	—	—	—	—	—	—	137	—	*

The species composition of the Chironomidae inhabiting the Sault Ste. Marie lakes was similar to other small Canadian Shield and oligohumic lakes (Hamilton, 1971; Mossberg, 1979). Typical of shield lakes were the <u>Polypedilum</u>, <u>Psectrocladius</u>, <u>Dicrotendipes</u>, and <u>Chironomus salinarius</u> group; while the <u>Heterotanytarsus</u>, <u>Zalutschia</u>, and <u>Chironomus anthracinus</u> group were more typical of humic lakes with a pH between 4.8 and 5.7. In Swedish lakes exhibiting a chronological increase in acidity, the <u>Chironomus anthracinus</u> group and <u>Limnochironomus</u> (<u>Dicrotendipes</u>) displayed an increase in abundance at pH < 5 (Mossberg, 1979). <u>Chironomus</u>, tolerant of a wide range of environmental conditions, is apparently resistant to very acidic conditions (Harp and Campbell, 1967). Mossberg (1979) noted that a few of the small Orthocladiinae, particularly <u>Heterotrissocladius</u> sp and the Tanytarsini, <u>Cladotanytarsus</u> and <u>Stempellina</u> exhibited a reduction in abundance at pH < 5, possibly due to changes in microbial activity on food sources.

The abundance and composition of the benthic community is greatly influenced by lake depth, temperature, substrate type, and dissolved oxygen regime. Crawford (1942) suggested that substrate composition was as important as pH in determining the composition of benthos in lakes receiving acid mine drainage. The substrates in the Sault Ste. Marie lakes were all similar, and the range of pH measured 2 cm below the mud-water interface (Table 2) was narrow. Unlike the plankton, the lentic benthos inhabit the boundary layer of the sediment, which to some degree has a capacity for buffering changes in pH.

The abundance of the Corixidae, <u>Sialis</u>, the Odonates, <u>Leucorrhinia</u> and <u>Enallegma</u>, as well as the large population of <u>Crangonyx</u> in the Upper Headwater lake is consistent (Henrikson and Oscarson, 1978) with the absence of fish. Ericksson <u>et al</u>. (1980) also postulated that tolerance to acidified soft waters permits these large invertebrates and other aquatic insects sensitive to fish predation, notably <u>Chaoborus</u>, to increase in both acidified lakes as well as in near neutral lakes devoid of fish. Excluding the amphipods, mollusks, and infrequent Ephemeroptera, few of the benthic invertebrates normally inhabiting lakes at depths greater than 3 m displayed notable change when pH levels were low. Below pH 5, Mossberg and Nyberg (1979) found a decrease in the number of benthic taxa but not in numbers of individuals.

Fisheries

Lakes in the Sault Ste. Marie district are principally salmonid producers (lake trout, <u>Salvelinus namaycush</u>, and brook trout, <u>Salvelinus fontinalis</u>) with other sport fish, particularly percids, being limited (Johnson <u>et al</u>., 1977).

Our sampled lakes supported simple fish communities, as only eight species were captured, including white sucker (<u>Catostomus commersoni</u>), northern hog sucker (<u>Hypentelium nigricans</u>), smelt (<u>Osmerus mordax</u>), walleye (<u>Stizostedion vitreum vitreum</u>), lake chub (<u>Hybopsis plumbea</u>), and creek chub (<u>Semotilus atromaculatus</u>) in addition to lake trout and brook trout.

The two most common species in the headwater lakes, brook trout and white sucker, occurred in lakes that ranged in size from 1.6 ha upward (Table 5). Although pH ranged only between 5.8 and 6.9 for brook trout and 5.2 and 7.0 for white sucker, alkalinities (bicarbonate) were as low as 18 µeq/l for both species. Metal concentrations (Table 5), particularly aluminum and lead, were only about one quarter the maximum detected in the study lakes. Brook trout generally existed in lakes with slightly higher pH values and alkalinities than did common white sucker.

The incidence of lakes devoid of fish increased with decreasing alkalinity and pH (Figure 11). At alkalinities below 75 µeq/l, barren lakes were common and greater than half of these lakes supported no fish communities. Although pH was less reliable than alkalinity as an indication of chemical status, lakes with pH < 6 more frequently supported no or limited fish communities. The increasing incidence of impoverished populations with decreasing pH and alkalinity has been well documented in Scandinavia (Leivestad, 1976) although it is unclear whether minimum pH, metal concentrations, or the chemical instability of the system is the limiting factor.

We made no examination of fish morphometric characteristics with lake condition (in part because of the small number of lakes fished - only 31), but we did examine the relationships of fish contaminant body burdens and total body lipid to aquatic conditions. Also, because of the difference in both body burden and morphometric characteristics among species as demonstrated through principal component analysis, we will deal only with brook trout and white sucker.

Body size and/or age significantly (P < 0.05 level) influenced few body components (Mg, Zn, Cu, DDE) in brook trout and many in white sucker (total body lipid, Hg, Ca, Mg, Zn, Cu, PCB, DDE, and Pb). Consequently, the interpretation of the relation between body burden and lake conditions (Table 6) is particularly difficult. It is obvious that the influence of lake features upon individual fish attributes varied drastically with species. Few lake factors have a common influence on characteristics of both species, exceptions being

Table 5

Mean and Range of Lake Characteristics for Brook Trout and White Sucker

Lake Characteristic	Brook Trout			White Sucker		
	Mean	Description Min	Max	Mean	Description Min	Max
Area (ha)	14.3	1.6	37.7	12.3	4.9	34.3
pH (std units)	6.3	5.8	6.9	6.2	5.2	7.0
Conductivity (μmhos/cm)	27	0	45	28	21	39
Alkalinity (μeq/ℓ)	151	18	597	109	17	180
Calcium (mg/ℓ)	4.0	2.1	8.1	3.4	1.9	4.4
Sulfate (mg/ℓ)	6.3	5.7	7.8	6.2	5.3	7.3
Chloride (mg/ℓ)	0.3	0.2	0.4	0.4	0.2	0.5
Magnesium (mg/ℓ)	0.7	0.5	0.9	0.8	0.5	1.4
Sodium (mg/ℓ)	0.7	0.6	0.7	0.8	0.8	1.1
Copper (μg/ℓ)	2	1	5	3	1	6
Lead (μg/ℓ)	14	1	25	13	1	31
Nickel (μg/ℓ)	2	1	4	1	1	3
Zinc (μg/ℓ)	7	3	13	7	1	18
Aluminum (μg/ℓ)	56	32	84	49	10	100

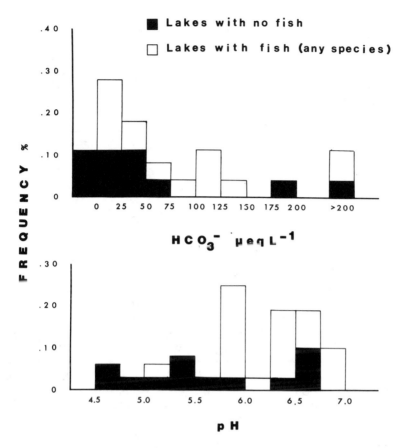

*Figure 11.   The presence and/or absence of fish, all vulner-
able species, in relation to pH and alkalinity
of Sault Ste. Marie district.*

apparent correlations between body zinc and lake pH, body weight
and lake aluminum, and body nickel and lake sulfate.

We examined the influence of lake features, particularly
those related to acidic conditions, to differences in life
style, between the species. The pelagic-littoral life style
of the brook trout contrasts strongly to the essentially
obligate benthic existence of the white sucker. We expected
water quality as reflected by epilimnetic water samples to
influence characteristics of brook trout and such is the case
(Table 6, Figure 12). When lakes are more acidic, sulfate
increases (Figure 7); and under these conditions, total lipid

Table 6

Correlation (Significant Values P<0.05 Only are
Metals and Ions and Lake Characteristics

Lake Character- istics	Brook Trout				Fish Attributes	
	FL	Wt	Lipid	Hg	Ca	Mg
Area	--	--	-0.38	--	--	--
DBA	0.37	0.41	--	--	--	--
pH	--	--	0.42	--	-0.47	--
Conductivity	--	--	--	--	--	--
Alkalinity	--	--	--	--	--	--
Cu	--	--	--	--	--	--
Pb	--	--	-0.35	--	--	--
Ni	--	--	--	--	--	--
Zn	-0.35	-0.39	--	--	--	--
Al	--	-0.42	-0.54	--	--	--
Ca	--	--	--	--	-0.41	--
Na	--	--	-0.41	-0.71	--	--
Mg	--	--	0.52	--	-0.44	--
SO_4	0.35	--	-0.57	--	--	--
Cl	--	--	0.38	--	-0.39	--
					White Sucker	
Area	0.63	0.77	0.76	0.45	-0.65	-0.63
DBA	--	--	--	--	-0.42	--
pH	0.39	0.38	--	--	--	--
Conductivity	--	--	--	--	--	--
Alkalinity	--	--	--	--	--	--
Cu	--	--	--	-0.52	--	--
Pb	-0.41	-0.40	--	--	--	--
Ni	--	--	--	-0.40	--	--
Zn	--	--	--	-0.42	--	--
Al	-0.39	-0.41	--	--	--	--
Ca	--	--	--	-0.34	--	--
Na	0.57	0.56	0.56	--	-0.71	-0.56
Mg	--	--	--	--	--	--
SO_4	--	--	--	--	--	--
Cl	-0.49	-0.60	-0.41	--	--	--

Presented) Between Fish Attributes Including
for Brook Trout and Common White Sucker

Zn	Cu	Na	PCB	DDE	Pb	Ni	K
--	--	0.40	--	--	--	0.41	--
--	--	--	--	--	--	--	--
-0.36	--	-0.43	0.37	0.35	--	-0.43	--
--	--	--	--	--	--	--	--
--	--	--	--	--	--	--	--
--	--	--	--	--	--	--	--
--	--	--	--	--	--	--	--
--	--	--	--	--	--	--	--
--	--	0.42	-0.39	-0.50	--	0.38	--
--	--	--	0.35	--	--	--	--
--	0.40	--	--	--	--	0.42	--
--	--	-0.44	--	-.45	--	-0.45	--
--	--	0.69	--	--	--	0.52	--
-0.40	-0.36	-0.44	--	--	--	-0.40	--
--	--	-0.37	0.64	0.43	--	-0.33	--
--	--	--	--	--	--	--	0.40
-0.43	--	--	--	--	-0.55	--	--
-0.53	--	--	--	--	--	--	--
-0.48	--	--	--	--	-0.48	--	--
--	--	--	--	--	-0.34	--	--
0.42	--	--	--	--	0.53	--	--
--	--	--	--	--	--	--	--
--	--	--	--	--	--	--	--
0.37	--	--	--	--	0.51	--	--
-0.41	--	--	--	--	-0.49	--	--
-0.54	-0.36	--	--	0.54	-0.35	-0.42	--
-0.40	--	0.35		--	--	--	--
--	--	--	--	--	--	0.33	--
--	--	--	-0.49	--	0.54	--	0.36

decreases in both brook trout and white sucker (Figure 12) although far more drastically in the former species.

Both species were confined to lakes with higher pH and alkalinity (Table 5) but metals, particularly lead and aluminum, were still elevated at lower alkalinities. Whether acidic conditions cause mobilization of lead, like mercury, from sediments or the drainage basin is unknown.

Fish body burdens of PCB (Figure 12), DDE, copper, and nickel (Table 6) do not appear influenced by acidic conditions. Both PCBs and DDT metabolites were low in both species (Figure 12) in these remote undeveloped lakes, suggesting a pervasive airborne contribution of at least these two organics of concern.

Mercury concentrations in walleye have been shown to increase (Scheider et al., 1979) with decreased pH. This also appears to be the case for brook trout and white sucker (Table 7) but the interrelated influence of fish size, fish body lipid, and lake characteristics demands that data be from a large number of lakes and better reflect the factors influencing body burden. The bottom feeding white sucker accumulates greater concentrations of both mercury and lead, suggesting that these metals may indeed be mobilized from the sediment. However, the ubiquitous presence of Hg and its reasonably high burdens in whole fish (0.01 to 0.22 in white sucker and 0.08 to 0.79 mg/gm wet weight in brook trout) indicates that this situation is a complex compelling question that must be clarified. Analysis by multiple regression (Table 7) sheds some light on the situation. Mercury content of brook trout was influenced by alkalinity, contrasting sharply to the influence of weight and lake chemistry on mercury content of white sucker. The remaining interrelations strongly support conclusions drawn from Figure 12.

SUMMARY

The fish, benthos, and phytoplankton reflected to varying degrees the acidic status of their aquatic system. Although lakes in the Sault Ste. Marie district range from acidic to well buffered, the link of acidic state to atmospheric acid addition is circumstantial, as historical data is limited. Nevertheless, a significant proportion of lakes in the district are acidic, high in aluminum and lead concentrations, and poorly buffered.

Biota in these lakes are to some extent protected by their life style and species tolerance against further change in

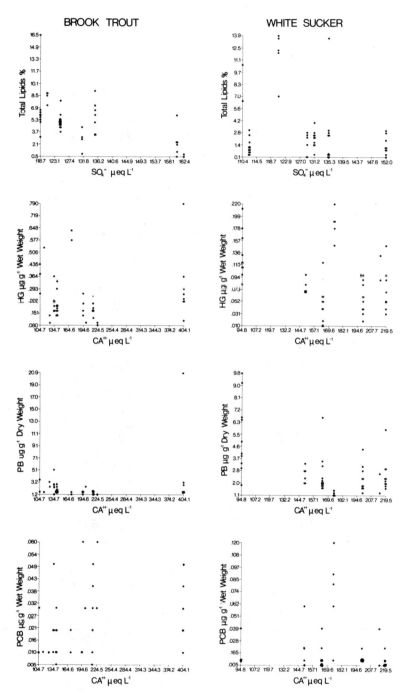

Figure 12.   Body burdens of lipid, mercury, lead, and PCB
              for brook trout and white sucker in Sault Ste.
              Marie district lakes.

Table 7

Significance of Independent Variables (F Statistic,
Significance at $P < 0.05$ Indicated by Asterisk) for Lipid,
Mercury, Lead, and PCB in Brook Trout and White
Sucker - Degrees of Freedom are in Parentheses

Chemical Parameter	Brook Trout	White Sucker
Dependent Variable	Independent Variable	Independent Variable
Total Body	(4/42) Wt 4.574* HCO_3 1.106 SO_4 15.197* Al 0.254	(4/48) Wt 58.921* HCO_3 1.238 SO_4 2.206 Al 2.260
Mercury	(5/41) Ca 0.037 Wt 1.243 HCO_3 2.547* Lipid 0.055 SO_4 0.014	(5/47) Ca 1.473 Wt 17.642* HCO_3 21.552* Lipid 0.522 SO_4 7.397*
Lead	(4/42) Wt 0.458 HCO_3 0.470 Ca 0.003 Al 0.352	(4/48) Wt 2.351 HCO_3 0.164 Ca 4.012* Al 2.859*
PCB	(4/42) Wt 2.045 Lipid 4.355* Ca 3.563* Al 0.054	(4/48) Wt 11.386* Lipid 4.106* Ca 1.244 Al 0.223

acidic status. Benthos are notably harbored against changes occurring in the water column by the ability of their microhabitat to counteract changes in lake pH. On the other hand, when behavior or development of the benthos removes them from the sediment, they then are prone to effects of predators, competitors, and lake chemistry. The phytoplankton of these lakes, while responding directly to lake conditions by shifts in community, do not appear to alter dramatically in community composition. Fish, on the other hand, operate much as the benthos with the response to acid state tempered by species tolerance, life style, metabolism of contaminants, and their compensation to changes induced by acidification in other trophic levels. In natural systems, although white sucker are generally more sensitive to low pH than brook trout (Baker and Schofield, 1980), difference in life style alters exposure to and uptake of contaminants and may favor existence of white sucker. In total, it seems that although benthos and phytoplankton respond to changes in their habitat, the effect upon fish is more blatant as seen through population success or metabolism of contaminants.

ACKNOWLEDGMENTS

The helicopter pilot, J. MacKenzie, of the Department of Transport, was particularly helpful. R.H. Collins, Great Lakes Biolimnology Laboratory, was continuously involved in the field sampling and consistently helpful. Our thanks are also extended to V. Proulx and the Ontario Ministry of Natural Resources for their help in aspects of laboratory and field work, respectively.

LITERATURE CITED

Almer, B., W. Dickson, C. Ekstrom, and E. Hornstrom.   1974. Effects of Acidification on Swedish Lakes. Ambio 3:30-36.

Andersen, A., M.F. Hovmand, and I. Johnson. 1978. Atmospheric Heavy Metal Deposition in the Copenhagen Area. Environ. Pollut. 17:133-151.

Baker, J.P. and C.L. Schofield. 1980. Aluminum Toxicity to Fish as Related to Acid Precipitation and Adirondack Surface Water Quality. In: Drabløs and Tollan, eds., "Ecological Impact of Acid Precipitation," March 11-14, 1980. Conf. Proc., Sandefjord, Norway. 383 pp.

Barrett, E. and G. Brodin. 1955. The Acidity of Scandinavian Precipitation. Tellus VII:251-257.

Beamish, R.J. and H.H. Harvey.   1972.   Acidification of the LaCloche Mountain Lakes, Ontario, and Resulting Effects on Fishes.   J. Fish. Res. Bd. Can. 29:1131-1143.

Beamish, R.J. and J.C. VanLoon.   1977.   Precipitation Loading of Acid and Heavy Metals to a Small Acid Lake Near Sudbury, Ontario.   J. Fish. Res. Bd. Can. 34:649-658.

Boissonneau, A.N.   1968.   Glacial History of Northeastern Ontario II.   The Temiskaming-Algoma Area.   Can. J. Earth Sci. 5:97-109.

Bousfield, E.L.   1958.   Freshwater Amphipod Crustaceans of Glaciated North America.   Can. Field Nat. 72:55-113.

Brinkhurst, R.O. and B.G. Jamieson.   1971.   Aquatic Oligochaeta of the World.   Oliver and Boyd, Edinburgh.   860 pp.

Clark, A.H.   1973.   The Freshwater Molluscs of the Canadian Interior Basin.   Malacologia 13:1-509.

Conroy, N., D.S. Jeffries, and J.R. Kramer.   1974. Acid Shield Lakes in the Sudbury, Ontario, Region.   Proc. 9th Can. Symp. on Water Poll. Res. 9:45-61.

Conroy, N., K. Hawley, W. Keller, and C. LaFrance.   1976. Influence of the Atmosphere on Lakes in the Sudbury Area.   Proc. 1st Symp. on Atmospheric Contributions to the Chemistry of Lake Waters.   Great Lakes Res. 2 (Suppl. 1):146-165.

Conroy, N. and W. Keller.   1976. Geological Factors Affecting Biological Activity in Precambrian Shield Lakes.   Can. Mineral. 14:62-72.

Cox, E.T.   1978.   Counts and Measures of Ontario Lakes.   Ont. Min. Nat. Res. Manuscript Report.   114 pp.

Crawford, B.T.   1942.   Ecological Succession in a Series of Strip-Mine Lakes in Central Missouri.   M.A. Thesis, University of Missouri, Columbia, Missouri.

Culp, J.M. and R.W. Davies.   1980.   Reciprocal Averaging and Polar Ordination as Techniques for Analyzing Macroinvertebrate Communities.   Can. J. Fish. Aquatic Sci. 37:1358-1364.

Dillon, P.J., D.S. Jeffries, W. Snyder, R. Reid, N.D. Yan, D. Evans, J. Moss, and W.A. Scheider.   1977. Acidic Precipitation in South-central Ontario: Recent Observations.   Ont. Min. Env. Tech. Report.   24 pp.

Dillon, P.J., D.S. Jeffries, W. Snyder, R. Reid, N.D. Yan, D. Evans, J. Moss, and W. Scheider. 1978. Acidic Precipitation in South-central Ontario: Recent Observations. J. Fish. Res. Board Can. 35:809-815.

Environment Canada. 1974. Analytical Methods. Inland Waters Directorate, Water Quality Branch, Ottawa, Ontario, Canada.

Environment Canada. 1979. The LRTAP Problem in North America: A Preliminary Overview. Prepared by US-Can. Res. Cons. Group. Manuscript Report. 49 pp.

Erikssen, E. 1955. Air Borne Salts and the Chemical Composition of River Waters. Tellus VII:243-250.

Eriksson, M.O., L. Henrikson, B.I. Nilsson, G. Nyman, H.G. Oscarson, and A.E. Stenson. 1980. Predatory Prey Relations Important for the Biotic Changes in Acidified Lakes. Ambio 9:248-249.

Findlay, D.L. and H.J. Kling. 1975. Seasonal Successions of Phytoplankton in Seven Lake Basins in the Experimental Lakes Area, Northwestern Ontario, Following Artificial Eutrophication. Env. Can., Fish. and Mar. Serv. Tech. Report No. 513. 53 pp.

Fisher, D.W., A.W. Gambell, G.E. Likens, and F.H. Bormann. 1968. Atmospheric Contributions to Water Quality of Streams in the Hubbard Brook Experimental Forest, New Hampshire. Water Res. 4:1115-1126.

Giblin, P.E. and E.J. Leahy. 1967. Sault Ste. Marie - Elliott Lake Geological Compilation Series. Ont. Dept. Mines, Geol. Comp. Ser. Map 2419.

Gorham, E. and A.G. Gordon. 1960. The Influence of Smelter Fumes Upon the Chemical Composition of Lake Waters Near Sudbury, Ontario, and Upon the Surrounding Vegetation. Can. J. Bot. 38:477-487.

Gorham, E. and A.G. Gordon. 1963. Some Effects of Smelter Pollution Upon Aquatic Vegetation Near Sudbury, Ontario. Can. J. Bot. 41:371-378.

Haapala, H., P. Sepponen, and E. Meskus. 1975. Effect of Spring Floods on Water Acidity in the Kiiminkijoki Area, Finland. Oikos 26:26-31.

Hamilton, A.L.   1971.   Zoobenthos of Fifteen Lakes in the Experimental Lakes Area, Northwesteren Ontario.   J. Fish. Res. Bd. Can. 28:257-263.

Harp, G.L. and R.S. Campbell.   1967.   The Distribution of Tendipes plumosus (L) in Mineral Acid Water.   Limnol. Oceanogr. 12;260-263.

Hendrey, G.R. and R.F. Wright.   1976.   Acid Precipitation in Norway: Effects on Aquatic Fauna. Proc. 1st Special Symp. on Atmos. Contribution to the Chemistry of Lake Waters. Int. Asso. Great Lakes Res., September 1975, pp. 192-207.

Henrikson, L. and H.G. Oscarson.   1978. Fish Predation Limiting Abundance and Distribution of Glacnocorisa p. propinqua. Oikos 31:102-105.

Henriksen, A.   1979.   A Simple Approach for Identifying and Measuring Acidification of Freshwater.   Nature 278:542-545.

Hill, M.O. 1973. Reciprocal Averaging: An Eigenvector Method for Ordination. J. Ecol. 61:237-249.

Hutchinson, G.E.   1957.   A Treatise on Limnology.   Vol. 1. Geography, Physics, and Chemistry.   J. Wiley and Sons, Inc., London.   1015 pp.

Johnson, M.G., M.F.P. Michalski, and A.E. Christie.   1970. Effects of Acid Mine Wastes on Phytoplankton Communities of Two Northern Ontario Lakes.   J. Fish. Res. Bd. Can. 27:425-444.

Johnson, M.G., J.H. Leach, C.K. Minns, and C.H. Olver. 1977. Limnological Characteristics on Ontario Lakes in Relation to Associations of Walleye (Stizostedion vitreum vitreum), Northern Pike (Esox lucius), Lake Trout (Salvelinus namaycush), and Smallmouth Bass (Micropterus dolomieui). J. Fish. Res. Bd. Can. 34:1592-1601.

Kling, H.J. and S.K. Holmgren. 1972. Species composition and Seasonal Distribution in the Experimental Lakes Area, Northwestern Ontario. Fish. Res. Board Can. Tech. Rep. #337. 51 pp.

Kwiatkowski, R.E. and J.C. Roff. 1976. Effects of Acidity on the Phytoplankton and Primary Producitivity of Selected Northern Ontario Lakes. Can. J. Bot. 54:2546-2561.

Leivestad, H., G. Hendry, I.P. Muniz, and E. Snekvik. 1976. Effects of Acid Precipitation on Freshwater Organisms. In: "Impact of Acid Precipitation on Forest and Freshwater Ecosystems in Norway," pp. 87-111. S.N.S.F. Res. Rep. No. FR6-76.

Likens, G.R., R.F. Wright, J.N. Galloway, and T.J. Butler. 1979. Acid Rain. Sci. Amer. 241:43-49.

MOE. 1979. Determination of the Susceptibility to Acidification of Poorly Buffered Surface Waters. Ont. Min. of the Env. Tech. Rept. 21 pp.

Mossberg, P. 1979. Bottenfannans sammonsattning i sura oligotrofa sjoar. Summary: Benthos of Oligotrophic and Acid Lakes. Information fran Sotvattens laboratoriet, Drottningholm 11. 40 pp.

Mossberg, P. and P. Nyberg. 1979. Bottom Fauna of Small Acidic Forest Lakes. Inst. of Freshwater Res., Drottningholm 58:77-87.

Okland, K. 1969. On the Distribution and Ecology of Gammarus lacustris. G.O. Sars in Norway, with notes on its morphology and biology. Nors. J. Zool. 17:111-152.

Oliver, D.R., D. McClymont, and M.E. Roussel. 1978. A Key to Some Larvae of Chironomidae (Diptera) from the Mackenzie and Porcupine River Watersheds. J. Fish. Res. Board Can. Fisheries & Marine Service Tech. Rept. 791.

Ostrofsky, M.L. and H.C. Duthie. 1975. Primary Productivity and Phytoplankton of Lakes on the Eastern Canadian Shield. Verh. Intern. Verein. Limnol. 19:732-738.

Parsons, J.D. 1968. The Effects of Acid Stripmine Effluents on the Ecology of a Stream. Arch. Hydrobiol. 65:25-50.

Pennak, R.W. 1978. Freshwater Invertebrates of the United States, 2nd ed. John Wiley & Sons, New York. 803 pp.

Pielou, E.C. 1969. An Introduction to Mathematical Ecology. Wiley Interscience, Toronto. 286 pp.

Pope, G.F., J.C. Carter, and G. Power. 1973. The Influence of Fish on the Distribution of Chaoborus spp (Diptera) and Density of Larvae in the Matamek River System, Quebec. Trans. Amer. Fish. Soc. 102:707-714.

Rawson, D.S. 1960. A Limnological Comparison of Twelve Large Lakes in Northern Saskatchewan. Limnol. Oceanogr. 5:195-211.

Roff, J.C. and R.E. Kwiatkowski. 1977. Zooplankton and Zoobenthos Communities of Selected Northern Ontario Lakes of Different Acidities. Can. J. Zool. 55:899-911.

Sanderson, M. and P.D. LaValle. 1979. Surface Loading From Pollutants in Precipitation in Southern Ontario: Some Climatic and Statistical Aspects. J. Fish. Res. Bd. Can. 5:52-60.

Scheider, W.A., B. Cave, and J. Jones. 1975. Reclamation of Acidified Lakes Near Sudbury, Ontario, by Neutralization and Fertilization. Ont. Min. Env. Tech. Rep. 48 pp.

Scheider, W.A. and P.J. Dillon. 1976. Neutralization and Fertilization of Acidified Lakes Near Sudbury, Ontario. Proc. 11th Cdn. Symp. Water Poll. Res. Can., pp. 93-100.

Scheider, W.A., D.S. Jeffries, and P.J. Dillon. 1979. Effects of Acidic Precipitation on Precambrian Freshwaters in Southern Ontario. J. Great Lakes Res. 5:45-51.

Schindler, D.W. and S.K. Holmgren. 1971. Primary Production and Phytoplankton in the Experimental Lakes Area, North-Western Ontario, and Other Low Carbonate Waters, and a Liquid Scintillation Method for Determining ^{14}C Activity in Photosynthesis. J. Fish. Res. Bd. Can. 28:189-201.

Schindler, D.W., H. Kling, R.V. Schmidt, J. Prokopowich, V.E. Frost, R.A. Reid, and M. Capel. 1973. Eutrophication of Lake 227 by Addition of Phosphate and Nitrate: the Second, Third, and Fourth Years of Enrichment, 1970, 1971, and 1972. J. Fish. Res. Board Can. 30:1415-1440.

Schindler, D.W., R. Wageman, R.D. Cook, T. Ruszczynski, and J. Prokopowich. 1980. Experimental Acidification of Lake 223, Experimental Lakes Area: Background Data and the First Three Years of Acidification. Can. J. Fish. Aquatic Sci. 37:342-354.

Stumm, W. and J.J. Morgan. 1970. Aquatic Chemistry. Wiley Interscience, Inc., London and New York. 583 pp.

Sutcliffe, D.W. and T.R. Carrick. 1973. Studies on Mountain Streams in the English Lake District, pH, Calcium, and Distribution of Invertebrates in River Duddon. Freshwat. Biol. 3:437-462.

Vollenweider, R.A. 1969. A Manual on Methods for Measuring Primary Production in Aquatic Environments. International Biological Programme. Blackwell Scientific Publications, Oxford.

Welch, P.S. 1952. Limnology. McGraw-Hill Book Company, Inc., New York. 538 pp.

Wiederholm, T. and L. Ericksson. 1977. Benthos of an Acid Lake. Oikos 29:261-267.

Wright, R.F., T. Dale, E.T. Gjessing, G. Hendry, A. Henriksen, M. Johannessen, and I.P. Muniz. 1976. Impact of Acid Precipitation on Freshwater Ecosystems in Norway. Water Air Soil Pollut. 6:483-499.

Wright, R.F. and E.T. Gjessing. 1976. Acid Precipitation: Changes in the Chemical Composition of Lakes. Ambio 5:219-223.

Wright, R.F. and E. Snekvik. 1978. Acid Precipitation: Chemistry and Fish Populations in 700 Lakes in Southern-Most Norway. Verh. Internat. Verein. Limnol. 20:765-775.

Yan, N.D. and P. Stokes. 1978. Phytoplankton of an Acidic Lake and Its Response to Experimental Alterations of pH. Environ. Conserv. 5:93-100.

CHAPTER 9

THE EFFECTS OF ACIDIFICATION ON STREAM ECOSYSTEMS

Thomas M. Burton, Richard M. Stanford,
and Jon W. Allan
Department of Zoology, Department of Fisheries
and Wildlife, and Institute of Water Research
Michigan State University
East Lansing, Michigan 48824

INTRODUCTION

A previous paper in this conference by O.L. Loucks (see Chapter 2) has delineated the acid precipitation problem on a worldwide basis with emphasis on the Great Lakes region of North America. The purpose of this paper is to summarize known effects of acid precipitation on stream ecosystems, again with emphasis on the Great Lakes region. Unfortunately, very little research has been conducted on streams in much of the Great Lakes region. Thus, data will be utilized from other regions of the United States as well as from Norway, Sweden, and elsewhere; from laboratory studies on effects of pH on biota; and from our own research on effects of pH on soft water streams in Michigan to summarize probable effects on stream ecosystems within the region. Data from studies of acid mine drainage will be used to some extent. However, these data will be used with caution, since many of the effects of acid mine drainage on biota are likely the result of toxic levels of heavy metals and are not comparable to acid precipitation effects. Thus, much of this summary will rely on the limited data available on acid precipitation effects. All components of the stream ecosystem will be emphasized except fish, which are covered in detail elsewhere by Schofield (1976).

This chapter consists of a general review of the literature followed by a description of preliminary results from our stream research.

EXTENT OF THE PROBLEM IN THE GREAT LAKES REGION

The sensitivity of streams to acid precipitation is coupled closely with the type of soil, the thickness of the soil, and bedrock geology of the regions they drain. Basically, this sensitivity is linked to the buffering capacity of the soils and waters of the watershed-stream ecosystem. In general, the higher the calcium carbonate content and the finer textured the soil, the less sensitive the soil is to acidification and the less likely it is that streams draining such watersheds will be acid. Streams draining areas of exposed, non-calcareous, granitic bedrock of low solubility and with thin soils of little buffering capacity, such as the Precambrian Shield area of Canada and some of the northern and mountainous portions of the United States, are especially sensitive to acid precipitation. The sensitive areas of the western Great Lakes region in the U.S. are concentrated in northern Wisconsin, northeastern Minnesota including the Boundary Waters Canoe Area, and the western portion of the upper peninsula of Michigan (Galloway and Cowling, 1978; McFee, 1980). Within this region of young, coarse textured glacial deposits, highly sensitive areas are interspersed with regions with little sensitivity at all (McFee, 1980). The streams reflect this, with soft water streams often being in close proximity to relatively hard water streams (Zimmerman, 1968).

Much of the western portion of the Great Lakes drainage in Ontario, with the exception of a few small areas of calcareous bedrock, is sensitive to acid precipitation, with the sensitivity extending to the boundary of the Precambrian Shield just south of the Muskoka-Haliburton region of Ontario south and east of the southern end of Georgian Bay (Dillon et al., 1978; Galloway and Cowling, 1978; Scheider et al., 1979). The carbonate regions south of this area in Ontario are not sensitive to acid precipitation inputs.

The sensitive areas of New York are concentrated in the Adirondack region or further east, primarily east of the Great Lakes drainage (McFee, 1980). Likewise, the Lake Erie dainage of Pennsylvania is not sensitive to acid rain inputs, with the most sensitive regions being in the Alleghany uplands and ridges further to the east (Arnold et al., 1980; McFee, 1980). With the exception of slight sensitivity of portions of southern Ohio and Indiana, Ohio, Indiana, Illinois, the lower peninsula of Michigan, and southern Wisconsin are not sensitive to acid precipitation (Galloway and Cowling, 1978; McFee, 1980).

The general statement that an overall region has soils and waters with high buffering capacity and low sensitivity to acid precipitation should not be taken to mean that no problem streams can exist in these areas. For example, streams on isolated areas of coarse textured soils or exposed bedrock or in sandy soils near the Great Lakes can have low alkalinities and little buffering capacity. However, such streams will be rare and, in general, the areas outlined above are most likely to contain acid-sensitive streams.

The size of the stream also influences its sensitivity to acid precipitation inputs. Even in areas of igneous bedrock, larger streams are essentially not affected by acid precipitation because of a two-step neutralization process (Johnson, 1979). The first step of this process consists of neutralization of acid rain in the upper soil by dissolution of pre-existing aluminum hydroxide compounds and by leaching of bases from biologic matter so that rain has already been neutralized to a great extent before it reaches the stream. This neutralization is accompanied by high leaching losses of aluminum to the stream (Cronan and Schofield, 1979; Johnson, 1979). As this water moves downstream, the second step occurs. This step consists of chemical weathering reactions with loss of aluminum and gain of strong bases and silica by the water. Carbonic acid replaces the strong acids (sulfuric and nitric) as the major acid present even though total alkalinity is still very low. These changes seem to occur as a consequence of stream order regardless of elevation, with major streams not being susceptible to acidification (Johnson, 1979). Neutralization occurs rapidly in even the smallest streams in carbonate areas. Fairly large streams can be sensitive if much of the basin is exposed granitic bedrock. One of the largest streams known to have been affected is the Tovdal River in Norway with peak discharge in excess of 300 m^3/sec (Overrein et al., 1980). The effects of acid precipitation on streams of the Great Lakes region will probably be significant only in small, low-order (unbranched or with only one or two branches), headwater streams in areas draining thin, coarse-textured, non-calcareous soils or in areas of exposed hard-to-weather silicate bedrock such as parts of northern Wisconsin, northeastern Minnesota, the western portion of the upper peninsula of Michigan, and most of Ontario.

Fortunately, many of the western areas of the Great Lakes are not yet receiving rain as high in acidity as are areas further east. Thus, the most sensitive areas of the western Great Lakes region in the United States are receiving the least inputs. Even so, increased construction of power plants to burn western coal are likely to increase exposure of this region to acid rain during the next few years. South central Ontario already has a significant problem with acidification of streams (Dillon et al., 1978; Jeffries et al., 1979; Scheider et al., 1979) as does much of the high-altitude areas east and south of the Great Lakes in the eastern United States.

EFFECTS ON STREAM CHEMISTRY

Acid precipitation contains highly elevated levels of hydrogen ions, ammonium, nitrate, and sulfate, and decreased levels of bicarbonate compared to unpolluted rain. For the eastern United States, about 70 percent of increased acidity is sulfuric acid while about 30 percent is nitric acid (Likens et al., 1979). These chemical constituents are highly modified by watershed processes with combined watershed-stream interactions resulting in major changes in stream chemistry. These changes in streams draining watersheds with low buffering capacity include increased levels of aluminum, manganese, calcium, magnesium, and potassium in streamwater and can include short-term increases in sodium as a result of replacement of these elements in the soil by hydrogen ions as part of cation exchange processes and weathering reactions (Cronan and Schofield, 1979; Hall et al., 1980; Gjessing et al., 1976; Overrein et al., 1980; Scheider et al., 1979). Increases in aluminum are especially important, with concentrations reaching toxic levels (>200 ug/l) for many aquatic species as acidification results in replacement of aluminum by hydrogen ions and also increases solubility of aluminum in water (Cronan and Schofield, 1979; Hall et al., 1980; Gjessing et al., 1976; Overrein et al., 1980).

The two major anions received from acid precipitation are sulfate (about 64%) and nitrate (about 29%)(Likens et al., 1979). These anions are received both as wet and dry deposition (rain, snow, dustfall, aerosol impacts, etc.). Most of the sulfate is not retained by the watershed and ultimately is leached to streams (Overrein et al., 1980). Therefore, elevated sulfate levels are characteristics of acidified streams with sulfate replacing the bicarbonate-carbonate complex as the major anion present. Leaching of sulfate is directly correlated to aluminum plus calcium plus magnesium losses, and Norwegian scientists have developed predictive equations for this relationship (Gjessing et al., 1976; Overrein et al., 1980).

Ammonium and nitrate-nitrogen inputs are usually retained by the watershed since nitrogen is often in short supply for terrestrial ecosystems. Thus, nitrate (the easily leached form of nitrogen) concentrations in acidified streams are not normally elevated over background levels for non-acidified streams. There is a normal cycle of increased nitrate in streams during the winter since plant and microbial uptake within the watershed is at a minimum. It is possible, therefore, that snowmelt events associated with mid-winter thaws could result in short-term elevations of nitrate in stream water. However, we know of no reports of such elevated levels.

In addition to aluminum, manganese, and the major nutrients described above, other metals may also increase in stream water as a consequence of acidification. The solubility of many metals increases substantially as pH is lowered from alkaline to acid conditions. This increased solubility should result in substantial increases in many of these metals in acidified stream water. For example, metals solubilized from average crustal rock in water should increase from 0.03 to 500 µg/l for iron as pH is reduced from 8.1 to 4.5, from 0.0003 to 25 µg/l for copper, from 0.006 to 170 µg/l for nickel, and from 0.004 to 80 µg/l for zinc (Kramer, 1976). Zinc and lead concentrations are elevated in acidified lakes in Norway (Overrein et al., 1980), but these increases could either be the result of increased deposition or increased leaching. Increased atmospheric deposition rates during the past 30 yr of silver, gold, cadmium, copper, lead, antimony, vanadium, and zinc were described for lakes in New York and New Hampshire by Galloway and Likens (1979). Many of these same elements plus arsenic and selenium were also elevated in deposition in the Norwegian studies (Overrein et al., 1980). Whether from leaching or from increased deposition rates, trace metals may be elevated in streams receiving acid precipitation.

Experimental acidification of Norris Brook, a third-order stream in New Hampshire, resulted in the increases in aluminum, manganese, calcium, magnesium, and potassium already discussed; in decreases or increases in iron and cadmium depending on time of sampling, but no significant change in copper, nickel, zinc, or lead (Hall et al., 1980). This experiment consisted of acidification to pH 4 of the stream channel only and may not reflect changes that would have occurred if the entire watershed had been acidified. Increased manganese levels as well as aluminum are well documented. Scheider et al. (1979) reported increases of manganese from 3 to > 200 µg/l. They also reported elevated levels of mercury in fish tissue from acidified lakes. Thus, elevated trace metal levels other than aluminum may contribute to biotic stress in acidified streams.

Many of the soft water streams in the western Great Lakes region are brown water streams draining swampy areas. These streams contain large amounts of organic matter which can alter the toxicity of aluminum and other metals due to complexation of these metals with the organic matter (Dickson, 1978; Driscoll et al., 1980; Patrick et al., 1981). Thus, results from other areas may not be totally germane to the western Great Lakes. More studies of the interaction between pH, trace metal toxicity, and organic content of the water are needed for prediction of effects in this region.

Increased aluminum concentrations can also result in increased flocculation of organic matter from the water column (Dickson, 1978). Thus, acid lakes are often highly transparent with little suspended sediment. This same process could result in less organic matter in stream water.

Episodic acidification events may be especially important in the response of biota to changes in stream chemistry resulting from acidification (Overrein et al., 1980). Changes in pH may be as great as one to two pH units following snow-melt (e.g., see review of Jeffries et al., 1979) or large storms and can result in loss of organisms even if the stream has a pH during most of the year that sensitive organisms can tolerate.

Changes in stream chemistry as a result of acid precipitation in poorly buffered streams can be summarized as follows. Changes will generally include elevated concentrations of aluminum, calcium, magnesium, potassium, and manganese and will often include elevated concentrations of zinc, lead, and other trace metals. Elevated aluminum is often toxic to sensitive species of invertebrates and fish. It also leads to increased sedimentation of organic matter, a primary food source for many stream organisms. Thus, acidification effects on stream chemistry other than just hydrogen ion changes may have significant effects on the biota. It is difficult to determine whether biotic changes resulting from acidification are linked to sensitivity to pH, to aluminum, or to some other chemical change. The overall effects of acidification on various trophic levels of the aquatic community will be described below.

EFFECTS ON PLANTS

The effects of acidification on plant communities in streams have not been extensively studied. Most of the work has come from studies conducted in Norway. These studies are summarized by Hendrey (1976), Leivestad et al. (1976), and Overrein et al. (1980). The Norwegian studies were initiated because of field observations that suggested that acidified streams often contained heavy growths of filamentous algae. To document this observation, Hendrey (1976) diverted flow from Ramse Brook, an already acidified stream (pH 4.3 to 5.5), through artificial stream channels and adjusted the pH to 4.0 in one channel, to 6.1 in another, and left it as ambient (4.3 to 5.5) for the third. It is important to note that ambient conditions were acid with acid-tolerant plants already present. Dominating species in all three channels were the filamentous green algae, Mougeotia sp., and the diatom, Tabellaria flocculosa. These two species accounted for greater than 50 percent of total cells for all periods of study and usually for more than 70 percent of total cells. At low pH, Mougeotia sp. was more common, while Tabellaria flocculosa was more common at pH 6.1. These are the same two species found to be dominant in most acidified streams of Norway. Acidification to pH 4.0 led to increased accumulation of algae, to increased chlorophyll levels, and to higher proportions of organic matter existing as algal cells. Carbon uptake per unit of chlorophyll was lowest at pH 4.0.

Hall et al. (1980) acidified Norris Brook, New Hampshire, from background levels of 5.7 to 6.4 to a pH of 3.9 to 4.5. This acidification resulted in 2- to 5.6-fold increases in chlorophyll-a and obvious visual increases in algal biomass in the stream. However, periphyton production rate was lower in the acidified stream despite the increased biomass.

Patrick et al. (1968) experimentally determined the effects of pH on diatom communities on glass slides in boxes with water diverted from Darby Creek, Pennsylvania, with a control pH of 7.3 to 10.1 and with a low pH of 4.7 to 5.7. They reported that low pH resulted in greatly reduced biomass with maximum reduction occurring in October and November and least reduction occurring in May and June. Likewise, community structure as measured by diversity was reduced more in October and November than in May and June by lowered pH, but overall species differences were not very great, with only a few species present at low pH that were not present in controls and vice versa.

Warner (1971) reported that sections of Roaring Creek, Virginia, polluted by acid mine drainage with a pH of 2.3 to 3.8, had fewer species of periphytic algae than did unpolluted sections of the stream (27 or more species in unpolluted sections, 10 to 19 in polluted sections). Dominant forms in acid waters included Ulothrix tennerima, Pinnularia termitina, Eunotia exigna, and Euglena mutabilis.

Muller (1980) conducted experiments on the effects of acidification on periphyton in lakes. Diatoms were dominant at pH levels above 6 but were replaced by green algae below 6 with Mougeotia becoming the dominant species at pH 4. Fewer species and lower species diversity were observed at lower pH. Standing crops increased at low pH, but production did not. Thus, results were very similar to results from acidification of streams.

While data are limited, benthic algal communities in streams appear to respond to acidification by:

(1) species shifts to acid-tolerant forms, often the filamentous green algae <u>Mougeotia</u> sp.;
(2) lower diversity;
(3) increased standing crop of plants overall, but decreased standing crop of many diatoms; and
(4) no change or lower production rates by plants.

Increased standing crop with same or reduced production rates is often attributed to one or a combination of the following:

(1) lower rates of decomposition,
(2) lower grazing rates by herbivores, and
(3) species shifts to a few acid-tolerant species (Hall <u>et</u> <u>al</u>., 1980; Hendrey, 1976).

There are few data on other types of plant communities in streams. In preliminary results of experiments at Monticello, Minnesota, acidification from pH 7.9-8.2 to 5.0 of outdoor, non-shaded artificial channels with low flow rates resulted in decreased diversity of macrophytes, invasion of terrestrial plants into the stream and changes in the stream from an autotrophic to a heterotrophic system, perhaps because of shading by the emergent plants and the expanded cover of duckweed (<u>Lemna</u> <u>minor</u>)(U.S. EPA, 1979). These results cannot readily be extrapolated to other systems, since the background levels of alkalinity would make this type of stream fairly insensitive to acid rain inputs.

There are some data on responses of macrophytes to acidification in lakes that can give insight into possible effects in streams. In general, effects on macrophyte communities tend to be the result of indirect factors rather than actual intolerance to increased hydrogen ions (Hultberg and Grahn, 1976). These indirect effects include the existence of more free carbon dioxide as the lake becomes acid, which favors plants with high free carbon dioxide requirements such as many of the mosses. <u>Sphagnum</u>, the peat moss of bogs, forms mats on the bottom of many acidified Swedish lakes and has replaced <u>Lobelia</u> and <u>Isoetes</u> as a result of over-growing them. Likewise, decomposition rates in lakes are slower, and dense felt-like mats of algae and fungi form over the sediments of many acidified lakes. This mat competes with macrophytes for sediment nutrients, reduces flow of nutrients into overlying water, and shades out <u>Lobelia</u> and <u>Isoetes</u>. The reduced decomposition rate also results in increased organic content of the sediments. Increased organic content leads to increases in the acidophilic, high organic-requiring species, <u>Juncus</u> <u>bulbosus</u> var. <u>fluitans</u>. Other changes described as a consequence of acidificaton of Swedish lakes include decreases

in <u>Phragmites</u> <u>communis</u> and <u>Myriophyllum</u> sp. but no effects on
<u>Nymphaea</u> <u>alba</u> or <u>Nuphar</u> <u>luteum</u> (Hultberg and Grahn, 1976).

Since many of these lake macrophytes also occur in and
along streams, similar changes mght be expected in the stream
macrophyte communities.

EFFECTS ON DECOMPOSITION

Acidification of streams leads to decreased rates of
decomposition of organic matter (Leivestad <u>et</u> <u>al</u>., 1976;
Overrein <u>et</u> <u>al</u>., 1980). Decreased rates of decomposition in
small, acid-sensitive headwater streams can have major impact
since most energy used by organisms in these streams is derived
from organic matter which enters the streams from adjacent
terrestrial ecosystems.

Experimental evidence for decreased rates of decomposition
is somewhat limited and is derived primarily from experiments
conducted in Norway (see summaries by Leivestad <u>et</u> <u>al</u>., 1976;
Overrein <u>et</u> <u>al</u>., 1980). One set of these experiments was
conducted on microbial activity on peptone in the laboratory.
As pH decreased, numbers of species of ciliated protozoans and
total bacterial cell counts decreased. Below pH 5,
nitrification was reduced, oxidation of ammonia ceased, and
decomposition rates were reduced. Below pH 4, numbers of fungi
increased. In experiments with glucose and glutamic acid, lag
time before start of rapid decomposition was increased at lower
pH but decomposition rates after this lag were not appreciably
affected down to pH 3.5, perhaps as a result of selection and
adaptation of microbes. There was a shift to fungal populations
as pH was lowered. Similar experiments with homogenized wilted
birch (<u>Betula</u> <u>verrucosa</u>) leaves did show reduced decomposition
rates during the three-week period of these studies in the
laboratory.

Experiments were conducted on the effects of pH on
decomposition rates of birch leaves using artificial channels
and Ramse Brook water in Norway (Leivestad <u>et</u> <u>al</u>., 1976). At
pH 4, decomposition after one year was lowest (about 46% weight
loss) and was significantly different at the $P < 0.001$ level
from decomposition at pH 6.0 (about 52% weight loss). Similar
experiments with aspen wood chips yielded similar results as
did experiments carried out in natural streams.

Hall <u>et</u> <u>al</u>. (1980) reported that acidification of a New
Hampshire stream resulted in growth of a basidiomycete fungus
on tree roots and mosses in the stream and resulted in a thin
layer of fungal hyphae over 70 percent of the stream bottom.

Maple and birch leaves had similar genera of aquatic hyphomycetes as did those in non-acidified stream sections nearby but species diversity and spore numbers were reduced in the acidified section of stream.

Preliminary data from open artificial channels in Minnesota suggested that decomposition rate was slowed substantially by acidification (61% weight loss at pH 7.9-8.2 to 48% at pH 5.0) in litter bags open to detritivores while there was no significant difference when detritivores were excluded (29% weight loss at pH 7.9-8.2 versus 26% weight loss at pH 5.0). This study suggested that the microbial rates were not substantially reduced but that decomposition rates were reduced by reduction in species diversity and numbers of dominant macroinvertebrates (U.S. EPA, 1979).

Decreased decomposition, accumulation of organic matter on the bottom, and development of thick gelatinous mats of fungal hyphae on lake bottoms are known to occur widely as a result of acidification. These processes are also known to reduce remineralization rates of nutrients such as phosphorus and can lead to self-accelerating oligotrophication of such acid lakes (Dickson, 1978; Leivestad et al., 1976; Overrein et al., 1980). Such changes in stream chemistry could also lead to major losses of stream productivity since organic matter forms the major food resource for animals in small, headwater streams.

EFFECTS ON INVERTEBRATES

Acidification can have major impacts on the invertebrate fauna of a stream. Certain species appear to be much more sensitive than others. Thus, community structure is likely to be altered as sensitive species are replaced by more tolerant forms. This reorganization of community structure has not been studied in detail. Instead, studies have been conducted on the species level or on a macrodistribution level (stream surveys of acid versus non-acid streams). The studies of Hall et al. (1980) did examine differential drift rates among different functional groups of invertebrates as a consequence of acidification.

Surveys of streams to ascertain which organisms occur within a region in rivers with low versus high pH, which species are tolerant of lower pH, and which species appear to be favored by lower pH are useful in predicting sensitivity to acidification of streams. However, these data must be used with caution since many other factors are linked to low pH. For example, soft waters with low levels of calcium, magnesium, sodium, and potassium can limit the distribution of certain species with high requirements for any of these elements. Snails and clams, for example, have a high requirement for calcium carbonate for shell building and may not exist in soft water. In fact, most snails cannot survive in streams that are more acid than pH 6.2, with most but not all species requiring more than 15 mg/l of carbonates (Pennak, 1978). Soft water is highly sensitive to acidification. Thus, absence of species may be correlated with low pH, but these species may be absent due to lack of calcium or carbonate for shell building or due to lack of the proper balance of calcium, sodium, and potassium ions necessary for osmotic regulation, etc. Thus, field survey work needs to be coupled to laboratory studies before cause and effect can be determined.

On the basis of distribution within the River Duddon in England, Sutcliffe and Carrick (1973) suggested that all mayflies (Ephemeroptera), certain species of caddisflies (Trichoptera) in the genera Wormaldia and Hydropsyche, the river limpet, Ancylus fluviatilis, and the amphipod, Gammarus pulex, were limited to waters with pH > 5.7, while stoneflies (Plecoptera), caddisflies in the genera Plectrocnemia and Rhyacophila and the family Limnephilidae, midges (Chironomidae), blackflies (Simulium) and craneflies (Tipulidae) were all present in acid waters (ph < 5.7) and appeared to be insensitive to pH. Minshall and Minshall (1978), while agreeing that there was a correlation with pH, disagreed that pH was the prime determinant of distribution in this river and suggested other water quality factors were more important.

The sensitivity of mayfly larvae and the relative insensitivity of stonefly larvae to low pH were also suggested by surveys of 25 stream sites in Norway (Leivestad et al., 1976). Sixteen of these sites had a pH between 4.0 and 5.0 and never contained more than 2 species (mean of 1.6) of mayflies. Above pH 5.0, numbers of species present varied between 1 and 7 (mean of 4.5) with all but one of the sites having at least 3 species present (Leivestad et al., 1976). Mean numbers of mayfly species are about 3 to 4 times higher in streams with a pH of 6.5 to 7.0 than in streams with a pH of 4.0 to 4.5 (Overrein et al., 1980). This relationship was not found for stream stoneflies (Plecoptera), even though it was found for lake stonefly species (Leivestad et al., 1976; Overrein et al., 1980).

Drift rates for mayflies increased as a result of experimental acidification (Hall et al., 1980; Herricks and Cairns, 1976-76; Overrein et al., 1980). Ephemerella and Epeorus showed largest drift densities as a result of acidification of a New Hampshire stream (Hall et al., 1980).

Drift rates of Baetis sp., Ephemerella sp., Isonychia sp., and Stenonema sp. increased substantially following a 15-minute shock treatment where pH of Mill Creek, a hardwater stream in Virginia, was reduced from pH 8.0 to 4.0 by addition of sulfuric acid. Drift remained elevated for 48 hr (Herricks and Cairns, 1974-76).

Other evidence for sensitivity of mayfly larvae (Ephemeroptera) includes laboratory studies of Ephemerella subvaria by Bell (1971) and Bell and Nebeker (1969). Their studies indicated that Ephemerella subvaria was the most sensitive of 10 insect species tested. Fifty percent of test species died after 30 days exposure to pH 5.38, while 50 percent successfully emerged at pH 5.9. However, not all members of the genus are as sensitive. Fiance (1978) found that experimental acidification of Norris Brook, New Hampshire, for six months from pH 5.85-6.45 to pH 3.9-4.3 with sulfuric acid did not affect emergence rates of Ephemerella funeralis but did cause a decrease in growth and nearly eliminated recruitment of the new cohort.

Other known acid-sensitive species of mayflies include Baetis rhodani, the most common species in rivers in Norway with a pH > 6.0. In rivers below pH 6, these species are eliminated. Experiments indicated that at pH of 4.5 to 4.7 and low conductivity, all species died in less than 48 hr while 10 percent survived the same pH with somewhat higher conductivity. Field observations indicated that Baetis rhodani does not reproduce and probably dies from physiological stress at pH below 5.0 (Overrein et al., 1980). Baetis sp. also had the highest drift rates as a result of a 15-minute shock treatment with pH reduced from pH 8 to 4 in Virginia (Herricks and Cairns, 1974-76).

The studies by Hall et al. (1980) suggested that stonefly larvae (Plecoptera) were relatively insensitive to low pH. Drift rates of the dominant species (Nemoura and Leuctra) did not increase during acidification. Malirekus/Isoperla did increase but numbers were too small to conclude much (from 0 to 8 individuals). Thus, the stream surveys in England and Norway, results from experimental acidification of streams, and laboratory tests (Bell, 1971; Bell and Nebeker, 1969) indicate that stoneflies (Plecoptera) in streams are fairly tolerant of low pH.

Stream insects are often grouped according to their ecological role in the stream into four functional trophic groups (Cummins, 1975). These groups are:

(1) grazers and scrapers--herbivores feeding on attached algae,

(2) shredders--large particle feeding detritivores,
(3) collectors--fine particle feeding detritivores which filter particles from the water or gather them from surfaces, and
(4) predators.

Hall et al. (1980) found that acidification of Norris Brook, New Hampshire, from pH 5.7-6.4 to pH 3.9-4.5 resulted in a 37 percent decrease in emergence of insects but that shredders and predators were unaffected, with all the decrease ascribed to collector organisms. These reductions in abundance were primarily due to decreased emergence of midges (Chironomidae, Subfamily Orthocladinae) and mayflies (Ephemeroptera). No scrapers emerged during the time of their study from either the control or acidified streams.

Hall et al. (1980) also studied the effects of acidification on drift of invertebrates. There was a 3.9-fold increase in overall macroinvertebrate drift on the first day of acidification followed by a peak 13-fold increase on the second day. Drift rates decreased to those typical of the control section one week after acid addition. There was a differential effect on functional groups. Drift of shredders was unaffected. Collectors were affected most with up to 17-fold increases in drift on the second day. Scrapers had a peak 9-fold increase on the first day, while drift of macroinvertebrate predators increased 4-fold.

Hall et al. (1980) also determined that there was a 75 percent reduction in total numbers of benthic invertebrates as a result of acidification. Aquatic Diptera formed about 68 percent of the numbers in the stream and were reduced by 87 percent by acidification. Cranefly larvae (Tipulidae) were reduced by 91 percent, biting midge larvae (Ceratopogonidae) by 88 percent, and the midge larvae (Chironomidae) by 86 percent. Chironomidae abundance in acidified lakes in Norway was also reduced by 60 to 80 percent (Overrein et al., 1980).

The 15-minute shock treatment of lowering pH from 8 to 4 in a hardwater stream in Virginia reduced total numbers of benthic invertebrates by 42 percent and reduced diversity by 28 percent. The stream took 28 days to recover from this stress (Herricks and Cairns, 1974-76).

Invertebrates other than insects are also known to be sensitive to low pH. Much of the data on these species come from lake studies in Norway. These lake studies give some insight into potential sensitivity of stream invertebrates. Amphipods (Gammarus) are known to be acid-sensitive in both lakes and streams. Gammarus lacustris does not tolerate pH less than 6.0 in lakes in Norway, while <u>Gammarus</u> <u>pulex</u> is not found below pH 5.7 in the River Duddon in England (Overrein <u>et</u> <u>al</u>., 1980; Sutcliffe and Carrick, 1973).

Freshwater snail species start to be eliminated from lakes in Norway at pH 6.0 and never occur below pH 5.2. This sensitivity is to be expected in North America as well since snails rarely occur in streams below pH 6.2 (Pennak, 1978). Most small clams (Sphaeriidae) are eliminated below pH 6.0 with only 2 of 20 species occurring below pH 5.0 in Norway (Overrein <u>et al</u>., 1980). The two major North American families of bivalves (Unionidae and Sphaeriidae) are known to be acid-sensitive, with Unionidae rarely found below pH 7.0 and Sphaeriidae found at pH as low as 6.0 (Pennak, 1978). Thus, no bivalves are likely to occur in acidified streams.

Isopods (<u>Asellus</u> <u>aquaticus</u>) are less acid-sensitive in lakes and can tolerate pH levels as low as 4.8 (Overrein <u>et</u> <u>al</u>., 1980). Preliminary data from our investigations suggest <u>Asellus</u> <u>intermedius</u> will be eliminated from streams acidified to pH 4.0 (see results below).

Segmented worms (Oligochaetes) are also less abundant in acid lakes (3 to 4 times higher in non-acidified lakes) in Norway (Overrein <u>et</u> <u>al</u>., 1980) so reductions could occur in acidified streams.

The crayfish (<u>Orconectes</u> <u>virilis</u>) is able to survive pH 4.0 in lakes for short periods but eventually dies due to interference with uptake of calcium leading to an inability to successfully molt (Malley, 1980). This species is also widespread in streams, so acidification to pH below 5.0 could lead to its demise.

On the other hand, surface-dwelling organisms such as water-boatmen (Corixidae) do not seem to be affected by acidification, and become a dominant group in acidified lakes (Overrein <u>et</u> <u>al</u>., 1980). Stream surface-dwelling insects might also increase as a consequence of acidification.

While this coverage of effects on stream invertebrates is not all inclusive, it does indicate that major impacts on the invertebrate fauna are to be expected. Since these invertebrates are important to processing of organic matter in streams (Cummins, 1975) and as the basis of fish production (Brocksen <u>et</u> <u>al</u>., 1968), effects on them can lead to major changes in nutrient cycling rates, processing of organic matter, and fish production.

STUDIES OF POTENTIAL EFFECTS OF ACIDIFICATION IN MICHIGAN

Introduction

We are using artificial stream channels at the Kellogg Biological Station to study the potential effects of acid precipitation on soft water streams in Michigan. These studies have been underway for about 10 months but actual acidification has only been in progress for four months. The purpose of this part of the chapter is to briefly describe our studies and to report preliminary results.

Materials and Methods

Two small, first-order streams that drain directly into Whitefish Bay near Paradise, Michigan, were selected as a source of organisms for use in simulation. The alkalinity of these streams varies from 8 to 19 mg $CaCO_3$/l and hardness varies from 14 to 18 mg $CaCO_3$/l. They are slightly higher in nitrogen and phosphorus than some of the least perturbed streams of acid-sensitive regions such as those in the Huron Mountains of Marquette County, Michigan (Table 1).

The temperature- and light-controlled concrete channels used for simulation of these streams were described in detail by Cummins (1971)(Figure 1). These channels are 12 m long, 1.5 m wide, and vary from 0.6 m deep at the upstream reservoir to 0.8 m deep at the downstream reservoir. In actual operation, water in the channels is maintained at a depth of 14 to 20 cm just below the upstream reservoir to 31 to 37 cm just in front of the downstream reservoir. The deepening of the channel in a downstream direction results in simulation of a riffle and pool environment. The channels were cleaned thoroughly and repainted with epoxy paint prior to the start of these experiments.

The channels were lined with a mixture of the large gravel to stone sediments taken from the Peshekee River in the Huron Mountains, Marquette County, Michigan, and fine sand sediments typical of the streams that enter Whitefish Bay near Paradise, Chippewa County, Michigan. The sand forms a layer over the gravel and rock except in the faster current within 3 m of the upstream reservoir. The water source is deionized well water. This deionized water was allowed to equilibrate with the sediment and then adjusted to levels typical of upper peninsula soft water streams by adding appropriate amounts of various constituents. The stream water equilibrated with sediments so that few chemical additions were needed. Enough potassium

Table 1

Comparison of Chemistry of Soft Water Streams in
Michigan Versus Artificial Channels Just Prior to Acidification

	Unnamed Streams[a]	Peshekee River[b]	Artificial Streams	
			Control	Acidified
		----mg/l[c]----		
Inorganic-nitrogen	0.16+.04	0.03	0.03	0.03
Organic-nitrogen	0.53+.09	0.04	0.07	0.12
Molybdate reactive phosphorus	0.038+.023	0.002	0.013	0.012
Total phosphorus	0.064+0.15	0.019	0.043	0.030
Chloride	1.75+.50	1	1	1
Sulfate	<1.0	2-3[b]	<1.0	<1.0
Hardness (as CaCO$_3$)	16.25+2.06	15	15	11
Alkalinity (as CaCO$_3$)	12.4+3.9	10-14[b]	17	13
pH (range)	6.2-6.8	7.3-7.7[b]	7.3	7.4
Calcium	5.24+1.22	4.8-6.0[b]	6.35	4.47
Magnesium	1.40+.45	1.0-1.4[b]	0.35	0.44
Sodium	0.80+.02	0.90-1.30[b]	0.63	1.40
Potassium	0.75+.26	0.30-0.40[b]	0.80	1.31
Aluminum (µg/l)	78+26	--	33	20
Manganese (µg/l)	6+7	--	1	1
Iron (µg/l)	287+146	400-500[b]	6	14
Copper (µg/l)	51+26	--	44	24

[a]Mean of samples from two unnamed streams near Paradise, Chippewa County, during October and November, 1980.

[b]Peshekee River at US 41 Bridge, Michigamme Township, Section 25, Marquette County, Michigan. Data marked by b are taken from STORET, other data collected in the field in June, 1980. Alkalinity varies from 10 to 14 mg/l for this site. Our data from further upstream indicate alkalinity as low as 3 mg $CaCO_3$/l.

[c]Except where indicated.

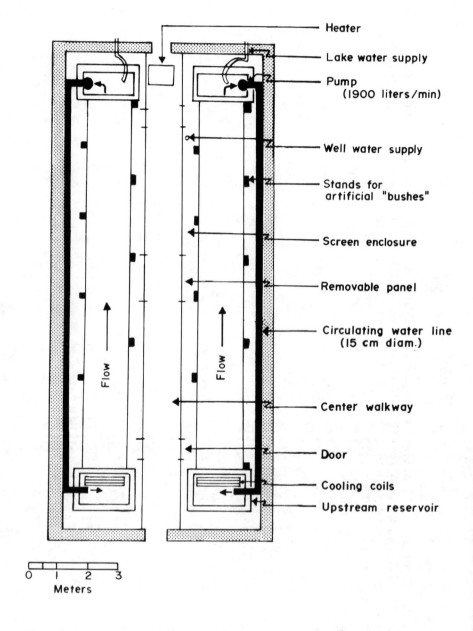

Figure 1.   Diagrams of artificial stream channels used for
            simulation of soft water streams in Michigan
            (Cummings, 1971).

nitrate was added to add 0.10 mg NO_3-N/l and 0.28 mg K/l to
each stream. Enough sodium monobasic phosphate was added to
add 0.05 mg P/l and 0.04 mg Na/l to each stream. The nitrate
and phosphate did not stay in solution long and were likely
taken up by the biota and/or the sediments. Thus, levels in
stream water just prior to acidification were somewhat low for
the streams near Paradise but approximated concentrations
typical of the Peshekee River (Table 1). Iron was very low
compared to the natural streams and magnesium, aluminum, and
manganese were somewhat low (Table 1). Otherwise, water
chemistry appeared to reasonably simulate chemistry of upper
peninsula streams.

    Organisms were collected from the two small streams near
Paradise, Chippewa County, Michigan, and were transported back
to the laboratory in well moistened, chilled leaf packs. These
organisms were used to stock the streams. Even though efforts
were made to evenly divide these organisms, population levels
for the two streams prior to experimentation were different.
Sampling of loose leaf material deposited in the stream
indicated that there were 14.5 + 4.4 (mean + 95% C.I.) organisms
per 5 gm of leaf material in the acidified channel prior to
acidification while there were only 6.3 + 2.1 organisms per 5
gm of leaf material in the control stream. Sampling of leaf
pack material prior to acidification yielded 19.5 + 6.3
organisms per leaf pack in the channel that was to be acidified
versus 9.0 + 4.8 per leaf pack in the control. Thus, organisms
appeared to be more concentrated in leaf packs than on the loose
organic matter, but in each case there were 2.2 to 2.3 times
more organisms per leaf pack in the stream that was to be
acidified than there were in the control stream. These
differences have to be taken into account in subsequent data
analysis.

    Five-gram leaf packs of white birch (<u>Betula</u> <u>papyrifera</u>)
and sugar maple (<u>Acer</u> <u>saccharum</u>), the dominant tree species of
the forest along the streams near Paradise, Michigan, were
placed in the streams on the leading edge of bricks using
techniques of Merritt <u>et al</u>., (1979). During initial stocking
of the streams with invertebrates, these leaf packs were made
up of leaves from terrestrial areas adjacent to the streams
near Paradise, Michigan, since all leaf material in the stream
had been decomposed by the start of the study in late May, 1980.
Prior to acidification, part of these leaf packs were removed
and all organisms collected and sorted to species while the
other half were removed 38 days after acidification began.
These "old" leaf packs were used, therefore, as a method of
screening losses of invertebrates during initial acidification.

    Newly fallen leaf material was collected in October, 1980,
dried, and used to construct new leaf packs following the
techniques of Merritt <u>et al</u>. (1979). Twenty 5 gm leaf packs

of each species (white birch and sugar maple) were placed in each stream on November 4, 1980. This was two weeks prior to removal of "old" leaf packs described above and 24 days prior to start of acidification. Since the start of acidification, these leaf packs have been removed from the stream, weighed, and all organisms identified and counted on a monthly basis. This process will continue until leaf material is degraded or until the experiment ceases in September, 1981.

Acidification began on November 28, 1980. The pH was dropped from 7.4 to 4.0 by slowly dripping 0.1 $\underline{N}$ sulfuric acid into the stream over a 6-hr period. The pH of 4.0 is maintained by dripping in appropriate amounts of acid on a daily basis. Thus, pH is maintained at levels between 4.0 and 4.3.

Some seepage of surface runoff through the high carbonate soils near the greenhouse enters the artificial streams despite our best efforts to stop it. Thus, hardness and alkalinity slowly increase over time and have to be reduced by deionization of stream water. Initial leaching of materials from sediments by acidification during the first 8 days increased hardness from 16 to 30 mg $CaCO_3$/l. The acidified stream is kept at this hardness (30 mg/l) by deionization of stream water. Alkalinity and hardness of the control stream are maintained between 15 and 20 mg $CaCO_3$/l, the level typical of both streams prior to acidification.

Preliminary Results

Acidification reduced the total number of benthic invertebrates on the "old" or conditioned leaf packs which had been in the stream for several months from 19.5 $\pm$ 6.3 to 7.2 $\pm$ 7.3 organisms per leaf pack or a 63 percent decease in total numbers after 38 days of acidification ($p < 0.05$, t test). Population levels in the control channel were not significantly different after 38 days (t test)(from 9.0 $\pm$ 4.8 to 7.8 $\pm$ 2.3 organisms per leaf pack) compared to the 63 percent reduction in the acidified stream over the same time period. The reduction in the acidified stream compares well to the 75 percent reduction in total numbers reported from acidification of Norris Brook, New Hampshire (Hall $\underline{et}$ $\underline{al}$., 1980). In fact, Hall $\underline{et}$ $\underline{al}$. (1980) reported an 84 percent reduction in total numbers of invertebrates from debris dams.

The most common invertebrate present was a shredder, $\underline{Lepidostoma}$ $\underline{liba}$ (Trichoptera). It appears to be relatively unaffected by acidification. Numbers in the old leaf packs were reduced by 44 percent in the acidified stream from 8.8 $\pm$ 4.5 to 4.9 $\pm$ 6.2 organisms per leaf pack, but they were also reduced by 34 percent in the control stream from 3.5 $\pm$ 3.0 to

2.3 $\pm$ 1.0 organisms per leaf pack, but neither reduction was significantly different (p 0.05, t test). Likewise, numbers on the "new" leaf packs (those put in the stream from newly fallen leaves in November, 1980) were reduced from 9.2 organisms per leaf pack based on collections of loose leaf debris prior to acidification to 7.8 after one month of acidification (a 15% reduction) while the control stream was reduced from 4.5 to 3.8 organisms per leaf pack (a 16% reduction; no significant difference in either case). Thus, natural mortality seems to be reducing population levels but no significant reduction can be ascribed to acidification effects. Both downstream drift rates and emergence of shredders were unaffected by acidification in New Hampshire as well (Hall et al., 1980).

The isopod, *Asellus intermedius*, is also an important component of the macroinvertebrate fauna in the natural streams near Paradise, Michigan. It was markedly reduced in numbers by acidification from 6.8 $\pm$ 4.5 to 0.9 $\pm$ 0.8 organisms per leaf pack on the "old" leaf packs after 38 days of acidification, or an 87 percent reduction in total numbers. At the same time, populations in the control stream increased slightly from 0.8 $\pm$ 0.7 to 0.9 $\pm$ 1.1 organisms per leaf pack. This same trend was characteristic of the new leaf pack with numbers reduced from 2.4 $\pm$ 1.2 per leaf pack to 0.25 $\pm$ 0.26 by acidification (a 90% reduction), while populations in the control stream expanded from 0.43 $\pm$ 0.28 organisms per leaf pack to 1.7 $\pm$ 0.61. *Asellus aquaticus* is eliminated from lakes in Norway at pH 4.8 to 5.2 (Overrein et al., 1980), and our findings suggest that *A. intermedius* in streams is also sensitive to acidification.

The snail *Physa heterostropha* had low population levels in the acidified stream prior to acidification and has apparently been eliminated as a result of acidification. Population of this species was larger in the control stream and population levels appear to have remained stable there (there was an apparent decrease from 2.8 $\pm$ 0.7 organisms per leaf pack to 0.9 $\pm$ 1.6 per leaf pack on the old leaf packs but an apparent increase from 0.4 to 1.5 organisms per leaf pack on the new leaf pack material).

Decomposition rates of leaf material have not been significantly affected by acidification after two months of exposure to pH 4.0. For example, sugar maple leaves retain 73.7 $\pm$ 5.0 percent of their original weight in the upper half of the acidified channel compared to 74.3 $\pm$ 1.1 percent in the upper half of the control channel and 78.2 $\pm$ 5.6 versus 74.7 $\pm$ 2.2 percent for the lower half of the acidified versus the control channel, respectively. Since decomposition rates are very low at low winter temperatures (Suberkropp et al.,

1975)(the streams have been at 2-3°C over the winter), differences due to acidification are not likely to show up until decomposition rates increase during spring. We expect these differences to occur based on studies summarized above, but they have not occurred as yet in our study.

The effects on stream chemistry due to leaching of material from the substrate by acidification can be summarized as follows. After 8 days of acidification, calcium increased 2.2-fold, magnesium increased 1.7-fold, potassium dropped slightly (13%), sodium increased slightly (1.3-fold), aluminum concentrations were over 6 times higher than pre-acidification levels, with copper, iron, zinc, and manganese all showing significant increases. The stream channel results may not be totally applicable to natural systems, and we are in the process of conducting leaching studies of stream sediments from the upper peninsula streams using acidified stream water in order to better predict these responses. As expected, sulfate levels increased due to addition of sulfuric acid, while carbonate-bicarbonate alkalinity was reduced to negligible levels.

The above summary represents preliminary results from our studies. We expect to continue these studies and to expand them to include individual growth, feeding rates, and mortality rates of <u>Lepidostoma</u> <u>liba,</u> <u>Physa</u> <u>heterostropha,</u> and <u>Asellus</u> <u>intermedius</u> as well as the collector, <u>Parapsyche</u> sp. under acidified and non-acidified conditions.

CONCLUSIONS

Acidification significantly alters the chemical and biological processes in soft water streams. These changes include increases in aluminum, manganese, calcium, magnesium, and potassium concentrations in stream water with aluminum concentrations increasing to potentially toxic levels.

Effects on the biota include increases in standing crop of benthic algae with a shift toward acid-tolerant filamentous green algae such as <u>Mougeotia</u> sp. However, productivity is often unaffected or may be reduced with the increased standing crop resulting from lower grazing pressure, species shifts, or decreased decomposition rates.

Decomposition rates are decreased in acidified streams, resulting in reduced energy flow to detritivores, the primary base of food support for fish, and also resulting in lowered rates of remineralization and recycling of nutrients.

Invertebrates vary in sensitivity. Mayflies (Ephemeroptera) appear to be very sensitive to acid waters as a group, while stoneflies (Plecoptera) are not. Caddisflies (Trichoptera) appear to contain both sensitive and tolerant species. Snails, clams, and many crustaceans appear to be very sensitive to acidification. Based on limited field and laboratory studies, acidification to pH 4.0 will reduce invertebrate numbers by 60 to 80 percent, will result in increased drift of collector and predator species, and will result in decreased emergence of certain groups, especially the aquatic Diptera (midges, craneflies, and biting midges) and Ephemeroptera (mayflies).

Finally, preliminary results from our studies in Michigan were summarized. Our results for invertebrates are consistent with those summarized above. However, no reduction in degradation of leaf packs is yet apparent, perhaps because of the need for conditioning of the leaves before invertebrate feeding rates are expected to increase (Cummins, 1975), low winter temperature exposures to date and the short time the experiment has been in progress.

ACKNOWLEDGMENTS

This work was supported by a subcontract from the U.S. EPA-NADP Acid Precipitation Program (U.S. EPA Cooperative Agreement Number CR806912-01-0 with North Carolina State University). It was conducted using the facilities of the Kellogg Biological Station and the Institute of Water Research at Michigan State University. We thank C. Annett, P. Little, C. Piening, and V. Kelley for technical assistance. Kellogg Biological Station Contribution Number 463.

LITERATURE CITED

Arnold, D.E., R.W. Light, and V.J. Dymond. 1980. Probable Effects of Acid Precipitation on Pennsylvania. EPA-600/3-80-012, U.S. Environmental Protection Agency, Environmental Research Laboratory, Corvallis, Oregon. 20 pp.

Bell, H.L. 1971. Effect of Low pH on the Survival and Emergence of Aquatic Insects. Water Res. 5:313-319.

Bell, H.L. and A.V. Nebeker. 1969. Preliminary Studies on the Tolerance of Aquatic Insects to Low pH. J. Kansas Ent. Soc. 42:230-236.

Brocksen, R.W., G.E. Davis, and C.E. Warren.   1968.
Competition, Food Consumption, and Production of Sculpins and
Trout in Laboratory Stream Communities.   J. Wildlife Mgmt.
32:51-75.

Cronan, C.S. and C.L. Schofield.   1979.   Aluminum Leaching
Response to Acid Precipitation: Effects on High-Elevation
Watersheds in the Northeast. Science 204(4390):304-306.

Cummins, K.W.   1971.   Predicting Variations in Energy Flow
Through a Semi-Controlled Lotic Ecosystem.   Tech. Rept. 19,
Institute of Water Research, Michigan State Univ., East Lansing,
Michigan.   21 pp.

Cummins, K.W. 1975. Macroinvertebrates. In: B.A. Whitton,
ed., "River Ecology," pp. 170-198. Studies in Ecology, Vol.
2, Blackwell Scientific Publ., Oxford, England.

Dickson, W. 1978. Some Effects of the Acidification of Swedish
Lakes. Verh. Internat. Verein. Limnol. 20:851-856.

Dillon, P.J., D.S. Jeffries, W. Snyder, R. Reid, N.D. Yan, D.
Evans, J. Moss, and W.A. Scheider. 1978. Acidic Precipitation
in South-Central Ontario: Recent Observations. J. Fish. Res.
Bd. Can. 35:809-815.

Driscoll, C.T., Jr., J.P. Baker, J.J. Bisogni, Jr., and C.L.
Schofield.   1980.   Effect of Aluminum Speciation on Fish in
Dilute Acidified Waters. Nature 284:161-164.

Fiance, S.B.   1978.   Effects of pH on the Biology and
Distribution of Ephemerella funeralis (Ephemeroptera). Oikos
31:332-339.

Galloway, J.N. and E.B. Cowling.   1978.   The Effects of
Precipitation on Aquatic and Terrestrial Ecosystems - A Proposed
Precipitation Network. J. Air Pollut. Control Assoc. 28(3):229-
235.

Galloway, J.N. and G.E. Likens. 1979. Atmospheric Enhancement
of Metal Deposition in Adirondack Lake Sediments.   Limnol.
Oceanogr. 24(3):427-433.

Gjessing, E.T., A. Henriksen, M. Johannessen, and R.F. Wright.
1976. Effects of Acid Precipitation on Freshwater Chemistry.
In: F.H. Braekke, ed., "Impact of Acid Precipitation on Forest
and Freshwater Ecosystems in Norway," pp. 64-85. SNSF Project,
Research Report FR6. Agricultural Research Council of Norway,
Norwegian Council for Scientific and Industrial Research, and
Norwegian Ministry of Environment, Oslo, Norway.

Hall, R.J., G.E. Likens, S.B. Fiance, and G.R. Hendrey. 1980. Experimental Acidification of a Stream in the Hubbard Brook Experimental Forest, New Hampshire. Ecology 61(4):976-989.

Hendrey, G.R. 1976. Effects of pH on the Growth of Periphytic Algae in Artificial Stream Channels. Report IR25, SNSF Project, NISK, 1432 Oslo-As, Norway. 50 pp.

Herricks, E.E. and J. Cairns, Jr. 1974-76. The Recovery of Stream Macrobenthos from Low pH Stress. Revista de Biologia 10(1-4):1-11.

Hultberg, H. and O. Grahn. 1976. Effects of Acid Precipitation on Macrophytes in Oligotrophic Swedish Lakes. In: D.H. Matheson and F.C. Elder, eds., "Atmospheric Contributions to the Chemistry of Lake Waters," pp. 208-217. Proc. First Specialty Symp., Int. Assoc. Great Lakes Res., Vol. 2, Suppl. 1.

Jeffries, D.S., C.M. Cox, and P.J. Dillon. 1979. Depression of pH in Lakes and Streams in Central Ontario During Snowmelt. J. Fish. Res. Bd. Can. 36:640-646.

Johnson, N.M. 1979. Acid Rain: Neutralization Within the Hubbard Brook Ecosystem and Regional Implications. Science 204(4392):497-499.

Kramer, J.R. 1976. Geochemical and Lithological Factors in Acid Precipitation. In: L.S. Dochinger and T.A. Seliga, eds., "Proc. First Int. Symp. on Acid Precipitation and the Forest Ecosystem," pp. 611-618. USDA Forest Service General Tech. Rept. NE-23.

Leivestad, H., G. Hendrey, I.P. Muniz, and E. Snekvik. 1976. Effects of Acid Precipitation on Freshwater Organisms. In: F.H. Braekke, ed., "Impact of Acid Precipitation on Forest and Freshwater Ecosystems in Norway," pp. 87-111. SNSF Project, Research Report FR6. Agricultural Research Council of Norway, Norwegian Council for Scientific and Industrial Research, and Norwegian Ministry of Environment, Oslo, Norway.

Likens, G.E., R.F. Wright, J.N. Galloway, and T.J. Butler. 1979. Acid Rain. Sci. Amer. 241(4):43-51.

Malley, D.F. 1980. Decreased Survival and Calcium Uptake by the Crayfish Orconectes virilis in Low pH. Can. J. Fish. Aquat. Sci. 37:364-372.

McFee, W.W. 1980. Sensitivity of Soil Regions to Acid Precipitation. EPA-600/3-80-013, U.S. Environmental Protection Agency, Environmental Research Laboratory, Corvallis, Oregon. 178 pp.

Merritt, R.W., K.W. Cummins, and J.R. Barnes. 1979. Demonstration of Stream Watershed Community Processes with Some Simple Bioassay Techniques. In: V.H. Resh and D.M. Rosenberg, eds., "Innovative Teaching in Aquatic Entomology," pp. 101-113. Can. Spec. Publ. Fish. Aquat. Sci. 43:1-118.

Minshall, G.W. and J.N. Minshall. 1978. Further Evidence on the Role of Chemical Factors in Determining the Distribution of Benthic Invertebrates in the River Duddon. Arch. Hydrobiol. 83(3):324-355.

Muller, P. 1980. Effects of Artificial Acidification on the Growth of Periphyton. Can. J. Fish. Aquat. Sci. 37:355-363.

Overrein, L.N., H.M. Seip, and A. Tollan. 1980. Acid Precipitation - Effects on Forest and Fish. Final Report of the SNSF Project 1972-1980. Research Report FR19. Norwegian Council for Industrial and Scientific Research, Agricultural Research Council of Norway, Norwegian Ministry of Environment, Oslo, Norway.

Patrick, R., N.A. Roberts, and B. Davis. 1968. The Effect of Changes in pH on the Structure of Diatom Communities. Notulae Naturae, Acad. Natural Sci. Philadelphia 416:1-16.

Patrick, R., V.P. Binetti, and S.G. Halterman. 1981. Acid Lakes from Natural and Anthropogenic Causes. Science 211(4481):446-448.

Pennak, R.W. 1978. Fresh-Water Invertebrates of the United States. John Wiley & Sons, New York. 803 pp.

Scheider, W.A., D.S. Jeffries, and P.J. Dillon. 1979. Effects of Acidic Precipitation on Precambrian Freshwaters in Southern Ontario. J. Great Lakes Res., Int. Assoc. Great Lakes Res. 5(1):45-51.

Schofield, C.L. 1976. Effects of Acid Precipitation on Fish. Ambio 5:228-230.

Suberkropp, K., M.J. Klug, and K.W. Cummins. 1975. Community Processing of Leaf Litter in Woodland Streams. Verh. Internat. Verein. Limnol. 19:1653-1658.

Sutcliffe, D.W. and T.R. Carrick. 1973. Studies on Mountain Streams in the English Lake District. Freshwater Biol. 3:437-462.

U.S. EPA. 1979. Quarterly Report, U.S. Environmental Protection Agency, Environmental Research Laboratory-Duluth, July-Sept., 1979, Duluth, Minnesota.

Warner, R.W. 1971. Distribution of Biota in a Stream Polluted by Acid Mine Drainage. Ohio J. Sci. 71(4):202-215.

Zimmerman, J.W. 1968. Water Quality of Streams Tributary to Lakes Superior and Michigan. U.S. Fish and Wildlife Service, Special Scientific Report - Fisheries No. 559, Washington, D.C. 41 pp.

CHAPTER 10

A WATERSHED APPROACH - THE EPRI INTEGRATED
LAKE WATERSHED ACIDIFICATION STUDY (ILWAS)

Arland H. Johannes
Department of Chemical and Environmental Engineering
Rensselaer Polytechnic Institute
Troy, New York 12181

Robert A. Goldstein
Environmental Assessment Department
EPRI, 3412 Hillview Avenue
Palo Alto, California 94303

Carl W. Chen
Tetra Tech, Inc.
3746 Mt. Diablo Boulevard
Lafayette, California 94549

INTRODUCTION

Acidic deposition is currently recognized as a problem of national and international dimensions. Since the first reports of acidic precipitation several decades ago, our understanding of the problem and the processes involved has increased substantially. However, characterization of long-term effects to the biosphere and source-effect relationships require increased attention to quantify and detail the sequences and mechanisms involved. Atmospheric removal processes are generally characterized as wet and dry deposition of chemical species and are at the end of a complex series of atmospheric transformations. This in turn marks the beginning of another chain of events whereby these same inputs into a receptor such as a watershed trigger changes within the underlying ecosystem. Biogeochemical processes acting in series and in parallel produce or consume acids and release chemicals that shift the pH equilibrium during transport through the system to the lake. Such changes can occur in the forest canopies, the litter zones and soils, the ground water and lake tributaries, and finally, the lake itself. The results of one process may be modified by others in a real system to yield answers which may be conflicting. Scientific investigation of a single process is, therefore, not sufficient to understand the chemical behavior of atmospheric burdens as they eventually reach and become an integral part of surface waters.

This chapter gives a brief overview of the philosophy and organization of the Integrated Lake Watershed Acidification Study (ILWAS).   In addition, two years of atmospheric input data are reviewed and analyzed on a seasonal basis.   These data are mandatory for detailed watershed budgets and establish a benchmark for future ecological effects studies in the Adirondacks.

APPROACH AND OBJECTIVES

An intensive, integrated, five-year study of three forested watersheds was established to determine how lake waters become acidified and to quantify the train of events occurring as acid precipitation becomes lake water.   The study integrated management questions into the scientific research at the planning stage (Goldstein et al., 1980).   The total system was divided into compartments for detailed scientific analysis with a model being developed to reassemble (i.e., integrate) the data from each subsection to represent the overall behavior of the system.   The model can then be used to:

(1) determine a general quantitative relationship between atmospheric inputs and acidification of surface waters,
(2) provide a link between cause-effect relationships,
(3) predict and provide input for future watershed studies,
(4) identify key variables for long-term monitoring,
(5) check on the consistency of hypotheses and identify data gaps,
(6) predict future environmental problems,
(7) evaluate future control strategies, and
(8) predict water quality in other lake basins.

The watersheds chosen for study are all located within a 15 km radius of the town of Eagle Bay, New York, in the Adirondack Mountains.   All three watersheds are densely forested and are of comparable evaluation.   However, the three lakes have substantially different pH and alkalinity values. Panther Lake is a neutral pH lake with a yearly mean alkalinity of 160 μeq/l; Woods Lake is very acidic and has a yearly mean alkalinity of -20 μeq/l; and Sagamore Lake is circumneutral (near neutral, but variable) with a year mean alkalinity of 10 μeq/l (Galloway et al., 1980).

The watersheds are divided into compartments that the acid rain must pass through before reaching the lake outlet. At selected locations along this path, intensive measurements are made, with frequency of data collection dictated by temporal variability. Figure 1 (Chen and Goldstein, 1980) illustrates major field program components and lists the key elements of data collection. The overall model under development represents a watershed as an aggregate of subcatchments to account for horizontal and vertical flow routing. Specific details of the mathematical model and its development have been reported elsewhere (Chen et al., 1978; Goldstein et al. 1980; Chen and Goldstein, 1980).

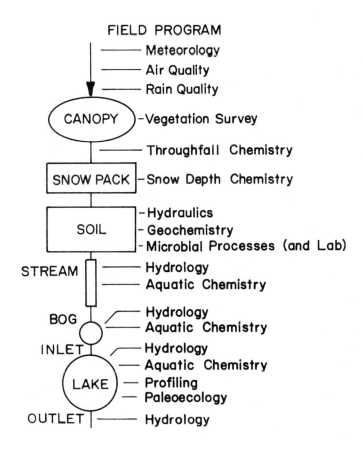

Figure 1. Field program components (after Chen and Gold-
*   stein, 1980).*

The project is sponsored by the Electric Power Research Institute (EPRI), Palo Alto, California. The study team includes twelve principal investigators from six universities and two federal agencies, with the overall modeling and project management done by Tetra Tech, Inc., Lafayette, California.

When the ILWAS project (Phase I) is completed in late December, 1981, a comprehensive data base for three Adirondack watersheds covering a period of almost four years will be available. In addition, a calibrated and verified mathematical model, as well as a series of interpretive reports and publications, will have been generated which should be readily applicable to other lake basins.

WET DEPOSITION

Atmospheric inputs were measured in each of the three watersheds since May, 1978. Precipitation, in the form of rain and snow, was collected at 4 to 7 sites within these watersheds on an event basis (defined as daily collection). Details of the network and sampling design have been reported elsewhere (Johannes and Altwicker, 1980; Altwicker et al., 1980; Johannes et al., 1981). Dry deposition has also been collected on a weekly basis. Wet and dry deposition samples were analyzed for pH, sulfate, nitrate, chloride, ammonium, calcium, magnesium, potassium, and sodium. More recently, ambient air concentrations of sulfate and nitrate aerosols, and sulfur dioxide, were measured in these watersheds (Altwicker and Johannes, 1981).

Mean monthly weighted ion concentrations for wet deposition during the period May, 1978, to August, 1980, are given in Table 1 for hydrogen, sulfate, nitrate, and ammonium. Weighted ion concentration is defined as $c_i v_i / v_i$; where c_i = concentration of a given ion in the ith event and v_i = rain volume for that event. Beginning in September, 1979, the total number of network sites was reduced from seven to four. The variabilities did not significantly change after the reduction. These results and linear correlation studies indicate that this reduction did not invalidate an earlier conclusion that the chemical composition of rain and snow was nearly identical (on a monthly basis) at all network sites. This was not true with precipitation quantity, where one watershed site (Sagamore) received approximately 20 percent less precipitation than the other two sites.

These observations lead to the conclusion that over a region such as the Adirondacks, precipitation quality could be accurately estimated using a small number of collectors.

However, to accurately predict loadings (concentration times amount), a much larger number of rain gauges would be required to account for the larger variability of precipitation quantity.

The mean monthly weighted ion concentrations in Table 1 can be viewed in another way to gain insight into ion dosages and trends. The log of each value was taken and the quantity ($\log c_i - [\Sigma \log c_i]/n$) was evaluated for $n = 24$ (May, 1978, to April, 1980). These values are plotted in Figure 2 for hydrogen, sulfate, and nitrate concentrations. The zero line is the two-year average (hydrogen, 71.1 µeq/l; sulfate, 63.6 µeq/l; nitrate, 37.3 µeq/l). Positive intervals, those above the line, can be viewed as "excess" concentrations; negative intervals, those below the line, can be viewed as "deficit" <u>vis a vis</u> the average. From May through September, 1978, hydrogen and sulfate tracked each other quite well, while nitrate appears to have been largely decoupled from the others. For the last three months of 1978, nitrate gradually increased relative to the other two species. With the exception of February and August, the behavior of all three species for the first ten

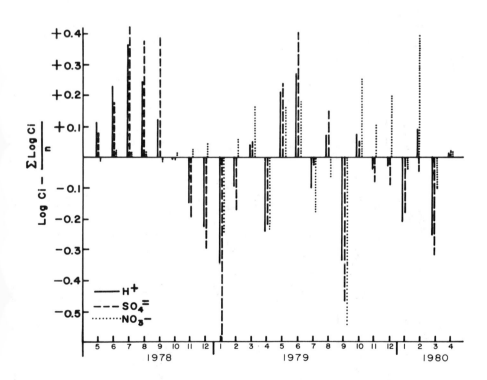

Figure 2. Net ion concentration.

Table 1

Mean Monthly Weighted Ion Concentrations (µeq/l), Standard Deviations, and Variability
(Basis: Seven Network Sites, May, 1978–August, 1979, Four Network Sites After August, 1979)

Month	H^+	S.D.	$\dfrac{S.D.}{X}$	SO_4^{-2}	S.D.	$\dfrac{S.D.}{X}$	NO_3^-	S.D.	$\dfrac{S.D.}{X}$	NH_4^+	S.D.	$\dfrac{S.D.}{X}$
1978												
May	85	12.8	0.15	66	11.9	0.18	33	1.9	0.06	20	1.4	0.07
June	112	7.6	0.07	84	4.3	0.05	35	1.1	0.03	25	2.1	0.08
July	151	29.8	0.20	145	16.6	0.11	34	7.5	0.22	22	3.2	0.37
August	115	7.6	0.07	133	7.6	0.06	35	4.2	0.12	21	3.6	0.17
September	87	2.8	0.03	136	4.0	0.03	32	5.0	0.16	41	2.5	0.06
October	64	6.2	0.10	54	4.6	0.09	34	2.3	0.07	18	2.5	0.14
November	47	4.6	0.10	35	4.6	0.13	35	5.4	0.15	14	3.7	0.26
December	39	2.7	0.07	28	3.3	0.12	36	4.5	0.13	9	2.4	0.27
1979												
January	30	0.9	0.03	14	1.6	0.11	19	1.6	0.08	2	0.4	0.20
February	53	8.0	0.15	37	6.2	0.17	40	4.1	0.10	9	3.0	0.33
March	72	8.1	0.11	62	10.8	0.17	50	6.8	0.14	19	3.5	0.18
April	38	1.5	0.04	33	5.1	0.15	19	2.0	0.11	8	1.4	0.16
May	106	7.9	0.07	95	7.7	0.08	48	3.7	0.08	31	11.3	0.36
June	123	10.5	0.09	141	23.5	0.17	55	6.1	0.11	46	8.7	0.19
July	53	11.6	0.22	53	14.6	0.28	22	4.9	0.22	14	4.6	0.33
August	77	7.3	0.09	77	4.2	0.05	29	3.4	0.12	20	4.2	0.21
September	29	3.9	0.13	18	3.0	0.16	10	1.1	0.11	6	0.5	0.10

October	78	4.2	0.05	61	1.7	0.03	57	4.3	0.08	31	2.4	0.08
November	60	5.6	0.09	45	7.0	0.15	41	4.2	0.10	16	3.6	0.23
December	59	7.4	0.13	45	2.3	0.05	53	5.8	0.11	18	1.4	0.08
1980												
January	41	1.9	0.05	36	1.3	0.04	31	1.2	0.04	10	1.1	0.11
February	86	4.2	0.05	50	10.5	0.21	35	5.0	0.07	18	3.1	0.17
March	36	4.4	0.12	26	3.6	0.14	26	2.9	0.11	6	1.5	0.25
April	66	4.5	0.07	52	5.2	0.10	36	4.1	0.11	15	2.3	0.15
May	69	5.4	0.08	87	9.3	0.11	43	7.2	0.17	27	1.9	0.07
June	65	7.4	0.11	66	4.8	0.07	25	2.1	0.08	22	1.8	0.08
July	83	10.8	0.13	82	11.4	0.14	31	2.1	0.07	25	5.5	0.22
August	107	7.4	0.07	105	7.3	0.07	42	3.0	0.07	30	2.6	0.09

months of 1979 could be described as coupled (i.e., increases and decreases were in phase).   These observations are of potential significance when one attempts to relate observed concentrations to source regions and transport behavior.   In addition, these plots can be used in effects studies to pinpoint possible synergistic interactions.   These plots can also be made using event data and limited to the growth season.   Plots of this type are valuable when investigating acute effects of acid rain compared to the average dose.

A   similar   transformation   was   made   for   loadings (concentrations   times   precipitation   quantity   in   units   of depth); for example, the expression (Log L_i - [Σ Log L_i] /n) was graphed for hydrogen, sulfate, and nitrate and is shown in Figure 3.   The zero line (two-year average) was 720 eq/ha for hydrogen, 666 eq/ha for sulfate, and 377 eq/ha for nitrate. Net sulfate and hydrogen ion loadings for the summer months of 1978 were considerably above the average.   By contrast, June and July, 1979, were below average.

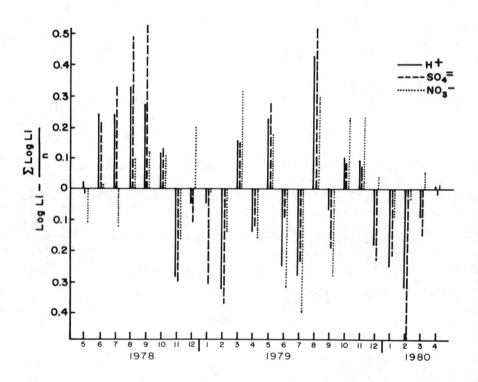

Figure 3.   Net ion loadings.

Net ion loading graphs similar to Figure 3 must be interpreted with care since values below the line (deficit compared to the average) can be caused by either low ion concentrations or small quantities of precipitation.   Three summer growth periods (June, July, and August, 1978 to 1980) were investigated in more detail to study the summer sulfate and hydrogen loading peaks. Table 2 lists the three-watershed monthly precipitation averages (cm) for 1978, 1979, and 1980. In 1978, similar quantities of rain fell each month; in 1979, June was extremely dry, whereas August was particularly wet. The three-month totals for 1978 and 1979, however, were nearly identical.   In June and July, 1980, precipitation was considerable, whereas August was quite dry. The 1980 total, however, was considerably above those of the previous two years. The total number of summer events in 1978 and 1979 was nearly identical (24 vs 22), but totaled 32 for 1980. A larger number of events in 1980 accounted for the observed difference, rather than a larger average amount per event.

Hydrogen, sulfate, and nitrate loadings corresponding to the same period are shown in Table 3. Loadings decreased from 1978 to 1979 by about one-third, although precipitation quantities were nearly identical. The 1979 loadings were lower due to a decrease in concentration. The increase in loadings from 1979 to 1980 was the result of the large quantities of 1980 rain relative to 1979. Concentration changes between 1979 and 1980 were slight. Put in terms of dosage shown in Table 4, one can see that all ions decreased in concentration from 1978 to 1980.

Table 2

Monthly Summer Rainfall ILWAS -- Three-Basin Averages

	Year		
Month	1978	1979	1980
June	8.4	2.1	13.8
July	6.5	7.3	15.0
August	9.7	16.0	6.2
Total	24.6	25.4	35.0

Table 3

Hydrogen, Sulfate, and Nitrate Loadings

		1978			1979			1980	
Month	H^+	SO_4^{-2}	NO_3^-	H^+	SO_4^{-2}	NO_3^-	H^+	SO_4^{-2}	NO_3^-
					eq/ha				
June	94	71	29	30	35	12	89	90	35
July	98	94	22	28	28	16	124	123	47
August	111	129	34	146	146	46	65	64	27
Total	303	294	85	204	209	74	278	277	109

Table 4

Mean Weighted Ion Concentration

		1978			1979			1980	
Month	H^+	SO_4^{-2}	NO_3^-	H^+	SO_4^{-2}	NO_3^-	H^+	SO_4^{-2}	NO_3^-
					eq/ha				
June	112	84	35	123	141	55	65	66	25
July	151	145	34	53	53	22	83	82	31
August	115	133	35	77	77	29	107	105	42
Average	123	120	35	80	82	29	78	78	30

Individual event concentrations were compared for the three summers. Figure 4 shows sulfate concentrations (μmol/l) for all summer rain events during 1978 to 1980. It is readily apparent that with three exceptions (one in 1979 and two in 1980) sulfate concentrations were considerably lower in 1979 and 1980 than in 1978. Similar results were noted for all other ions.

One must be careful not to jump to the conclusion that acid rain is becoming less acidic. Precipitation quality is highly variable and many years of data would be required to show a definite trend.

DRY DEPOSITION

Weekly dry deposition data were collected using a water extraction of the dryfall bucket. Dryfall data were taken for each watershed from March to December, 1978; therefore, dryfall collection was eliminated in all watersheds except one (Woods). In September, 1979 (to the present), dryfall collection was again expanded to all three watersheds.

Originally, dry deposition collection was interrupted because it was felt that the dry data were of little quantitative

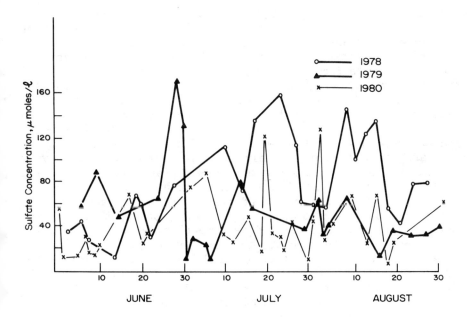

Figure 4.   Mean event summer sulfate concentrations.

value (since collection on a surrogate surface, i.e., a bucket, did not model deposition on natural surfaces). However, it was decided that dry deposition measured in a dryfall bucket may well have a use as a lower limit of what could be expected to deposit in a watershed. This point will be discussed further below.

Dryfall loadings to watersheds were calculated based on the laboratory analyses of the water extraction and the surface area of the bucket opening according to: Dryfall Ion Loading (μeq/m^2) = ($_iEc_i$)/A where E = volume of extraction water (0.5 l), c_i = ion concentration (μeq/l), A = area of collection bucket opening (0.064 m^2). Loadings were calculated for each ion for each collection and summed to give monthly values. Variabilities (S.D./$\overline{X}$, where $\overline{X}$ = mean dryfall loading) were considerably greater than variabilities found for wet deposition.

Dry deposition makes an important contribution to the total flux of ions into these watersheds. This is readily apparent from an inspection of Table 5. The key question that is not answered by these comparisons is the representativeness of the numbers, obtained from dry bucket collection. First, the values are only time-averaged values and not "event" values; second, the interpretation of the values is not straightforward, since the collection efficiencies of the surfaces employed are in reality unknown; third, dryfall contributions arise from many sources; therefore, it becomes difficult to relate the bucket measurements to pollutants transported in the ambient air. Such sources are local terrestrial soil inputs, bird droppings, leaves, needles from conifers, and twigs. Errors can be caused by dry light snow and certain dew, mist, and fog events. In such cases, the sensor on the wet/dry collector may not be activated at the beginning of the event (if at all). As a consequence, such precipitation is collected (at least in part) in the dry bucket and becomes part of the analysis. It would appear, however, that a likely interpretation would be to view these dry loadings as minimum values. Gravitational settling and diffusion are probably the major mechanisms and these are known not to be the most efficient for small particle size ranges. The dry bucket offers a relatively stagnant environment in a sheltered watershed and the approach velocity with which air masses move through forests is virtually absent inside a bucket. The forest surface (trees) and the lake surface are much more efficient collectors. In the future, as the data base increases, it may become possible to make some general statements about the relationship between dry deposition and ambient air concentrations.

Table 5

Comparison of Wet/Dry Fluxes for Selected Major Ions

Watershed

-------eq/ha/yr-------

Ion	Woods				Panther				Sagamore			
	Year 1		Year 2		Year 1		Year 2		Year 1		Year 2	
	Wet	Dry	Wet	Dry	Wet	Dry	Wet	Dry	Wet	Dry	Wet	Dry
Sulfate	808	106	612	143	819	124	643	240	647	165	466	198
Nitrate	418	51	420	82	397	110	420	58	302	83	304	128
$\Sigma(-)$	1226	157	1032	225	1216	234	1063	358	949	248	770	326
Ammonium	216	24	192	46	205	45	201	65	146	33	157	78
Calcium	128	68	143	82	132	120	169	96	97	87	110	79
Hydrogen	859	55	671	114	855	96	718	139	698	76	521	211
$\Sigma(+)$	1203	147	1006	242	1192	261	1088	300	941	196	788	368
$\frac{\Sigma(+)}{\Sigma(-)}$	0.98	0.94	0.97	1.08	0.98	1.12	1.02	0.89	0.99	0.79	1.02	1.13

Year 1 - May, 1978 to April, 1979
Year 2 - May, 1979 to April, 1980

ACKNOWLEDGMENT

The authors would like to thank the Electric Power Research Institute, Palo Alto, California, for support of this work.

LITERATURE CITED

Altwicker, E.R., A.H. Johannes, and N.L. Clesceri. 1980. Acid Deposition Measurements in the Northeastern USA. 73rd Annual APCA Meeting, Montreal, Canada, June, Paper 80-71.3.

Altwicker, E.R. and A.H. Johannes. 1981. Trends in Acid Precipitation in the Northeastern United States and the Role of Dry Deposition. 74th Annual APCA Meeting, Philadelphia, Pennsylvania, June, Paper 81-6.2.

Chen, C.W., S.A. Gherini, and R.A. Goldstein. 1978. Modeling the Lake Acidification Process. In: M.G. Wood, ed., "Ecological Effects of Acid Precipitation. Report of a Workshop, Galloway, United Kingdom," September 4-8. Electric Power Research Institute, Palo Alto, California, EA-79-6-LO.

Chen, C.W. and R.A. Goldstein. 1980. Integrated Lake-Watershed Acidifiction Study. In: "Proceedings of the Symposium of Effects of Air Pollutants on Mediterranean and Temperate Forest Ecosystems," June 22-28, University of California, Riverside, California.

Galloway, J.N., C.L. Schofield, G.R. Hendry, E.R. Altwicker, and D. Troutman. 1980. An Analysis of Lake Acidification Using Annual Budgets. In: "Proceedings of International Conference on the Ecological Impact of Acid Precipitation," March 11-14. Sandefjord, Norway.

Goldstein, R.A., C.W. Chen, S.A. Gherini, and J.D. Dean. 1980. A Framework for the Integrated Lake-Watershed Acidification Study. In: "Proceedings of International Conference on the Ecological Impact of Acid Precipitation," March 11-14. Sandefjord, Norway.

Johannes, A.H. and E.R. Altwicker. 1980. Atmospheric Inputs Into Three Adirondack Lake Watersheds. In: "Proceedings of International Conference on the Ecological Impact of Acid Precipitation," pp. 256-257, March 11-14. Sandefjord, Norway.

Johannes, A.H., E.R. Altwicker, and N.L. Clesceri. 1981. Characterization of Acidic Precipitation in the Adirondack Region. EPRI EA-1826, RP 1155-1, May.

CHAPTER 11

LIMING ACID PONDS IN NEW YORK

Leigh M. Blake
Associate Aquatic Biologist
New York State Department of
Environmental Conservation
Watertown, New York 13601

INTRODUCTION

Acid precipitation has been much in the news in recent years. Causes and effects have been described by Schofield (1976b) and others. In New York, the effects are most serious in ponds at higher elevations in the Adirondack region which historically are poorly buffered. Typically, pristine ponds have pH of 6.5 or less and alkalinities of 50 μeq/l or less. Brook trout, indigenous to Adirondack ponds, are somewhat tolerant of acid conditions, but reproductive success is limited when the pH drops below 5.0 to 5.5 and alkalinity below 5 μeq/l (Schofield, 1976a). As acid conditions worsen, brook trout and associated species no longer survive.

Of 2,109 Adirondack ponds (open to the public) totalling 74,000 ha (Pfeiffer, 1979), 170 (3,000 ha) were known casualties of acid precipitation in 1979 with the number increasing each year. Although public awareness and concern over the situation is keen, excellent fishing in hundreds of ponds is still available. Acid precipitation and subsequent loss of additional fishing opportunity will constitute a problem for future fisheries management in this region.

POND LIMING

Pond liming is not new; fisheries workers in New York have been liming ponds since the late 1950s. While some experiments were conducted with different forms of lime, hydrated lime, $Ca(OH)_2$, was used almost exclusively. Plosila (1966) studied the effects of various application rates on water chemistry and fish survival in 19 ponds in the late 1950s and early 1960s. He adopted rates of 3.65 to 9.17 kg of hydrated lime per 1,000 m^3 depending on the severity of acid conditions and the type of pond. Bog ponds received the higher rate; clear ponds, the lower.

Selecting candidate ponds for treatment is straight-forward. In addition to having an acid condition, a pond must meet the following criteria:

(1) have a flushing rate of more than one year,
(2) be a suitable habitat for trout survival and growth, and
(3) be open to public fishing.

Priority is given to ponds accessible by road and to ponds with a history or potential for natural reproduction of trout. Threatened ponds containing a heritage strain of brook trout receive the highest priority.

Application of hydrated lime is restricted to open water periods. If spread on ice, it "reverts" to calcium carbonate which is less potent.   Since it costs twice as much as agricultural lime ($CaCO_3$) loss of its greater neutralizing capacity makes it economically expedient to use it in its most active form.   The advantage of hydrated lime is its superior buffering capacity.   On a weight basis, compared with agricultural lime, approximately one-fifth as much is needed to treat a pond. For example, an 8 ha pond with an average depth of 6 m would be treated with 1.8 to 4.5 tonne of hydrated lime in contrast with 3.6 to 4.5 tonne of agricultural lime. On the negative side, hydrated lime dissolves instantly and its effects are generally short-lived and depend on the flush rate of the pond. Also, pH shock has been reported to be potentially harmful to certain aquatic fauna (Schneider et al., 1975).

To date, 50 ponds have been treated with hydrated lime, some more than once. Follow-up studies to evaluate results were not always as complete as might be desired. Nevertheless, 35 treatments resulted in measurable improvements in water quality and fish survival. Many of these ponds were completely barren of fish life prior to treatment, but after treatment they produced excellent fishing (Blake, 1977).

APPLICATION RATES AND METHODS

In recent years, a shift to the use of agricultural lime has taken place, primarily in an effort to extend the time between treatments. Commercially available limestone is a mixture of particle sizes. The very fine particles produce an immediate, but less sharp, increase in pH, while progressively larger particles dissolve more slowly. There has been too little time to thoroughly evaluate the use of agricultural lime. However, private ponds treated with it have proven to retain the effect longer than in the case of hydrated lime (Dwight Webster, personal communication). Agricultural lime is applied at a rate of 1.13 to 2.27 tonne/ha of surface area. A more precise application rate based on water chemistry is being developed. For the present, 2.24 tonne/ha is the rule for initial treatments.

In accessible ponds, application is the same for both types of lime; the lime is dumped from bags from a slowly moving boat. More lime is spread in shallow water areas (i.e., 1 to 6 m) than over deep water areas. The purpose here is to place the lime where the fish live and to expose it to wave action and currents as much as possible. Six ponds totalling 30 ha have been treated in this manner (Table 1).

Moving large quantities of lime to remote areas is more complicated. To date, the New York State Department of Environmental Conservation efforts have been limited to the use of a helicopter. However, fixed-wing aircraft have been used successfully elsewhere (Dwight Webster, personal communication). Four means of transporting lime by helicopter have been tested. Three proved impractical for a number of reasons, high labor costs being the principal factor.

The first and most straightforward method was simply to load bagged lime into a helicopter. Clear Lake was treated in this manner with hydrated lime in March, 1975 (Table 1). The method was expensive, requiring a crew of eight men, but it was rejected as a viable method when a bag broke while the helicopter was operating and the interior of the aircraft was filled with a cloud of lime dust.

In the second method, lime was loaded in a sling and transported by helicopter in 900 kg lots. In 1977, Clear Lake was retreated and Tamarack Lake was initially treated in this way, both with agricultural lime (Table 1). This method, too, was expensive and required considerable labor. A crew on the ice was needed to break open bags, spread the lime, and gather and dispose of the bags. These first two methods were limited to winter as it was necessary for the helicopter to land for unloading at a pond. Also, winter operation in itself was a problem due to swirling snow, which made flying difficult, and extremely low temperature (often -29°C).

Table 1

Summary of Data Concerning Ponds Limed in New York by the State Department of Environmental Conservation, 1975 to 1979

Name	County	Watershed	Pond No.[a]	Ha	Date	Liming Treatment Type	Amount (Tonne)	Delivery	Cost ($)	Justification
Clear Lake	Herkimer	Oswegatchie-Black	625	82.5	March, 1975	Hydrated	16.3	Helicopter[b]	2,700	History of high use
Tomkettle Lake	Oneida	Mohawk-Hudson	809	4.9	February, 1977	Hydrated	32	Helicopter[b]	2,340	History of high use
Townline Pond	St. Lawrence	St. Lawrence	371	14.6	September, 1975	Hydrated	0.8	Boat	125	History of high use
Pine Pond	St. Lawrence	St. Lawrence	368	6.5	October, 1975	Hydrated	5.9	Boat	682	History of high use
						Hydrated	1.4	Boat	245	
Long Pond	Lewis	Oswegatchie-Black	610	8.1	November, 1977	Agricultural	7.3	Boat	462	History of high use
Pitcher Pond	Lewis	Oswegatchie-Black	662a	2	October, 1975	Hydrated	1.1	Boat	165	History of high use
Payne Lake	Lewis	Oswegatchie-Black	620	4.4	October, 1975	Hydrated	0.7	Boat	95	History of high use
Evies Pond	Lewis	Oswegatchie-Black	608	2	October, 1975	Hydrated	1.8	Boat	195	History of high use
						Hydrated	0.7	Boat	98	
Horn Lake	Herkimer	Oswegatchie-Black	845	15.4	November, 1975	Hydrated	4.1	Helicopter[c]	2,370	Preserve heritage strain of brook trout
					October, 1978	Agricultural	18.1	Helicopter	3,786	
Falls Pond	Hamilton	Oswegatchie-Black	885	16.2	November, 1975	Hydrated	4.5	Helicopter[c]	2,495	History of high use
Duell Pond	Franklin	Lake Champlain	195	0.6	October, 1978	Agricultural	18.1	Helicopter	5,242	High use; popular roadside brook trout pond
					August, 1976	Hydrated	0.8	Boat	120[e]	
Florence Pond	Lewis	Oswegatchie-Black	664a	1.4	September, 1976	Agricultural	3.2	Boat	285	High use; one of few brook trout ponds in area
Cooler Pond	Franklin	St. Lawrence	205	0.4	September, 1976	Hydrated	0.3	Boat	40[e]	High use; roadside pond on Paul Smith's College campus
Duck (Echo) Pond	Franklin	Lake Champlain	136	4	September, 1976	Hydrated	2.9	Boat	450[e]	High use; on Fish Creek Ponds campsite
Nine Corner Lake	Fulton	Mohawk-Hudson	719	44.5	May, 1977	Hydrated	19.3	Boat	7,743	History of high use; preserve fishery in one of few brook trout ponds in county
					November, 1979[d]	Hydrated	19	Helicopter	15,000[e]	
						Agricultural	49.9			
Deer Pond	St. Lawrence	St. Lawrence	372	10.1	November, 1977	Agricultural	10.9	Boat	597	History of high use
Boottree Pond	St. Lawrence	St. Lawrence	374	8.1	November, 1977	Agricultural	9.1	Boat	502	History of high use
Buck Pond	Herkimer	Oswegatchie-Black	578	4.4	November, 1977	Agricultural	8.2	Boat	500	History of high use
Tamarack Lake	St. Lawrence	Racquette	171	5.3	February, 1978	Agricultural	11.8	Helicopter[b]	2,450	Preserve heritage strain of brook trout

Pond	County	Watershed[a]			Date			Method		Comments
Ice House Pond	Hamilton	Oswegatchie–Black	876	2.8	October, 1978	Agricultural	6.4	Boat	652	History of high use; natural brook trout spawning adequate
Brewer Lake	Herkimer	Oswegatchie–Black	967	5.3	November, 1979	Agricultural	18.1	Helicopter	2,550	Preserve heritage strain of brook trout
Livingston Pond	Essex*	Upper Hudson	705	2.4	November, 1979	Agricultural	5.4	Helicopter	1,783	High use potential on major trail maintained by State

[a] As designated in "Biological Survey" of watershed as published by New York State Conservation Department.
[b] Spread on ice.
[c] Liquid.
[d] Not complete.
[e] Estimated.

Ideally, lime should be spread directly by the helicopter (the third method), reducing crew size and allowing for year-round operation. This method was tried on Horn Lake and Falls Pond in November, 1975 (Table 1). Hydrated lime mixed with an equal amount of water by weight was transported in a fire-fighting water bucket and spread directly on the lake surface as the helicopter flew slowly at an elevation of about 60 m. The method proved too costly because, despite the lower amount of hydrated lime required, half of each load was water and the mixing required extra equipment and a crew of six.

While these experiments were taking place, the fourth method, dry dispersal, was being investigated. Commercial operators spreading fertilizer by helicopter were contacted and their methods evaluated. Ultimately, two lime "buckets" were purchased. Dispersal of lime by this method is straight-forward. Two buckets are used, the one on the ground being loaded while the second is transported to the pond by the helicopter. One-tonne loads are normally carried. The helicopter moves over the pond at a moderate speed at an altitude of about 60 m. A trap door is opened by the pilot and the load is dropped quickly (Figure 1).

RESULTS

To date, 5 ponds totalling 84 ha have been treated in this way. As with any new method, initial efforts involved additional costs in equipment and manpower. During the first trial in 1975, a bucket was inadvertently dropped and damaged. Then, in 1979, a bucket was accidentally dropped and damaged beyond repair. Thus, most work was done with a single bucket, which resulted in an inefficient operation.

Overall, 12 ponds have been treated with agricultural limestone. Only three follow-up studies have been made (on ponds treated in 1977); all demonstrated improved pH values and excellent trout survival. Three ponds were barren prior to treatment, and it is planned to monitor them closely over the next several years to measure effects on water quality and survival of initial stockings.

Recent reports, primarily from Scandinavia, have suggested that mobilization of mercury may cause increased levels in fish flesh following application of lime to acid ponds (Brosset et al., 1977). To ascertain if this phenomenon occurs in New York lakes, a study of mercury in brook trout from selected lakes, limed and unlimed, was undertaken in 1977 (Sloan et al., 1979). No increase in mercury levels was apparent in either limed or

*Figure 1.   Spreading dry lime on a remote pond from a bucket
slung under a helicopter and (insert) loading a
bucket with lime.*

control waters. In fact, in one pond the levels in trout taken
and analyzed prior to the initial liming were greater than those
in fish taken in 1977 after liming.

COSTS

Cost records were kept on 27 projects done in recent years.
Because much of the effort was experimental, costs were probably
higher than would be the case for "routine" projects.   For
example, assigning a crew to lime a series of accessible ponds
without interruption results in maximum efficiency and lower
costs. Liming 4 ponds in October, 1977, by a three-man crew
cost approximately $12/ha.   On the other hand, a single pond
limed in 1978 cost $21/ha.

Costs for liming remote ponds ranged as high as $120/ha.
However, it is felt that, on a routine basis, the average can

be substantially reduced. Experience with the dry dispersal method has shown that lime can be moved at a rate of 4.5 tonne/hr. At current rates, a helicopter costs $250/hr and a ground crew of eight men to load the buckets $35/day each. Added to this is the cost of the lime, presently $44/tonne delivered to the site, and travel expenses to the site which vary depending on the job but are assumed to average $100/day. Thus, the average costs should be about $40/ha as follows: lime ($16), helicopter ($20), labor ($3), and other expenses ($1).

BENEFITS

Benefits are based on angler use as derived from aerial censuses and the estimated value of an angler trip (Pfeiffer, 1979). Costs and benefits for 6 accessible ponds treated in 1977 and 1978, and for 4 remote ponds treated by the dry dispersal method in 1978 and 1979, are shown in Table 2. The assumed 3-yr life of the treatments was based on experience. This is likely a conservative estimate as most of that

Table 2

Relation of the Cost to Benefits for Accessible and Remote Ponds in New York That Were Limed to Abate Acidity

Item	Accessible Ponds	Remote Ponds
Ponds limed		
Number	6	4
Ha	80	40
Cost of liming		
Total	$10,456.00	$13,362.00
Per ha	$131.00	$334.00
Benefits		
Angler trips per year		
Total	3,528	388
Per ha	44	10
Valuation per year		
Per trip per ha	$23.90	$26.27
Total per ha	$1,051.60	$262.70
Cost-benefit ratio[a]	1:9.7	1:0.9

[a]Based on expected 3-yr life of treatment.

experience was with hydrated lime, and treatments with agricultural lime should be effective longer. Also, the cost-benefit ratios are based on treatments where the lime was applied at a maximum rate of 900 kg/ha. More recent data suggest that maintenance liming may require less than that for ponds with flushing rates of less than 90 percent of their volume per year. Thus, the cost-benefit ratio for remote ponds would improve to 1:1.3 if the cost of $40/ha for routine operations were accurate.

It was concluded that, for both accessible and remote ponds, liming is an effective and economically feasible management tool which can be used to counteract the adverse impact of acid precipitation and maintain selected fisheries.

LITERATURE CITED

Blake, L.M. 1977. A Review of Pond Liming in New York. New York State Department of Environmental Conservation, Watertown, New York 13601 (xerox).

Brosset, C. and J. Suedung. 1977. IVL Report B-378. Swedish Water and Air Pollution Res. Lab., Gothenburg.

Brown, T.L. 1975. The 1973 Statewide Angler Survey. Department of Natural Resources, Cornell University, Ithaca, New York 14853 (mimeo.).

Pfeiffer, M.H. 1979. A Comprehensive Plan for Fish Resource Management Within the Adirondack Zone. New York State Department of Environmental Conservation, Watertown, New York 13601 (mimeo.).

Plosila, D.S. 1966. Trout Water Chemistry Relationship in Soft Water Adirondack Pond. New York State Department of Conservation (mimeo.).

Schneider, W., J. Adamski, and M. Paylor. 1975. Reclamation of Acidified Lakes Near Sudbury, Ontario. Ontario Ministry Environment, Rexdale, Ontario (xerox).

Schofield, C.L. 1976a. Effects of Acid Precipitation on Fish. *Ambio* 5:228-230.

Schofield, C.L. 1976b. Acidification of Adirondack Lakes by Atmospheric Precipitation: Extent and Magnitude of the Problem. New York Federal Aid in Fish and Wildife Restoration Project F-28-R-4. New York State Department of Environmental Conservation, Watertown, New York 13601 (mimeo.).

Sloan, R.J., C.L. Schofield, and E. Harris. 1979. Mercury Levels in Brook Trout from Selected Acid and Limed Adirondack Lakes. Presented at 35th Annual Fish and Wildlife Conference, Providence, Rhode Island, April 1-4, 1979 (abstracted in Fisheries Abstracts of that conference).

CHAPTER 12

THE NEUTRALIZATION OF ACID PRECIPITATION IN WATERSHED
ECOSYSTEMS OF THE UPPER PENINSULA OF MICHIGAN

J. Robert Stottlemyer
Department of Biological Sciences
Michigan Technological University
Houghton, Michigan 49931

INTRODUCTION

In the absence of information regarding the relationships
and interactions among energy and nutrient inputs and outputs
of ecological systems, it is difficult to predict the impacts
of man's activities on natural systems. Decisions regarding
land use activities are often made with little or no basis to
assess the direct and especially the indirect effects of the
proposed action.

During the last decade, it has become increasingly evident
that the examination of entire interacting systems instead of
just a few of their components can be accomplished (Bormann and
Likens, 1979; Stottlemyer and Ralston, 1970; Cole et al., 1967).
The procedure requires the definition and comprehension of
biogeochemical cycling among the living and non-living
components of ecological systems. A large number of variables
such as climate, season, geological heterogeneity, and
biological structure and diversity affect the flow of water and
flux of nutrients through any given ecological system.
Therefore, data acquired over a number of years or possibly
decades provide the most effective baseline.

A series of studies was initiated in the summer of 1979
on watershed ecosystems located within Isle Royale National
Park, the Keweenaw Peninsula, and Pictured Rocks National
Lakeshore in Michigan's Upper Peninsula (Figure 1). The overall
goal is to obtain baseline data on the structure and function
of northern ecosystems, and how they vary as a result of
differences in vegetation, community composition, and
precipitation inputs. Precipitation quantity and quality in
the Lake Superior Basin are important ecological variables.

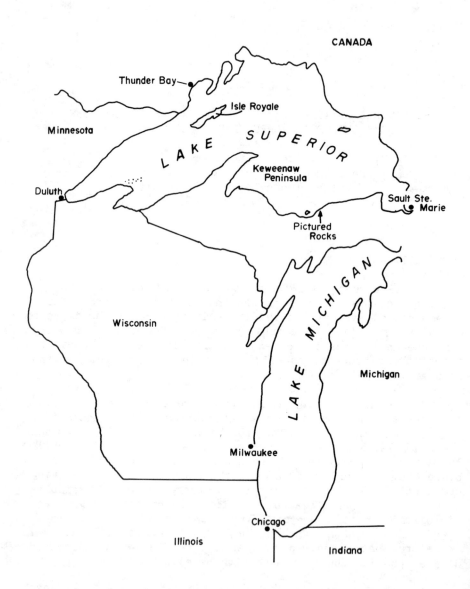

*Figure 1.   Regional map showing location of Isle Royale
            National Park, the Keweenaw Peninsula, and
            Pictured Rocks National Lakeshore, Michigan.*

This chapter summarizes the data collected to date on the
neutralization of acid precipitation in low-order streams in
the upper portions of watersheds with differing substrates, the
results of experimental acidification of one watershed, and

examines some historical watershed data which imply that acid loading and neutralization have occurred in the Lake Superior Basin for at least the last 15 yr.

SITE DESCRIPTIONS

Isle Royale, a 56,000 ha island in northern Lake Superior, is mostly Precambrian basaltic and andesitic lava flows of the Portage Lake Volcanics formation (Huber, 1975). Glacial till is much in evidence in the southwest third of the island. Scattered glacial debris is found over the remainder of the park. The island's vegetation consists of both boreal coniferous forest and northern hardwood forest (Janke et al., 1978). About one-third of the island contains mature steady-state forested ecosystems. The remaining successional forests are the result of fire. The Washington Creek watershed, the site of this research, is a moderate-sized (36 km^2) drainage. Its elevation varies from lake level (about 195 m) to about 440 m. It has a glacial till substrate. The dominant forest vegetation species are white spruce (<u>Picea</u> <u>glauca</u>), balsam fir (<u>Abies</u> <u>balsamea</u>), white birch (<u>Betula</u> <u>papyrifera</u>), yellow birch (<u>Betula</u> <u>alleghaniensis</u>), and sugar maple (<u>Acer</u> <u>saccharum</u>). The vegetation is over 200 years old.

Little Beaver Creek watershed, draining an area of 270 ha, is in the Pictured Rocks National Lakeshore on the south shore of Lake Superior. The watershed substrate is Early Cambrian sandstones overlain by Valders glacial till. This entire region has been disturbed by man, and the dominant vegetation in the watershed is pole-sized sugar maple about 60 yr old with some scattered beech (<u>Fagus</u> <u>grandifolia</u>) and white pine (<u>Pinus</u> <u>strobus</u>). Unlike Washington Creek, Little Beaver Creek first drains into a small lake before flowing into Lake Superior. The watershed's elevation varies from about 215 m to 330 m.

The Keweenaw Peninsula watershed is located about halfway between Isle Royale and Pictured Rocks. It is similar in size to the Beaver Creek watershed and, based on a preliminary assessment, has very similar vegetation. The geologic substrate is like that of Isle Royale, but the quality of its widespread surface glacial till has not been determined.

METHODS AND MATERIALS

Rainfall is collected in 20 cm diameter polyethylene rain gauges with qualitative filters placed in the collecting funnels to keep out insects. Multiple gauges are located in each watershed to determine spatial or elevational variation. Summer precipitation is collected on an event basis on Isle Royale and every ten days to two weeks at Pictured Rocks. During winter, precipitation is collected in large open wastecans. Winter elevation transects are established in both parks and the Keweenaw Peninsula with snow gauges recording the amount of precipitation with increased elevation at 25 m intervals above Lake Superior. Snow samples are collected monthly at Pictured Rocks, at varying time intervals on Isle Royale due to limited access, and bi-weekly on the Keweenaw Peninsula. On each collection date, 5 snow cores are taken at each station with depth and density recorded, and a snow core sample is retained for qualitative analysis. Snow samples are then melted and, as with the summer precipitation samples, pH and conductivity are recorded. The sample is then frozen in a 125 ml amber polyethylene bottle until analysis.

Stream samples are collected on a similar schedule except for Isle Royale where the mouth of Washington Creek is sampled every week in summer. Two small adjacent drainages within the Washington Creek watershed, WATR and WATRB, are sampled at two-week intervals during the summer (May 1 to November 1) at a series of 3 to 7 stations running the entire length of the drainages. The stream within WATR is first-order for most of its length (2.5 km), but flows through a series of 10 beaver ponds. WATRB also has a first-order stream for its entire length (1 km), but is without wetlands or beaver ponds. The purpose of sampling these small watersheds is to look more closely at what stream water chemical changes occur in the uppermost portions of the watershed, and to see if the presence of beaver ponds is a significant modifying factor.

Samples are collected in 500 ml amber polyethylene bottles. Stream flow and temperature are recorded. Alkalinity, pH, and conductivity are then determined and the remainder of the sample frozen until analysis.

An additional experiment was conducted in the headwaters of Little Beaver watershed during the summer of 1980. On three occasions, the pH of the stream near its headwater was lowered about two pH units and held there for a period of 1 hr with sulfuric acid. One upstream and three downstream stations were sampled before, during, and after each acidification to measure chemical change in stream water.

Atomic absorption techniques were used in the laboratory analyses for calcium, magnesium, sodium, potassium, iron, aluminum, and silicon. An ion chromatograph was used for the analysis of sulfate, phosphate, nitrate, nitrite, and chloride.

RESULTS

The precipitation at all stations is acidic (Figure 2). The weighted mean values ranged from 4.88 at the Windigo site adjacent Washington Creek on Isle Royale to 3.87 for Mott Island in the northeast portion of the park. This sampling period did not include the first two months of the year. Since most of the winter precipitation is snow, a higher pH value would be expected during winter conditions. This is true for the data from Pictured Rocks but not for Isle Royale. Winter precipitation increases with elevation above Lake Superior for all ecosystems studied, and it represents about one-third the annual precipitation input. On occasion, there is a lowering of pH with increased elevation, but the trend is not significant nor is it consistent from one sample date to the next. Summer precipitation does not vary significantly in quantity or quality at the elevations sampled.

Precipitation quality during 1980 was similar to 1979, but has not been fully analysed. The relatively strong acids in precipitation are quickly neutralized in the upper portions of first-order streams, and are generally significantly altered before reaching the stream. Table 1 summarizes results found in two small watersheds, WATR and WATRB, within the Washington Creek drainage on Isle Royale. Station WATR1 is separated from

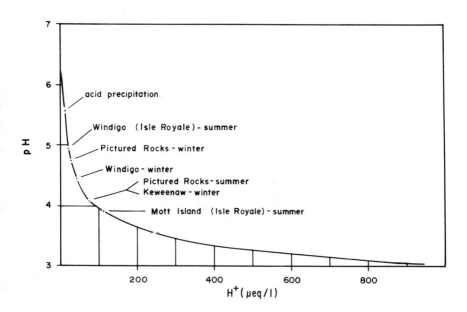

Figure 2. Concentration of hydrogen ions (weighted means for CY 1979).

Table 1

Average Bi-Weekly Concentration of Dissolved Substances for Two Tributaries to Washington Creek, Isle Royale National Park – Sampling Period was May 1 to November 1, 1980

Station[a]	pH	Conductivity	Alkalinity	Ca	Mg	Na	K
	--std units--	----μmho----			-mg/ℓ-		
WATR1	6.13	84	48	4.9	5.2	2.2	0.38
WATR2	6.43	86	61	4.7	5.4	2.2	0.28
WATR3	6.59	108	84	4.5	6.3	2.9	0.49
WATR6	7.12	127	101	2.6	6.6	2.9	0.67
WATRB1	6.41	69	28	2.4	3.3	1.6	0.55
WATRB2	7.00	129	75	2.6	7.4	3.4	0.57
WATRB3	7.16	134	78	2.7	7.3	3.0	0.60

[a]Station numbers increase downstream.

WATR2 by a distance of only 200 m, which includes a beaver pond. Similar conditions separate WATR2 from WATR3. About 2 km and a series of ponds separate WATR3 and WATR6. The changes in pH, conductivity, and alkalinity which occur between the first several stations equals or exceeds that found in the remainder of the watershed. Despite the fact summer precipitation pH values are less than 5, the very smallest first-order stream has an average pH of 6.13.

Similar results are found in watershed WATRB which is without beaver ponds or wetlands. The distance between WATRB1 and WATRB2 is only 250 m, yet the chemical alteration of stream water is greater than found in the upper stations of the WATR watershed. Soluble aluminum has never been found in amounts exceeding 1 mg/l except for one time at station WATRB1 immediately following a rain (2.50 cm at pH 4.5) when the station pH was 5.5 and soluble aluminum was 2 mg/l. Similarly, silicon has not been found in very high concentrations in stream water.

To find out more on the chemical processes involved in the observed neutralization and to see what effect different basement rock might have on the process, three acidification studies were conducted in the Beaver Creek watershed at Pictured Rocks (Table 2). The distance between the upstream LB1 and LB2 is about 1 km, and LBA and LBB are spaced about 200 m and 100 m above LB2. Station LB2 is about 2 km above LB5 at the mouth of Beaver Creek. The data collected one week before each treatment reveal that there is significant chemical change in the streamwater from LB1 to LB2, and that in general this change is greater in this relatively short distance than the change throughout the remainder of the drainage. pH increases on the average about 0.7 units, and alkalinity about 18 mg/l. The relatively high levels of magnesium probably reflect the presence of local deposits of dolomite in the watershed. Lowering the pH to about 4 at station LBB had relatively little effect on the cation concentrations observed downstream except for increases in calcium and some slight increase in potassium levels. Iron concentrations also increased 2- to 3-fold at LBB, but had assumed pre-treatment concentration by the time the water reached LB2. Samples collected at all stations one week after each acid treatment showed concentrations quite similar to pre-treatment data if one accounts for the fact that some time had elapsed.

DISCUSSION

Much of the concern over acid precipitation in the Lake Superior Basin involves the possible direct impact such contamination might have on the extensive surface aquatic

Table 2

Summary of Streamwater Chemistry Changes Before,
During Summer of 1980, Little Beaver Creek,

Station[a]	pH	Conductivity
	--std units--	----μmho----
Before		
LB1	6.5[b]	89
LBA	6.62	119
LBB	6.59	115
LB2	6.99	117
LB5	7.67	145
During		
LB1	--	--
LBA	7.17	122
LBB	3.95	270
LB2	5.98	153
LB5	--	--
After		
LB1	--	--
LBA	6.93	120
LBB	7.05	118
LB2	7.28	123
LB5	7.53	138

[a]Station numbers increase downstream. LB1 is
about 1 km above LB2 which is about 2 km above
LB5. LBA is about 100 m above LBB which is
about 100 m above LB2. Acid was added just
above LBB.

During, and Following Three Acid Treatments
Pictured Rocks National Lakeshore

Alkalinity	Ca	Mg	Na	K	Fe
----------------------------------mg/ℓ----------------------------						
48	11	4.3	0.85	0.32	--
64	14	5.0	0.93	0.31	0.12
65	14	5.0	0.94	0.32	0.12
66	15	5.5	0.94	0.32	0.12
76	18	6.2	0.99	0.5?	--
--	--	--	--	--	--
67	16	5.1	0.91	0.34	0.07
0	19	5.6	0.91	0.38	0.25
34	18	5.6	0.92	0.38	0.12
--	--	--	--	--	--
--	--	--	--	--	--
69	16	5.1	0.93	0.37	0.13
68	16	5.4	0.94	0.36	0.12
69	16	5.3	0.95	0.36	0.17
75	18	5.7	0.97	0.48	0.10

[b]1979 data, average of 8 readings.

resources. There is particular concern regarding those wetland
and stream systems found on Precambrian volcanic substrates
since such a substrate weathers very slowly and has a very
limited buffering capacity. Another potential concern, though
one not yet addressed to any degree in the Lake Superior Basin,
is change in precipitation quantity and quality with elevation.
This has been a topic of considerable interest in the
Adirondacks where the concentration of atmospheric contaminants
probably coming from the Ohio River Valley considerably
increases with increasing elevation. Though not presented here,
snowpack data taken in all the watershed ecosystems discussed
above reveal that there are pronounced increases in both
snowpack and snowfall with slight (100 m) elevation increases.
But the quality of precipitation does not appear to deteriorate.
These results must be qualified somewhat since the data
collected in this study cannot be directly compared to those
from the more sophisticated sampling devices in the Adirondacks.
Also, the elevation gradients found in and about the Lake
Superior Basin are small compared to the Adirondacks.

Nevertheless, the proximity of Lake Superior results in a
snowpack containing one-third of the annual precipitation, and
this occurs at the higher elevations where it can be expected
the natural system has the least capacity to mitigate
anthropogenic inputs such as acid precipitation (Johnson,
1979). The seasonal variation of precipitation quality,
particularly the low pH of snow samples, on Isle Royale
aggravates this situation. Based on the data gathered to date,
the seasonal variation is probably due to a local source, in
this instance the Thunder Bay area. This has been previously
discussed (Dohrenwend et al., 1980).

What happens to this load of nutrients and contaminants
when the snowpack ripens and melts in the spring? Snowpack
nutrient release is being closely followed both on the Keweenaw
Peninsula and on Isle Royale this winter and spring. Data
collected by the U.S. Geological Survey at the Washington Creek
Benchmark Hydrologic Station provide some implications. Table
3 summarizes the water quality data which have been collected
at this station 6 or 8 times each year since 1967. There is a
continuous water discharge record, but Table 3 only reports the
flow at the time water samples were collected for analysis.
About one-half the annual runoff from Washington Creek occurs
during April and May. The lack of access does not allow sampling
in April, but the early May sampling occurs when the stream is
still at about 70 percent maximum flow. Table 3 shows the
dilution of cation and anion concentrations expected with
increased runoff from a mature steady-state ecosystem. This
holds for everything except sulfate which is at maximum
concentration during high runoff. Converting volume and

Table 3

Mean Monthly Concentration of Dissolved Substances for 1967 to 1980, Washington Creek, Isle Royale National Park – Data Compiled from USGS Records for Washington Creek Gauging Station

Month	O	F	M	M	J	J	A	S
				--mg/ℓ[a]--				
Discharge (m^3/sec)	0.246	0.149	0.076	1.450	0.236	0.150	0.137	0.299
Conductivity	138	162	154	83	112	148	167	152
pH (std units)	7.28	7.47	7.41	7.20	7.33	7.57	7.38	7.19
Alkalinity	61	70	71	28	52	62	78	63
Calcium	18.8	20.5	20.7	9.8	15.9	19.4	23.1	19.2
Magnesium	5.5	6.2	6.3	3.1	4.8	5.2	6.2	5.8
Sodium	3.3	3.8	4.2	1.6	2.5	3.2	3.7	3.3
Potassium	0.7	0.5	0.6	0.4	0.5	0.6	0.7	0.6
Silicon	12.6	14.6	14.8	8.0	8.3	11.4	13.4	12.7
Sulfate	6.0	6.1	6.0	8.3	6.2	5.9	4.9	5.6
Nitrate[b]	0.20	0.25	0.13	0.26	0.39	0.47	0.60	0.16
Chloride	3.8	4.9	4.9	1.3	3.5	3.1	3.7	3.8

[a]Unless otherwise specified.

[b]Total nitrogen expressed as nitrate.

concentration to discharge of dissolved ions (gm/sec) and comparing the change from March to May we find that with a 19-fold increase in streamflow the discharge of sulfate increases 27 times; calcium, magnesium, and silicon about 9 times; sodium about 7 times; and potassium increases 13 times. The quality of precipitation on Isle Royale can account for this increased output of sulfate (precipitation concentration was 3.4 mg/l in 1979), but cannot account for the loss of cations.

The historical Washington Creek data also show just a slight dip in stream pH during spring runoff. Looking at the first 3 yr of USGS data (1967 to 1969) and the last 3 yr data (1978 to 1980) reveals that the discharge (gm/sec) of sulfate was as high in the period from 1967 to 1969 as it is today when determined for equal water discharge. This implies that sulfate loading was occuring then also. However, additional work is needed before any conclusions can be drawn.

The tributaries WATR and WATRB have calcium concentrations much lower (Tables 1 and 3) than those observed in the USGS data for the mouth of Washington Creek. Magnesium concentrations are as high, which is probably a local anomaly. At the upper stations, WATR3 and WATRB3 alkalinity values are equal to those at the mouth of Washington Creek. These data have been supported by data from two additional small tributaries to Washington Creek which also show very high buffering capacities at the uppermost stations.

Surprisingly, the concentrations of major cations in Washington Creek are as high as those observed in Little Beaver Creek despite the fact that Little Beaver watershed has a sedimentary basement rock with local deposits of limestone and dolomite. While it is risky to draw comparisons between the situation at Pictured Rocks and that at Isle Royale, it appears probable, at least for these two locations, that the presence or absence of glacial till is more important in determining buffering capacities and stream water chemistry than is the character of the basement rock. But the watersheds at Pictured Rocks have a different vegetation type and succession stage, which complicates direct comparisons.

This difference due to glacial till is further supported by comparing the data for Washington Creek or its small tributaries with that for the experimental watersheds at Hubbard Brook in New Hampshire (Likens et al., 1977). While there are general similarities in the resistence to erosion and low buffering capacity of the parent rock at both sites, there are striking differences in the chemical composition of stream water even when comparing the uppermost stations on WATR and WATRB with the mouths of the experimental watersheds at Hubbard Brook. With the possible exception of potassium, all major cations are much higher in concentration on Isle Royale. Obviously, this is a comparison of two seasons' data on Isle Royale with the average values for 15 yr at Hubbard Brook. If indeed the quality of precipitation has deteriorated in the 15-yr interval, more recent stream data from Hubbard Brook might be more comparable. For anions, sulfate concentrations in WATR, WATRB, and in Washington Creek are quite comparable to those at Hubbard Brook. Nitrate values on Isle Royale are much below those for Hubbard Brook. These comparisons can be explained in part by the quality of precipitation inputs.

The geologic substrates of Isle Royale and northeastern Minnesota are in general similar. A survey of 16 tributaries to Lake Superior from the Minnesota shore (Wagner and Lemire, 1976) shows that the stream water chemistry for Washington Creek is not greatly different from that of tributaries in the Duluth-Two Harbors area. Unfortunately, many of these Minnesota tributaries are disturbed by man and direct comparisons cannot be made. Calcium values for Washington Creek are much above all the Minnesota streams, however. When comparing less disturbed streams and watersheds in the Grand Marais-Grand Portage area, which is much closer to Isle Royale, stream water chemistry is more similar to that found in the very small watersheds on Isle Royale. However, notable differences occur here also. Alkalinity is on the average an order of magnitude greater at the upper stations of watersheds WATR and WATRB than at the mouth of the largest Lake Superior tributary in the Grand Marais-Grand Portage area.

While the results of this research certainly indicate that stream water chemistry cannot be easily generalized in this region, the deficiencies point out quite clearly the need for an ecosystem approach in the evaluation of ecological effects associated with anthropogenic impacts. As a necessary first step in this work, nutrient budgets will be computed for individual watersheds. These will be a little more valid with an additional year or two of precipitation data. Also, there is the need to look at the role forest vegetation plays in precipitation modification, and more closely examine what chemical changes occur as precipitation enters the soil. Some work along these lines is planned for Isle Royale this summer.

While these data show surprisingly high buffering capacities for some watershed ecosystems in this region, this should not be construed as being the general rule even where sedimentary substrates exist. For example, beginning last summer a survey of Isle Royale's streams and lakes was started.

A half-dozen streams were found to have average alkalinity values below 25 mg/l and a mean pH of 6.4 for samples collected at their mouths. Such a survey does not exist for Pictured Rocks.

ACKNOWLEDGMENTS

The National Park Service provided financial support for the conduct of this research. The author also wishes to thank the staff and seasonal employees of Isle Royale National Park for their logistical support and field assistance.

LITERATURE CITED

Bormann, F.H. and G.E. Likens. 1979. Pattern and Process in a Forested Ecosystem. Springer-Verlag, New York. 253 pp.

Cole, D.W., S.P. Gessel, and S.F. Dice. 1967. Distribution and Cycling of Nitrogen, Phosphorus, Potassium, and Calcium in a Second-Growth Douglas-Fir Ecosystem. Symposium on Primary Productivity and Mineral Cycling in Natural Ecosystems, New York City, December 27, 1967, University of Maine Press, Orono, Maine. 245 pp.

Dohrenwend, R.E., S.G. Shetron, J.R. Stottlemyer, and R.J. Olszewski. 1980. Acid Precipitation in the Keweenaw Peninsula of Michigan's Upper Peninsula. In: D. Drablφs and A. Tallon, eds., "Ecological Impact of Acid Precipitation," pp. 106-107. Proceedings of an International Conference, Sandefjord, Norway, March 11-14, 1980.

Huber, N.K. 1975. The Geologic Story of Isle Royale National Park. U.S. Geological Survey Bull. No. 1309. GPO, Washington, D.C. 66 pp.

Janke, R.A., D. McKaig, and R. Raymond. 1978. Comparison of Presettlement and Modern Upland Boreal Forests on Isle Royale National Park. Forest Sci. 24:115-121.

Johnson, N.M. 1979. Acid Rain: Neutralization Within the Hubbard Brook Ecosystem and Regional Implications. Science 204:497-499.

Likens, G.E., F.H. Bormann, R.S. Pierce, J.S. Eaton, and N.M. Johnson. 1977. Biogeochemistry of a Forested Ecosystem. Springer-Verlag, New York. 146 pp.

Stottlemyer, J.R. and C.W. Ralston. 1970. Nutrient Balance Relationships for Watersheds of the Frazer Experimental Forest. In: C.T. Youngberg and C.B. Davey, eds., "Tree Growth and Forest Soils," pp. 359-381. Oregon Univ. Press, Corvallis, Oregon.

Wagner, D.M., R.S. Lemire, and D.W. Anderson. 1976. Water Quality of Sixteen Minnesota Rivers Tributary to Lake Superior. J. Great Lakes Res. 2:111-123.

CHAPTER 13

ECOLOGICAL EFFECTS OF ACID PRECIPITATION
ON PRIMARY PRODUCERS

H. Lee Conway and George R. Hendrey
Department of Energy and Environment
Brookhaven National Laboratory
Upton, New York 11973

INTRODUCTION

Over the past few decades, the acidity of lakes and rivers has been increasing in several areas of the world. In southern Norway, western Sweden, the Canadian Shield, and the northeastern United States, acidification of freshwaters has become a major environmental problem. It has been postulated that acid precipitation is the cause of decreasing pH levels in waters of these affected areas (Conroy et al., 1975; Wright et al., 1976; Likens, 1976). Acidification results in the modification of communities of aquatic flora and fauna at all ecosystem levels (Almer et al., 1974; Beamish, 1976; Grahn, 1977; Hendrey and Wright, 1976; Hendrey et al., 1976; Schofield, 1976). The number of species is reduced and changes in the biomass of some groups of plants and animals have been observed. Decomposition of leaf litter and other organic substrates is hampered, nutrient recycling is retarded, and nitrification inhibited at pH levels frequently observed in acid-stressed waters.

Studies of the hydrobiological problems caused by acid precipitation have been primarily qualitative. Biological aspects of the synoptic lake surveys of Sweden (Almer et al., 1974), Norway (Hendrey and Wright, 1976), and the United States (Schofield, 1975) have concentrated on changes in the kinds and numbers of species. Very little quantitative information is available concerning the effects of acidification on primary production and biomass of pelagic and benthic algae.

PHYTOPLANKTON

Non-acidic, oligotrophic lakes are typically dominated by Chrysophyceae (golden-brown algae), Chlorophyceae (green algae), and Bacillariophyceae (diatoms), while Cyanophyceae (blue-green algae) are usually scarce in these lakes (Schindler and Holmgren, 1971; Duthie and Ostrofsky, 1974). With increasing acidity, the number of species in each algal group decreases and the species composition changes, with Dinophyceae (dinoflagellates) and Chrysophyceae dominating (Almer et al., 1978; Yan and Stokes, 1978; Hendrey et al., 1981). Several factors may contribute to this change in species composition. The intolerance of various species to low pH or to consequent chemical changes (Moss, 1973) will allow a few algal species to utilize the nutrients available in these predominantly oligotrophic waters. Many species of planktonic and benthic invertebrates are absent at low pH (Almer et al., 1974; Hendrey and Wright, 1976; Almer et al., 1978) and removal of algae by grazing is probably diminished. Microbial decomposition may be inhibited, thereby reducing the recycling of essential nutrients.

In the Swedish west coast region, 115 lakes were investigated with regard to phytoplankton (Almer et al., 1974) and lakes with pH values < 5 displayed a homogenous and limited composition consisting of about 10 species. The greatest changes in species composition were found in the pH interval 5 to 6. Similar observations were made in a regional survey of 55 lakes in southern Norway (Hendrey and Wright, 1976) and in a study of four lakes in Ontario, Canada (Conroy et al., 1976). Phytoplankton biomass was low in these acid lakes (< 1 mg/l) and it was correlated with the concentration of phosphorus in the water, which generally decreased at lower pH. In the Ontario lakes, phytoplankton biomass at pH near 4.5 was one-third to one-ninth as much as the pH near 6.5. Yan (1975) investigated the phytoplankton in the LaCloche Mountain lakes near Sudbury, Ontario, and found that in lakes with a pH <5.1 the phytoplankton communities were dominated by Dinophyceae while in lakes with pH > 6.1 domination was by Bacillariophyceae and Chrysophyceae (Figure 1). Yan concluded that if atmospheric input of acid to these lakes continued the number of algal species would decrease and the community structure would shift from Chrysophyceae and Bacillariophyceae to Dinophyceae. This shift in community structure predicted by Yan (1975) was supported by the results of other workers (Almer et al., 1978; Dickson et al., 1975; Brock, 1973). Brock (1973) found that blue-green algae were absent from habitats where the pH was less than 4.8.

Figure 2 shows the total numbers of species in three oligotrophic Adirondack Mountain lakes. Phytoplankton species

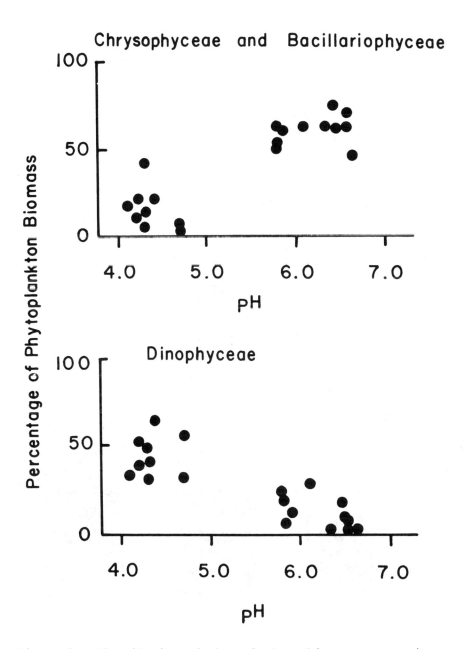

Figure 1. *Distribution of phytoplankton biomass among Dino-*
phyceae, Chrysophyceae, and Bacillariophyceae in
oligotrophic Sudbury area lakes (before manipula-
tion) and Haliburton area lakes. Each point is
monthly-weighted, ice-free period mean of biweekly
collected, morphometrically weighed euphotic zone
composite (Yan, 1979).

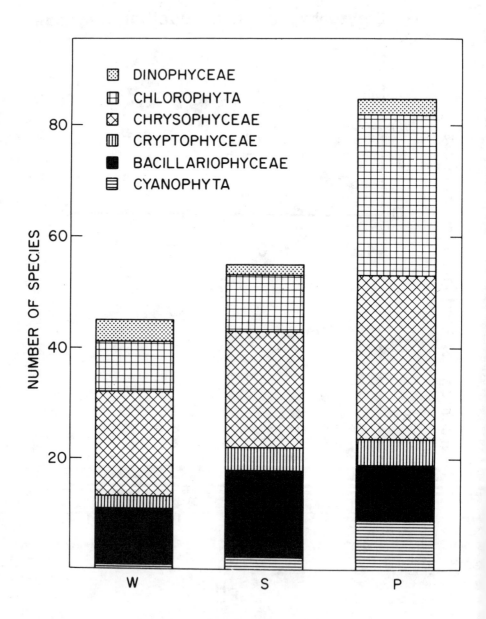

Figure 2.   *Total number of phytoplankton species observed in each of three Adirondack Mountain Lakes arranged by classes. Samples collected biweekly during the ice-free season, monthly during winter from 3 to 5 depths in each lake (1979). Lakes are Woods (W), Sagamore (S), and Panther (P)(Hendrey et al., 1981).*

diversity (Shannon index) values were Woods, 1.89; Sagamore, 1.90; and Panther, 2.11, with pH values of 5, 6, and 7, respectively. Chrysophyceae dominate both in terms of species number and biomass all year in Woods Lake, the most acidic of these lakes (Figure 3). The Dinophyceae, especially Peridinium inconspicuum, make up a significant portion of the biomass during the ice-free season. Dinophyceae become increasingly important contributors to the communities with increasing acidity, but do not dominate as often reported for other acidic lakes. At pH 4 to 5, Peridinium inconspicuum is frequently the dominant dinoflagellate in acidic, clearwater lakes (Almer et al., 1978; Yan, 1979; Hornström, 1979).

Analysis of sediments in acidified lakes shows a decrease in the number of planktonic diatoms while a few benthic forms increase, notably Eunotia, Tabellaria, and Amphicampa (Almer et al., 1978). The decrease in number of algal species is most striking in the pH range 4 to 5 and many species commonly found in oligotrophic lakes are absent from the upper sediment of acidified lakes (Almer et al., 1974).

Although the general rule is that Chrysophyceae and Dinophyceae dominate the plankton of acidic lakes, there are many exceptions to this rule. In Florence Lake (Sudbury area), with a pH range of 4.4 to 4.9, blue-green algae comprised 70 percent of the phytoplankton community (Conroy et al., 1976). In six other lakes in the Sudbury area, blue-green algae were found in large numbers in the acidic lakes and there was an apparent increase in the biomass of these species with decreasing pH (Kwiatkowski and Roff, 1976).

Acid precipitation may influence primary production by reducing the limiting nutrient for growth and reproduction. Two potential growth-limiting compounds in softwater, oligotrophic lakes are dissolved inorganic carbon (DIC) and dissolved inorganic phosphorus (DIP).

With an organic carbon content of approximately 50 percent by dry weight, algae require an abundant source of DIC to grow and reproduce in aquatic systems. However, the solubility of carbon dioxide decreases dramatically as hydrogen ion concentration increases (e.g., at pH < 5 the solubility of carbon dioxide is 0.24 mg C/l, but at pH 9 this value increases to 60 mg C/l)(Galloway et al., 1976). The lower DIC levels available in acid lakes (pH < 5) do not appear to be limiting to the growth of algae; however, the shift from bicarbonate ion to carbon dioxide (96.2% at pH 5) may influence the species composition of the algal community.

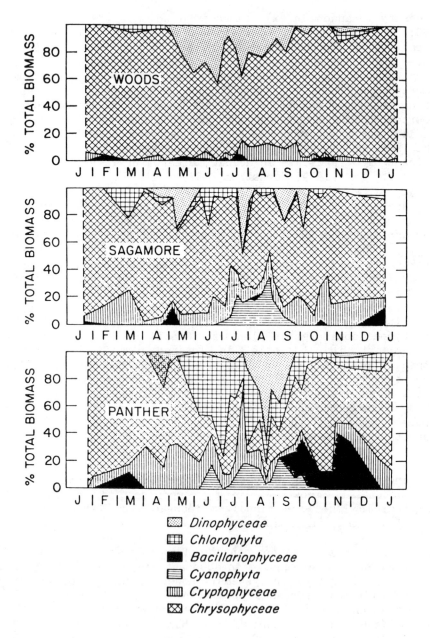

Figure 3.   Seasonal variation in the proportion of total
            phytoplankton biomass contributed by each taxo-
            nomic group in three Adirondack mountain lakes
            for 1979.   Biomass expressed as cell volume con-
            verted to wet weight assuming a unit density
            (Baumgartner, 1981).

In most cases, the acidic lakes are not carbon-limited. Atmospheric diffusion, bacterial regeneration, and benthic respiration supply adequate carbon dioxide for photosynthesis. For example, in Woods Lake, with DIC values ranging from 0.3 to 0.8 mg C/l, the maximal daily primary production rate never exceeded 6.4 percent of the available DIC. In Clearwater Lake (pH 4.2), the daily production rate (Yan, 1979) utilized approximately 14 percent of available DIC, as calculated from carbon dioxide solubility values.

Experiments in softwater Star Lake in Vermont suggested that open water bacterial regeneration provided sufficient carbon dioxide for peak productivity (Allen, 1972); however, equally important sources of carbon dioxide are diffusion from the air and regeneration in surficial sediments. Open water regeneration is probably more important in deep lakes (>25 m) and the latter two are more important in shallow lakes with strong mixing and a significant percentage of the total volume of the lake in contact with these interfaces.

The addition of sulfuric acid to ELA Lake 223 during 1976 and 1977 reduced the DIC levels from 1.4 to 0.8 mg C/l and the pH from 6.65 to 6.05; however, photosynthesis was not affected (Schindler et al., 1980). After comparing production rates in acidic (pH 4.1 to 4.3) and heavy metal contaminated Clearwater Lake (Sudbury) to uncontaminated Blue Chalk Lake (pH 8), Dillon et al. (1979) concluded that acidification had not reduced primary production in acidic Clearwater Lake. However, in acidic Woods Lake, primary productivity values were significantly lower (2 to 3 times) than those for Panther Lake (Table 1).

Based on the small amount of data available, it is not possible to determine conclusively that increasing acidity reduces pelagic primary productivity. Other factors besides pH play a vital role in regulating the productivity and ultimately the biomass of acidic lakes.

Phosphorus is the element that many researchers believe is limiting the growth of algae in freshwater systems (Schindler, 1971; Scheider and Dillon, 1976; Nicholls and Dillon, 1978; Yan, 1979). For example, the pH of acidic Middle Lake (pH 4.4) was increased to pH 7 in 1973, but the phytoplankton biomass did not increase and total phosphorus (TP) concentrations remained stable (Table 2; Hendrey et al., 1981). However, when the TP concentration was increased in 1975, the biomass increased as well. A similar effect was observed in Mountaintop Lake. However, a comparison of TP and biomass in 6 acidic and 15 non-acidic Canadian Shield lakes showed no significant differences between the two types of lakes

Table 1

The Mean and 95 Percent Confidence Interval ($\overline{X} \pm 95\%$ C.I.) And
Number of Observations (n) for pH, Dissolved Inorganic Carbon,
Productivity, and Chlorophyll for Panther and Woods Lakes,
1978 to 1980

Variable	n	$\overline{X} \pm 95\%$ C.I.
Panther Lake		
pH	286	6.58 ± 0.16
Dissolved inorganic carbon (mg/l)	286	3.31 ± 0.19
Net production (µg C/l/hr)	142	1.03 ± 0.18
Nanno production (µg C/l/hr)	141	0.44 ± 0.10
Ultra production (µg C/l/hr)	141	1.74 ± 0.30
Total production (µg C/l/hr)	141	4.79 ± 1.31
Net chlorophyll (µg/l)	206	0.67 ± 0.14
Nanno chlorophyll (µg/l)	204	0.26 ± 0.04
Ultra chlorophyll (µg/l)	206	1.17 ± 0.11
Total chlorophyll (µg/l)	227	2.10 ± 0.32
Woods Lake		
pH	294	4.94 ± 0.01
Dissolved inorganic carbon (mg/l)	147	0.81 ± 0.17
Net production (µg C/l/hr)	79	0.20 ± 0.05
Nanno production (µg C/l/hr)	79	0.07 ± 0.01
Ultra production (µg C/l/hr)	82	0.76 ± 0.12
Total production (µg C/l/hr)	92	1.79 ± 0.24
Net chlorophyll (µg/l)	116	0.11 ± 0.03
Nanno chlorophyll (µg/l)	115	0.09 ± 0.01
Ultra chlorophyll (µg/l)	122	0.80 ± 0.17
Total chlorophyll (µg/l)	154	0.82 ± 0.09

(1) The three size fractions are net > 48 µm, 48 µm > nanno >
20 µm and 20 µm > ultra > 0.45 µm

(Hendrey et al., 1981). Almer et al. (1978) found a minimum
in biomass at intermediate pH values for 58 Swedish lakes
(Figure 4). It was not possible to distinguish differences in
phosphorus concentrations in these 58 lakes. However, it is
possible that the highest TP concentration existed in the most
acidic lakes. Aluminum, which is found at elevated levels in
acid lakes, removes dissolved phosphorus from the water column
by flocculation and precipitation (Almer et al., 1978). This
removal of phosphorus is greatest in the pH interval between 5
and 6 (Dickson, 1978) and could result in higher phosphorus

Table 2

Total Phosphorus (TP), pH, and Biomass Observed in Two Sudbury Experimental Study Lakes (Data are for Ice-Free Period)(Hendrey et al., 1981)

Lake	Year	pH	TP	Phyto Biomass	Chlorophyll
		--std units--	--µg/l--	----mg/l-----	---µg/l----
Middle	1973	4.4	7.3	0.46	0.91
	1974	7.0	7.1	0.16	0.92
	1975-77	6.5	11.6	0.68	2.70
Mountaintop	1976	4.4	43	0.72	5.7
	1977	4.5	58	2.05	20.1
	1978	5.0	75	6.35	64.8

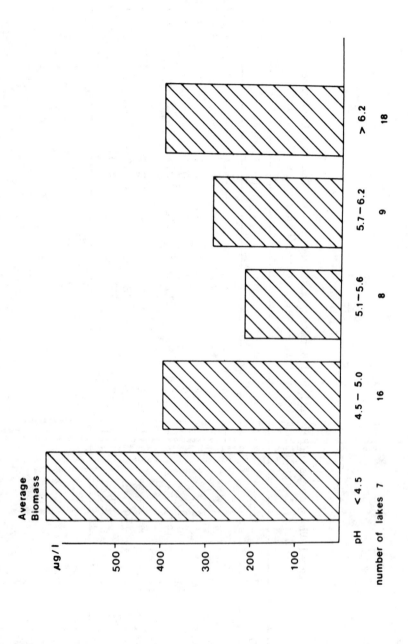

Figure 4.   Average biomass phytoplankton in 58 lakes having different pH values on the Swedish west coast (Almer et al., 1978).

concentrations at the lower pH levels. On the other hand, watershed acidification greatly increases the amount of aluminum in waters below pH 5 (Dickson, 1980). Thus, aluminum complexing of phosphorus may actually increase, overall, at pH 5.

Chlorophyll concentration, another measure of algal biomass, decreased with decreasing pH in six lakes of the Sudbury area (Kwiatkowski and Roff, 1976) and in two Adirondack Mountain lakes (Table 1). The acidification of ELA Lake 223 from pH 6.7 to 6.1 produced no apparent change in chlorophyll concentration (Schindler et al., 1980), but the further acidification to pH 5.6 resulted in an increase in the chlorophyll concentration even though TP concentration was not affected (Schindler, 1980).

Although pH and biomass data are often contradictory, it is apparent that hydrogen ion concentration does not control phytoplankton biomass directly. However, phosphorus, by virtue of its growth-limiting status, may control biomass in many acidic lakes, and phosphorus may be controlled by altered biogeochemical processes associated with acidification, including decomposition of organic matter, nutrient cycling and aluminum release from soils.

BENTHIC ALGAE AND MACROPHYTES

The majority of lakes are small and shallow and, as such, littoral flora can contribute a major fraction of the total primary productivity (Wetzel, 1975), (e.g., up to 83 percent of the mean annual production)(Hargrave, 1969). Although the littoral flora represent a vital component in aquatic ecosystems, few data exist on the effect of acidification on species composition, biomass, and primary production.

Acidification of large enclosures in ELA Lake 223 in northwest Ontario resulted in fewer periphyton species and lower species diversity at low pH values (Müller, 1980). In this study, diatoms dominated at pH 6.2 while green algae dominated at pH values between 4 and 6. The observed effect of pH on the number of periphyton species was similar to that observed for phytoplankton (Lind and Campbell, 1970; Almer et al., 1974).

In waters impacted by acid precipitation, major changes also occur within macrophyte communities. The plant communities of Lake Colden were surveyed in 1932 and again in 1979 and marked changes in the macrophyte community were observed (Table 3; Hendrey and Vertucci, 1980). Although no pH values are available for 1932, a decrease in pH was observed from 1960 to 1979 (5.4 to 4.9). In 5 lakes of the Swedish west coast, a

Table 3

Plant Communities in Lake Colden Observed
in 1932 and 1979 (Hendrey and Vertucci, 1980)

Plant	Colden 1932	Colden 1979
Potamogeton confervoides	a	a
Potamogeton oakesianus	a	-
Scirpus subterminalis	f	a
Eleocharis acicularis	f	r
Eriocaulon septangulare	r	a
Nuphar luteum	c	r
Myriophyllum tenellum	f	c
Utricularia sp.	r	a
Lobelia dortmanna	f	r
Isoetes muricata	c	c
Vallisneria americana	-	r
Drepanocladus exannulatus	-	f
Sphagnum pylaesii	?	a
Sphagnum cuspidatum	?	r

a = abundant; c = common; f = frequent; r = rare; - = not reported

region severely affected by acid precipitation, Grahn (1977) reported that in the past 3 to 5 decades the macrophytic communities dominated by Lobelia and Isoetes have regressed, while communities dominated by Sphagnum have expanded. In the 0 to 2 m depth zone of Lake Örvattnet, the bottom area covered by Sphagnum increased from 8 to 63 percent between 1967 and 1974. In the 4 to 6 m depth zone, the increase was from 4 to 30 percent. At the same time, the pH in Örvattnet decreased 0.8 units to ca. 4.8. At the pH of these acid lakes, essentially all of the available inorganic carbon is in the form of carbon dioxide or carbonic acid. Conditions are more favorable for Sphagnum, an acidophile which is not able to utilize bicarbonate ion as do many other aquatic plants. Similar growths of Sphagnum occur in Norwegian lakes as well (Halvorsen, personal communication). In some lakes, the moss appears to simply outgrow the flowering plants under acid conditions.

In developing their hypothesis on oligotrophication, Grahn et al. (1974) have stressed two biologically important consequences of this Sphagnum expansion. First, Sphagnum has an ion-exchange capacity which results in the withdrawal of essential ions from solution, thus reducing their availability

to other organisms. Secondly, dense growths of Sphagnum form
a distinct biotope which is unsuitable for many members of the
bottom fauna. Utricularia also forms dense stands in Lake
Colden and Woods Lake (300 and 107 gm dry wt/m^2, respectively)
in the Adirondack Mountains and in Lake Gårdsjon (Lazarek,
personal communication) in Sweden. In these three lakes,
Utricularia is found at depths between 2 to 4 m. The dense
stands of Sphagnum and Utricularia may inhibit the exchange of
essential regenerated elements between the sediment and
overlying water, which could also decrease nutrient
availability to other algae.

Only a few studies have been published on the effects of
pH on benthic primary production and biomass. The data that
have been published are somewhat contradictory without a
definite trend in production or biomass with increased acidity.

Intact lake sediment cores which included the rooted
macrophyte Lobelia dortmanna were incubated at three pH levels
(4.0, 4.3 to 5.5, and 6.0) at Tovdal in southern Norway. The
growth and productivity of L. dortmanna (oxygen production)
were reduced by 75 percent at pH 4 compared to the control (pH
4.3 to 5.5) and the period of flowering was delayed 10 days at
the low pH (Laake, 1976).

Acidification of artificial stream channels to pH 4
resulted in an increase in the biomass of periphytic algae and
a decrease in their productivity rates (Hendrey, 1976). This
was also observed in an experimentally acidified stream (pH 4)
in the Hubbard Brook watershed (Hall et al., 1980). However,
in the experimental acidification of large enclosures in ELA
Lake 223, Müller (1980) found that periphyton biomass was higher
at pH < 6.0 than at pH > 6.0; however, productivity did not change.
Our own data show that benthic primary production was
significantly higher in neutral Panther Lake (pH 7.0) than in
two more acidic lakes (Table 4). However, in contrast to the
results of Hendrey (1976) and Hall et al. (1980), the
preliminary data for algal biomass (chlorophyll a) imply that
the lowest values were found in acidic Woods Lake (pH 4.9).
Hendrey (1976) found little preference among algal species for
low pH, but suggested that a decrease in invertebrate grazing
pressure due to a decrease in their (invertebrate) population
may have contributed to increased algal biomass at pH 4. This
theme was developed into a hypothesis by Hendrey et al. (1981).
The preference of some algal species for low pH was found by
Müller (1980); however, he did not observe a decrease in the
number of invertebrates at low pH.

Although the overlying water may have a certain physical
controlling influence on benthic algae (e.g., light, T^{oC}, etc.),

Table 4

The Mean and 95 Percent Confidence Interval ($\overline{X} \pm 95$
C.I.) and the Number of Observations (n) for pH,
Percent Organic Content and Production for Panther,
Sagamore, and Woods Lakes

Variable	N	$\overline{X} \pm 95$ C.I.
Panther Lake		
pH	27	6.95 $\pm$ 0.17
Organic content of sediment (%)	56	47.6 $\pm$ 3.3
Production (mg C/m^2/day)	27	30.7 $\pm$ 15.0
Sagamore Lake		
pH	37	5.90 $\pm$ 0.10
Organic content of sediment (%)	58	36.6 $\pm$ 6.4
Production (mg C/m^2/day)	44	8.9 $\pm$ 1.0
Woods Lake		
pH	40	4.92 $\pm$ 0.04
Organic content of sediment (%)	57	46.9 $\pm$ 3.1
Production (mg C/m^2/day)	14	6.7 $\pm$ 3.7

the chemical factors required for their growth are, for the
most part, controlled by the interstitial water of the sediment,
with the exception of epilithic algae. This, of course, gives
benthic algae a competitive advantage over planktonic forms
since their supply of growth-limiting elements (P, C, Si, N)
is not strongly influenced by water column stratification and
the resulting depletion of these elements in surface waters.
These essential elements are found in the sediments at
concentrations much higher than those in the overlying waters
(Wium Andersen and Andersen, 1972; Carignan and Kalff, 1980).
Regenerated nutrients in the sediments and their flux out of
sediments are utilized by the microscopic benthic algae living
in the surficial sediments and by submergent macrophytes.
Lobelia dortmanna, a macrophyte commonly found in acid lakes,
is able to absorb the carbon dioxide required for photosynthesis
from the sediment through its roots (Wium Andersen, 1971). This
ability has enabled Lobelia to become the dominant macrophyte
in many acidic lakes in Sweden (Lazarek, personal
communication). This phenomenon is not limited to carbon
dioxide utilization; Carignan and Kalff (1980) found that nine
common macrophytes took all of their phosphorus from the

sediments in oligotrophic and mildly eutrophic lakes. The ability of certain species to utilize regenerated nutrients found in the sediment at high concentrations, but perhaps at critically low concentrations in the water column, may explain some of the changes in the structure of macrophyte communities in acid lakes.

CONCLUSIONS

One of the most apparent conclusions to be drawn from the data described above is the lack of agreement in trends of primary production and biomass with increasing acidity. In fact, the results of many studies are contradictory. Some of these contradictions can be partially explained by the differences between measurements taken in aquatic systems at a quasi-steady state (e.g., survey of many lakes at different pH) versus the experimental acidification of an aquatic system or a part thereof. In the former, inherent differences among lakes with different chemical and physical characteristics may produce variation on a scale similar to that observed for the chemical and physical variables. An ideal experiment would be to study a set of lakes for several decades to document acidification effects on a time scale sufficient for species to adapt. However, in most cases this may not be a practical approach in determining the ultimate consequences of acidification of natural waters. What must be done is to study the entire ecosystem from bacteria to fish, together with the chemical and physical properties of the lake and its watershed. To date, this has not been done; either the watershed has been intensively studied and the aquatic biota ignored or selected components of the food chain have been studied without attempting to study other aspects of the ecosystem. Of course, funding and special interests of the participants dictate the scope of any scientific study. However, without an integrated approach in the study of acidified ecosystems, our knowledge will remain fragmentary.

LITERATURE CITED

Allen, H.A. 1972. Phytoplankton Photosynthesis, Micronutrient Interactions, and Inorganic Carbon Availability in a Soft-Water Vermont Lake. In: G.E. Likens, eds., "Nutrients and Eutrophication," pp. 63-83. Amer. Soc. Limnol. Oceanogr. Special Sym., Vol. I. 328 pp.

Almer, B., W. Dickson, C. Ekström, E. Hörnström, and U. Miller. 1974. Effects of Acidification of Swedish Lakes. Ambio 3:30-36.

Almer, B., W. Dickson, C. Ekström, and E. Hörnström. 1978. Sulfur Pollution and the Aquatic Ecosystem. In: J.O. Nriagu, ed., "Sulfur in the Environment, Part II: Ecological Impacts," pp. 271-311. Wiley, New York.

Baumgartner, K.J. 1981. A Quantitative Study of the Phytoplankton of Three Adirondack Lakes With Differing pH. M.S. Thesis, Cornell University. 83 pp.

Beamish, R.J. 1976. Acidification of Lakes of Canada by Acid Precipitation and the Resulting Effects on Fishes. Water, Air Soil Pollut. 6:501-514.

Brock, T.D. 1973. Lower pH Limit for the Existence of Blue-Green Algae: Evolutionary and Ecological Implications. Science 179:480-483.

Carignan, R. and J. Kalff. 1980. Phosphorus Sources for Aquatic Weeds: Water or Sediments? Science 207:987-988.

Conroy, N., K. Hawley, W. Keller, and C. LaFrance. 1976. Influences of the Atmosphere on Lakes in the Sudbury Area. J. Great Lakes Res. 2(Suppl. 1):146-165.

Dickson, W. 1978. Some Effects of the Acidification of Swedish Lakes. Verh. Int. Ver. Limnol. 20:851-856.

Dickson, W. 1980. Properties of Acidified Water. In: A. Drabløs and A. Tollan, eds., "Ecological Impact of Acid Precipitation," pp. 75-83. SNSF Project, Oslo.

Dickson, W., E. Hörnström, C. Ekström, and B. Almer. 1975. Rodingsjoar soder om Dalalven. (Char Lakes South of the River Dalalven.) Freshwater Laboratory, Drottningholm, Information No. 7. 139 pp.

Dillon, P.S., N.D. Yan, W.A. Scheider, and N. Conroy. 1979. Acidic Lakes in Ontario, Canada: Characterization, Extent, and Response to Base and Nutrient Additions. Arch. Hydrob. Beih. Ergebn, Limnol. 13:317-336.

Duthie, H.C. and M.L. Ostrofsky. 1974. Plankton, Chemistry and Physics of Lakes in the Churchill Falls Region of Labrador. J. Fish. Res. Bd. Canada 31:1105-1117.

Galloway, J.N., G.E. Likens, and E.S. Edgerton. 1976. Acid Precipitation in the Northeastern United States: pH and Acidity. Science 194:722-724.

Grahn, O., H. Hultberg, and L. Landner. 1974. Oligo-trophication - A Self Accelerating Process in Lakes Subjected to Excessive Supply of Acid Substances. Ambio 3(2):93-94.

Grahn, O. 1977. Macrophyte Succession in Swedish Lakes Caused by Deposition of Airborne Acid Substances. Water, Air Soil Pollut. 7:295-306.

Hall, R.J., G.E. Likens, S.B. Fiance, and G.R. Hendrey. 1980. Experimental Acidification of a Stream in the Hubbard Brook Experimental Forest, New Hampshire. Ecology 61(4):976-989.

Hargrave, B.T. 1969. Epibenthic Algal Production and Community Respiration in the Sediments of Marion Lake. J. Fish. Res. Bd. Canada 26:2003-2026.

Hendrey, G.R. 1976. Effects of pH on the Growth of Periphytic Algae in Artificial Stream Channels. Internal Report IR 25/76. SNSF, Oslo, Norway. 50 pp.

Hendrey, G.R. and F.R. Wright. 1976. Acid Precipitation in Norway: Effects on Aquatic Fauna. J. Great Lakes Res. 2(Suppl. 1):192-207.

Hendrey, G.R. and F. Vertucci. 1980. Benthic Plant Communities in Acidic Lake Colden, New York: Sphagnum and the Algal Mat. In: D. Drabløs and A. Tolan, eds., "Ecological Impact of Acid Precipitation." Proc. Internat. Conf., March 11-14, Sandefjord, Norway.

Hendrey, G.R., K. Baalsrud, T.S. Traaen, M. Laake, and G. Raddum. 1976. Acid Precipitation: Some Hydrobiological Changes. Ambio 5:224-227.

Hendrey, G.R., N.D. Yan, and K.J. Baumgartner. 1981. Responses of Freshwater Plants and Invertebrates to Acidification. Internat. Sym. for Inland Waters and Lake Restoration, Portland, Maine.

Hornström, E. 1979. Kalkning och försurning-effekter på växtplankton i tre Västkustsjöar. Statensnaturvårdsverket Rapport SNV-PM 1220. Box 1312 171 25 Solna Sweden. 45 pp.

Kwiatkowski, R.E. and J.C. Roff. 1976. Effects of Acidity on the Phytoplankton and Primary Productivity of Selected Northern Ontario Lakes. Can. J. Bot. 54(22):2546-2561.

Laake, M. 1976. Effekter av lav pH på produksjon, nedbrytning og stoffkretsløp in littoralsonen. SNSF Project IR 29/76, Aas-NHL, Norway, 75 pp.

Likens, G.E.   1976.   Acid Precipitation.   Chem. Eng. News, 22 Nov.:29–44.

Lind, O. and R.S. Campbell. 1970. Community Metabolism in Acid and Alkaline Strip-Mine Lakes. Trans. Am. Fish. Soc. 90:577–582.

Moss, B.   1973.   The Influence of Environmental Factors on the Distribution of Freshwater Algae: An Experimental Study, II. The Role of pH and the Carbon Dioxide-Bicarbonate System.   J. Ecol. 61:157–177.

Müller, P.   1980.   Effects of Artificial Acidification on the Growth of Periphyton.   Can. J. Fish. Aquat. Sci. 37:355–363.

Nicholls, K.H. and P.J. Dillon.   1978.   An Evaluation of Phosphorus-Chlorophyll-Phytoplankton Relationships for Lakes. Int. Rev. Ges. Hydrobiol. 63:141–154.

Scheider, W. and P.J. Dillon.   1976.   Neutralization and Fertilization of Acidified Lakes Near Sudbury, Ontario. Proc. 11th Canadian Symp. Water Poll. Research Canada. 93 pp.

Schindler, D.W.   1971.   Carbon, Nitrogen, and Phosphorus and the Eutrophication of Freshwater Lakes.   J. Phycol. 7:321–329.

Schindler, D.W.   1980.   Experimental Acidification of a Whole Lake: A Test of the Oligotrophication Hypothesis.   In: D. Drabløs and A. Tollan, eds., "Ecological Impact of Acid Precipitation," pp. 370–374.   SNSF Project, Oslo, Norway.

Schindler, D.W. and S.K. Holmgren.   1971.   Primary Production and Phytoplankton in the Experimental Lakes Area, Northeastern Ontario, and Other Low Carbonate Waters, and a Liquid Scintillation Method for Determining C Activity in Photosynthesis.   J. Fish. Res. Bd. Canada 28:189–201.

Schindler, D.W., R. Wageman, R.B. Cook, T. Ruszczynski, and J. Prokopowich.   1980.   Experimental Acidification of Lake 223, Experimental Lakes Area: Background Data and the First Three Years of Acidification.   Can. J. Fish. Aquat. Sci. 37:342–354.

Schofield, C.L.   1975.   Lake Acidification in the Adirondack Mountains of New York: Causes and Consequences.   In: "Proc. First Intl. Symp. Acid Precipitation and the Forest Ecosystem," May 12–15, Ohio State Univ.   USDA Forest Service Gen. Tech. Rep. NE-23.

Schofield, C.L. 1976. Effects of Acid Precipitation on Fish. Ambio 5:228-230.

Wetzel, R.G. 1975. Limnology. Saunders, Philadelphia. 473 pp.

Wium Andersen, S. 1971. Photosynthetic Uptake of Free CO_2 by the Roots of Lobelia dortmanna. Physiol. Plant 25:245-248.

Wium Andersen, S. and J.M. Andersen. 1972. Carbon Dioxide Content of the Interstitial Water in the Sediment of Grane Langsø, a Danish Lobelia Lake. Limnol. and Oceanogr. 17(6):943-947.

Wright, R.F., T. Dale, E.T. Gjessing, G.R. Hendrey, A. Henriksen, M. Johannessen, and I.P. Muniz. 1976. Impact of Acid Precipitation on Freshwater Ecosystems in Norway. Water, Air Soil Pollut. 6:483-499.

Yan, N.D. 1975. Acid Precipitation and Its Effect on Phytoplankton Communities of Carlyle Lake, Ontario. M.S. Thesis, University of Toronto.

Yan, N.D. and P. Stokes. 1978. Phytoplankton of an Acidic Lake and Its Response to Experimental Alterations of pH. Environ. Conservation 5:93-100.

Yan, N.D. 1979. Phytoplankton Community of an Acidified, Heavy-Metal Contaminated Lake near Sudbury, Ontario: 1973-1977. Water, Air Soil Pollut. 11:43-55.

CHAPTER 14

ECOLOGICAL EFFECTS OF ACID PRECIPITATION ON ZOOPLANKTON

D. F. Malley
D.L. Findlay
P.S.S. Chang
Department of Fisheries and Oceans
Freshwater Institute
Winnipeg, Manitoba R3T 2N6
Canada

INTRODUCTION

Zooplankton occupy a central position in lakes funnelling energy from phytoplankton to higher levels of secondary production. Not only do the production, biomass, and species composition of algae influence the biomass and nature of the zooplankton community (Brooks, 1969; Hillbricht-Ilkowska and Weglenska, 1970; McNaught, 1975; Porter and Orcutt, 1980), but the nature of the zooplankton affects phytoplankton diversity and biomass as well (Porter, 1977; Lynch, 1980). Similarly, the zooplankton appear to be able to influence the standing crop of secondary producers such as fish (Hrbáček, 1969; Stockner, 1977) and, in turn, are themselves influenced with respect to species composition, body size and standing biomass by fish and other predators (Hrbáček et al., 1961; Brooks and Dodson,1965; Hall et al., 1970; Anderson, 1980; Threlkeld et al., 1980).

Acidification affects lakes in multiple, far-reaching ways, changing them physically, chemically, and biologically. Thus, we can expect a complex of effects of acidification on zooplankton directly from changes in the physical or chemical environment or indirectly from changes in food supply, competition, or predation. Reciprocally, changes to the zooplankton will affect other parts of the lake ecosystem.

Effects of acid precipitation on zooplankton populations have been described a number of times during the last decade (Almer et al., 1974; Sprules, 1975; Leivestad et al., 1976; Wright et al., 1976; Roff and Kwiatkowski, 1977; Chrisman et al., 1980; Raddum et al., 1980). A consistent conclusion is that progressive acidification causes a reduction in the number of species comprising the zooplankton community. Species begin to disappear as the pH falls below 6.0 but by pH 5.5 to 5.0 and below, rate of loss is much more rapid.

During the same decade, considerable advance has been made in understanding factors controlling composition of species or functional type in zooplankton communities, body size of species or of individuals within species, and species dominance (Lewis, 1979; Kerfoot, 1980a). Although the actual food and feeding rates of grazers, omnivores, and invertebrate predators are poorly known in field situations, testable hypotheses on or models of dynamics of zooplankton communities are available (Lynch, 1979; Sprules, 1980; Anderson, 1980; Kerfoot and DeMott, 1980; Zaret, 1980). Only recently have discussions of the effects of acidification on zooplankton considered ecological effects through changes in the food supply, competition, and predation (Ericksson et al., 1980; Henrickson et al., 1980; Raddum et al., 1980; Nilssen, 1980a; 1980b), but these have not yet been treated in any depth. Incorporation of such information may well be fruitful, explaining changes in zooplankton populations in acidic lakes. Conversely, by providing a perturbation to the zooplankton community and selectively removing certain species, low pH may shed more light on dynamics of zooplankton communities.

This paper suggests a number of factors which may be responsible for changes in zooplankton communities of lakes with acidification. Each factor is described using data from an experiment involving the artificial acidification of a Precambrian Shield lake, number 223, and from the literature. An attempt is made to judge the relative importance of each factor, primarily in explaining the response of the zooplankton community of Lake 223 to acidification. Lake 223 has been experimentally acidified from 1976 to 1980, during which time the average pH of the epilimnion was reduced from 6.79 to 5.37. Therefore, data from this experiment can clarify earlier stages of acidification, which often proceed relatively rapidly in lakes exposed to acidic precipitation. Information on moderate and extreme acidification is taken from the literature.

Possible factors affecting zooplankton communities in acid lakes are:

(1)   increased temperature in portions of the water column as a result of increased transparency and greater solar heating,

(2)   change in algal food abundance and/or quality,

(3)   increase in the hydrogen ion concentration to toxic levels,

(4) increase in concentrations of metals to levels toxic at low pH,

(5) decreased or increased predation by invertebrate predators (<u>Mysis relicta,</u> <u>Epischura,</u> <u>Chaoborus,</u> corixids),

(6) change in predation by minnows or larger fish species,

(7) change in competition among zooplankton due to elimination of competing species.

LAKE 223, DESCRIPTION AND EXPERIMENTAL METHODS

Lake 223 is an oligotrophic lake with surface area of 27.3 ha and maximum depth of 14.4 m situated in the Experimental Lakes Area (ELA), northwestern Ontario, at 49°42'N latitude and 93°43'W longitude. Initial alkalinity was about 80 meq/l. Mean annual pH of precipitation in ELA ranged from 4.9 to 5.0 over 10 yr. The lake was studied during two pre-acidification years, 1974 and 1975, and has been subject to experimental acidification with sulfuric acid beginning in 1976, causing the pH of the epilimnion to drop from its natural level of between 6.5 and 7.0 to an average pH of 5.37 in 1980 at a rate of about 0.25 pH units/yr. Acid additions were made directly to the lake and within hours mixed with the epilimnion. Acid mixed with the entire water column at fall turnover.

Method of acidification and results from the first 4 yr of acidification are described by Schindler <u>et al.</u> (1980b) and Schindler (1980). Methods for phytoplankton sampling and analysis are given by Findlay and Saesura (1980). Zooplankton sampling, identification, and enumeration methods are described by Chang <u>et al.</u> (1981). In 1974, vertical series of zooplankton samples were taken with a 28.7-1 transparent trap at the deepest part of the lake. In 1975 and 1976, only the top 2 m of the water column were sampled, and the data are not included here. In 1977 to 1980, vertical net hauls were made. During thermal stratification (May to September) zooplankton were separately sampled from epilimnion, metalimnion, and hypolimnion. Mesh size of nets, including that on the trap, was 53 μ. Samples were preserved upon collection with 4 percent formaldehyde. Identification of species and life stages and counts of individuals were performed on duplicate 1 ml aliquots of the sample, together representing 6 to 10 percent of the sample volume. Data are presented as number of individuals/m^2 lake surface area after weighting of numbers in each lake layer to reflect relative volume of that layer in the lake.

ZOOPLANKTON OF ACIDIFIED LAKES

Acidic lakes possess a lower diversity of zooplankton species than do neutral-pH, soft-water lakes (Sprules, 1975; Leivestad et al., 1976; Roff and Kwiatkowski, 1977; Raddum et al., 1980) and this decline is most clearly evident below pH 5.5 or 5.0. A reduction in biomass of zooplankton communities generally appears to accompany acidification (NRCC, 1981).

Daphnids disappeared first from Swedish lakes, with only a few specimens found below pH 6.0 (Almer et al., 1974), but the species were different from those found in Lake 223. Diaphanosoma brachyurum, Holopedium gibberum, and Leptodora kindtii were observed at pH greater than 4.9. The most tolerant crustaceans in the Swedish lakes were Bosmina coregoni and Diaptomus gracilis, both occurring at pH 4.6 (Almer et al., 1974). Daphnids were the least tolerant of crustaceans in acidified lakes in south Norway and disappeared at about pH 5.0 (Raddum et al., 1980).

Epischura lacustris was the first species to be lost from 47 acidified lakes under the influence of metal smelters in the Sudbury, Ontario, area (Sprules, 1975). It disappeared at about pH 5.9. Daphnia galeata mendotae and Tropocyclops prasinus mexicanus disappeared below pH 5.2. Mesocyclops edax, Cyclops bicuspidatus thomasi, Holopedium gibberum, and Diaphanosoma sp. were tolerant to about 4.1; only Diaptomus minutus and Bosmina were found in the most acidic lakes, below pH 4.0. Daphnia retrocurva and D. longiremis were present only in lakes between about pH 5.0 and 6.0. One species, D. catawba, was acid-tolerant to pH 4.1 (Sprules, 1975).

Six other lakes in the Sudbury area ranging in mean pH from 3.95 to 6.75 contained D. g. mendotae occasionally in all but the most acidic ones (Roff and Kwiatkowski, 1977), whereas E. lacustris was absent from lakes less than pH 6.0. C. b. thomasi and M. edax occurred most abundantly above pH 4.38. T. prasinus occurred at 5.6 and above. Most acid-resistant species were Bosmina longirostris, H. gibberum, and D. minutus.

In 32 ponds near the shores of Georgian Bay, Ontario, naturally ranging from pH 4.7 to alkaline pH, D. g. mendotae was a rare species and present only above pH 7.0. D. sicilis, also rare, was present only at pH 7.0 and above. E. lacustris was absent above pH 6.1. Most acidic ponds were dominated by B. longirostris, Diaphanosoma sp., and M. edax, as well as a few other species not common to Lake 223 (Carter, 1971).

In Lake 223, total number of cladocerans/m^2 remained relatively constant with acidification from mean epilimnetic

Table 1

Number of Cladocerans/m^2 (Entire Water Column) in Lake 223 Averaged Over the May to September Period During 1974 and 1977 to 1980 — Mean pH of the Epilimnion for Each Year is Given in Parentheses

Species	Year				
	1974	1977	1978	1979	1980
	(6.64)	(6.08)	(5.84)	(5.60)	(5.37)
			--pH--		
Bosmina longirostris	10,836	10,795	12,175	4,433	18,634
Daphnia galeata mendotae	5,250	1,390	785	241	1,916
Daphnia catawba x schoedleri[a]	0	0	0	0	4,164
Holopedium gibberum	111	713	1,654	821	2,642
Diaphanosoma brachyurum	3,941	5,261	2,232	4,185	6,176
Total	20,140	18 160	16,845	9,680	33,530

[a]This population possesses half of the taxonomic characters of D. catawba and half of those of D. schoedleri as described by Brooks (1957). Both D. catawba and D. schoedleri occur, rarely, in ELA (Patalas, 1971).

Table 2

Number of Copepods/m^2 (Entire Water Column) in Lake 223 Averaged Over the May to September Period During 1974 and 1977 to 1980

Species	1974	1977	1978	1979	1980
			CALANOIDS		
Diaptomus minutus adults	4,323	8,952	785	7,963	12,107
Diaptomus sicilis adults	628	86	0	0	0
Epischura lacustris adults	422	1,189	1,295	0	0
Calanoid nauplii N_1 to N_6	54,052	67,036	46,092	48,607	90,441
Calanoid copepodids C_1 to C_5	71,961	102,002	85,664	63,348	151,466
Total calanoid adults	5,370	10,230	2,080	7,960	12,110
			CYCLOPOIDS		
Cyclops bicuspidatus thom. adults	3,670	11,506	7,836	3,706	10,088
Tropocyclops prasinus mex. adults	1,346	5,049	6,564	1,230	34
Mesocyclops edax adults	1,021	3,630	1,194	1,141	894

Cyclopoid nauplii N_1 to N_6	51,843	102,010	155,763	58,350	77,083
Cyclopoid copepodids C_1 to C_5	44,470	128,709	51,695	58,215	102,030
Total cyclopoid adults	6,040	20,185	15,645	11,080	11,015

pH of 6.64 to 5.84, but appeared to decrease at mean pH 5.6 and
to increase at mean pH 5.37 (Table 1). Daphnia galeata mendotae
appeared to decline with acidification from 1977 (pH 6.08) to
1979 (pH 5.60), but in 1980 increased in numbers.   Average
number of Daphnia catawba x schoedleri, not previously recorded
in Lake 223, exceeded D. g. mendotae in 1980 (Table 1).   B.
longirostris and D. brachyurum were relatively unaffected by
acidification whereas acidification appeared to enhance numbers
of H. gibberum (Table 1).

Calanoid nauplii and copepodids and adults of D. minutus
showed no trend towards increase or decrease with acidification
until 1980 when each was more abundant than in previous years
(Table 2).   The first species to disappear from Lake 223, at
average epilimnetic pH 5.84, was the minor species, D. sicilis,
predominantly an inhabitant of deep, clear lakes (Carter et
al., 1980).   The following year, 1979, at mean pH 5.60, E.
lacustris declined to low numbers and was absent in 1980 (Table
2).

Cyclops bicuspidatus thomasi and M. edax appeared largely
unaffected by acidification but numbers of T. p. mexicanus
appeared to be reduced. Numbers of nauplii and copepodids of
cyclopoids showed no trend with acidification (Table 2).

The rotifers, Polyarthra vulgaris, P. remata, Keratella
taurocephala, and Kellicottia longispina, increased in numbers
with acidification (Malley and Chang, 1981; Malley et al., in
manuscript) and remained very abundant in 1980.   Dramatic
increases in abundances of Kellicottia bostoniensis,
Trichocerca cylindrica, and Keratella cochlearis occurred in
1980.   K. cochlearis increased with decreased pH in 1977 and
1978 but was relatively rare in 1979.   In 1980, this species
showed its highest abundance in the period between 1974 and
1980.

Another species, the opossum shrimp, Mysis relicta, was
lost from Lake 223 with acidification. This is an important
zooplanktonic predator by night. By day it is benthic. Prior
to acidification of Lake 223, this species was abundant (Nero,
1981), but it was not studied quantitatively until 1978. In
midsummer, 1978, population size was estimated to be 5.34×10^6
$\pm 1.06 \times 10^6$ (95% confidence interval). Acidification had not
yet affected the population in summer, 1978, because the
densities recorded were in the range expected for Mysis based
on studies of other lakes (Nero, 1981). By midsummer, 1979,
the population was reduced to less than 5 percent of its size
in 1978 and by October, 1979, no Mysis were captured from the
lake despite extensive sampling efforts. Its sharp decline

over the winter of 1978 to 1979 was associated with maximum pH within its vertical habitat of 5.9 to 5.6.

The major changes in the zooplankton community in Lake 223 with acidification to pH 5.37 were thus the loss of the predaceous copepod, E. lacustris; the opossum shrimp, M. relicta, and the minor calanoid, D. sicilis; and an increase in number of rotifers, particularly of species known to be acid-resistant. D. g. mendotae was well represented at mean pH 5.37, in contrast to its apparent decline at pH 5.60, and a second daphnid species, D. catawba x schoedleri, appeared for the first time. Average numbers of crustaceans at mean epilimnetic pH 5.37 were greater than they had been in 1974 before acidification.

Effects of Increased Solar Heating

Increased transparency is one of the primary and most visible effects of lake acidification (Almer et al., 1974; Sprules, 1975; Dickson, 1978). Dickson (1978) hypothesized that the precipitation of humic substances by increased levels of aluminum is responsible for most of the increase in transparency, particularly between pH 5.5 and 4.0. A pronounced increase in transparency measured either by Secchi-disc depth or extinction coefficient (0-10 m depth interval) occurred in Lake 223 by 1977 (Schindler et al., 1980b). Average Secchi-disc transparency in the ice-free season was greatest in 1977 and 1978 (Schindler, 1980) but was indistinguishable from pre-acidification values (4.25 m) during 1980. Nevertheless, more frequent and precise light extinction coefficient measurements indicated that the trend to increasing transparency with acidification continued in 1978, 1979, and 1980 (J. Shearer, unpublished data; Figure 1). Mean extinction coefficients in the post-acidification years were significantly less than in the 2 yr before acidification (P< 0.005, analysis of variance). This increase in transparency is thought to be due to a change in the color of dissolved matter (humic acids) in Lake 223 with decreased pH since there was no decrease in the quantity of dissolved organic matter (Schindler, 1980).

The thermal regime of Lake 223 has altered with acidification, presumably as a direct result of enhanced transparency. The depth of the epilimnion has increased in 3 out of 5 yr by 1 to 2 m. Fall turnover, consequently, may be earlier. Typical depths of the epilimnion in July of 1974 and 1975 were 3.5 to 4.5 m; whereas in July, 1980, epilimnetic depth was 6 to 7 (Figure 2). Epilimnetic depth in Lake 223 was compared with that in 2 non-acidified lakes, 224 and 239. Lake 224 (25.9 ha in surface area, 27.4 m maximum depth) is less

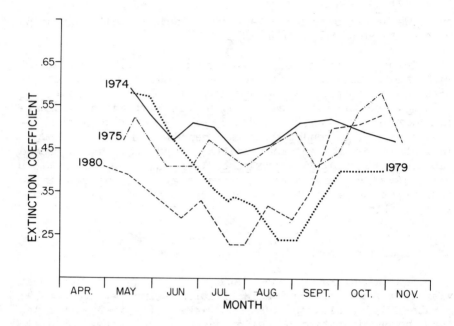

Figure 1.   *The vertical extinction coefficients for light in
            Lake 223 measured over the 0- to 8-m interval.
            Ice-free season averages for pre-acidification
            years 1974 and 1975 were 0.50 and 0.47. For 1976,
            1977, and 1978 (not shown), averages were 0.46,
            0.36, and 0.42. Averages for 1979 and 1980 were
            0.39 and 0.36. (Data supplied by ELA project
            personnel).*

productive, more transparent, and had an average epilimnetic
depth 1 m deeper than Lake 223. Epilimnetic depth in the 2
lakes sampled on the same dates in the years 1974 to 1978 were
highly correlated ($r = 0.771$ to 0.994) indicating that they
varied seasonally in a nearly identical manner. In 1977,
epilimnion depth in Lake 223 was significantly deeper in
relation to that in Lake 224 than it has been previously,
although in 1978 the relationship was statistically the same
as it had been in 1974 to 1976. Lake 239 was not sampled on
the same dates as Lake 223, but data are available for 1974 to
1980. Epilimnetic depth of Lake 239 (56.1 ha in surface area,
30.4 m maximum depth) in July was usually between 2.2 and 5.3
m. Epilimnetic depth in Lake 223 was deeper than that in Lake
239 on most dates in 1974 (Figure 2), but the epilimnion of
Lake 223 deepened further in the years from 1977 to 1980 relative
to Lake 239.

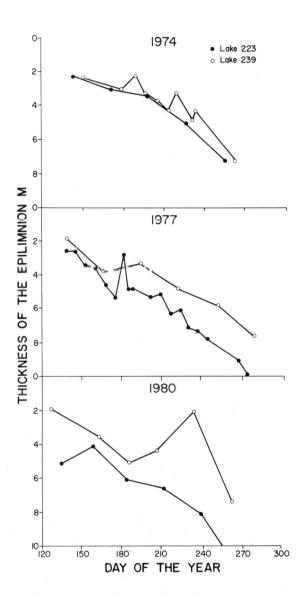

Figure 2. Comparison of the depths (thickness) of the epilimnion in Lakes 223 and 239 for three selected years (data supplied by ELA personnel).

As a consequence of epilimnetic deepening, Lake 223 contained more heat in 1980 than in 1975 (Figure 3). Seasonally, epilimnetic temperatures in Lake 223 and Lake 239 were similar to each other both before and after acidification (Turner and

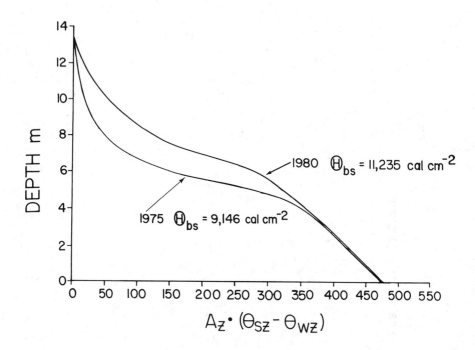

*Figure 3.   Heat in Lake 223 as a function of depth where A_z
is area (m^2) of the lake at depth z, θ_{sz} is
highest observed summer temperature (°C) at depth
z and θ_{wz} is 4°C. $\textcircled{H}_{bs}$ is summer heat income
calculated according to Hutchinson (1957).
(Temperature data supplied by ELA project personnel.)*

Cruikshank, unpublished data). Thus, with greater clarity,
Lake 223 contains more heat, not because of higher epilimnetic
temperatures but because of a deeper epilimnion. Thus, water
at 4-to 10-m depths is warmer than it was prior to acidification
either because it has been incorporated into the epilimnion or
because it is relatively higher in the thermocline (Figure 3).

Lakes in Ontario ranging from acidic (pH $\leq$ 4.3) near
Sudbury to circum-neutral in the Muskoka-Haliburton region give
similar data on transparency and solar heating. Lakes with
lowest pH had the greatest Secchi disc depths. An acidic lake,
Clearwater, had a deeper epilimnion than a comparable, non-
acidic lake, Blue Chalk Lake. The summer heat budget of
Clearwater Lake was greater than that of Blue Chalk Lake (NRCC,
1981).

Change in Algal Food Abundance or Quality

Analysis of 18 published measurements of phytoplankton production in lakes acidified by acid precipitation indicates that acidification does not reduce phytoplankton production (NRCC, 1981). Nevertheless, most studies report that phytoplankton diversity declines with decreasing pH (Almer et al., 1974; Kwiatkowski and Roff, 1976; Leivestad et al., 1976; Raddum et al., 1980; Chrisman et al., 1980; NRCC, 1981). In 115 Swedish lakes, those with pH 6 contained algal species equally from all groups characteristic of oligotrophic freshwaters. Lakes with pH lower than 5 had few species of diatoms and cyanophytes. Below pH 5.0 only about 10 species were present, including species of chlorophyceans, chrysophyceans, flagellates, and dinoflagellates. At pH 4.0, the populations were mainly dinoflagellates, Peridinium inconspicuum, and some species of Gymnodinium. Occasionally, one or two species of chlorophytes were dominant. The acid-resistant, filamentous chlorophyte, Mougeotia scalaris, was observed free floating in many acidified lakes (Almer et al., 1974). Yan and Stokes (1978) reported an apparent replacement of Chrysophyceae by Dinophyceae and to a lesser extent by Cryptophyceae in Carlyle Lake at pH 4.5 to 5.0.

The response of Lake 223 differed in several ways from these published results. First, phytoplankton production per unit area in Lake 223 increased from 1975 to 1979, although this trend also was shown to a lesser extent by Lake 239, probably due to a succession of warm, sunny years (Schindler, 1980). This led Schindler (1980) to reject the oligotrophication hypothesis as a general description of the acidification process, at least during acidification to pH 5.6. Consistent with published results, a mat of Mougeotia developed in the littoral zone of Lake 223 during 1979 but evidently did not yet affect nutrient availability (Schindler, 1980). Primary production in 1980 at pH 5.37 did not differ greatly from 1979 values (De Bruyn and Shearer, 1981). Secondly, there has been no decrease in diversity of algal species in Lake 223 up to 1980 (pH 5.37)(Findlay and Saesura, 1980; unpublished data).

With acidification of Lake 223, chrysophyceans have continued to dominate in the epilimnion but there have been proportionate increases in chlorophyceans and cyanophyceans in 1977 to 1980. In 1980, in addition, the dominance of chrysophyceans has been further eroded by a relative increase in the peridineans (dinoflagellates). Peridineae represented up to 30 percent of live standing biomass in the epilimnion in July. Nevertheless, over the whole water column peridineans were insignificant until 1980 when they increased in absolute biomass, but still represented a low proportion of biomass compared with total or edible phytoplankton (Figure 4A and 4B).

Total phytoplankton biomass in the euphotic zone has increased with acidification (Figure 4A). The percentage of that total phytoplankton biomass that is edible, arbitrarily considered to be those cells < 20 μ in any dimension, was at least as great during acidification years as prior to acidification. Therefore, total edible phytoplankton has increased with acidification (Figure 4B). Even though the largely inedible peridineans increased significantly in 1980, both in absolute biomass and as a proportion of total phytoplankton biomass, they are not replacing edible components (Figure 4B).

Increase in Hydrogen Ion Concentration

Depletion of bicarbonate alkalinity, depression of pH, and increase in sulfate concentration all occur with acidification by acid precipitation (Almer et al., 1974; Dillon et al., 1978; Dickson, 1978; Henriksen, 1980; Wright et al., 1980) and occurred as well in experimentally acidified Lake 223 (Schindler et al., 1980b). No changes were detected up to 1979 in depth distribution or average concentrations of calcium, manganese, sodium, potassium, chloride, suspended carbon, nitrogen and phosphorus, dissolved inorganic carbon, total dissolved phosphorus, or reactive silicate (Schindler et al., 1980b).

Increase in Metal Concentrations

Increased concentrations of certain metals are almost invariably associated with acidification because metals are deposited together with the acids and acid precursors (McFarlane et al., 1979; Franzin et al., 1979; Allen and Steinnes, 1980), or are leached from the watershed (Dickson, 1978; Cronan and Schofield, 1979). Concentrations of several metals, such as cadmium, manganese, zinc, and lead, increased as pH declined in 16 Swedish lakes (Dickson, 1980). In Lake 223, concentrations of copper, aluminum, zinc, manganese, and iron were elevated after acidification (Table 3; Schindler et al., 1980a; 1980b; Schindler and Turner, in press). The metals in this case did not come from precipitation, the acid that was added or from the watershed, but were mobilized from the sediments. Cadmium, chromium, and lead were undetectable in lake water of Lake 223 in 1976 and 1977. Cadmium was frequently below detection limits between 1976 and 1979. Not unexpectedly, the zinc and aluminum concentrations in Lake 223 in 1980 were considerably lower than the 15 to 20 μg/l and the 50 to 200 μg/l, respectively, in

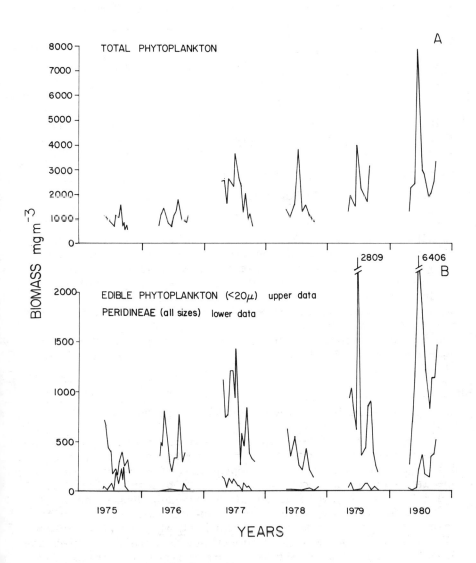

Figure 4. Live biomass of phytoplankton in Lake 223 during
ice-free seasons of pre-acidification year, 1975,
and acidification years 1976 to 1980. A. all
species of algae; B. upper:phytoplankton below
20 μ in all dimensions, lower:all species of
peridineans regardless of size.

Table 3

Concentrations of Metals (μg/l) in the Epilimnion of
Lake 223 Averaged Over May to September Periods

Year	Mean Ice-Free Season (pH)	Al	Zn	Cu	Mn
1976	6.79	15.5	2.5	<1.0	11.3
1979	5.60	24.5	3.0	<0.8	72.6
1980[a]	5.37	27	5.6	0.77	127

[a]Unpublished data, R. Wagemann.

Swedish lakes at pH 5.4 (Dickson, 1980). Lakes at pH 5.4 to 6.0 in Norway and Ontario contain 50 to 90 μg/l aluminum (NRCC, 1981). The watershed is the primary source of zinc and aluminum. In contrast, the source of manganese is primarily lake sediments. As expected, the concentrations of manganese in Lake 223 with acidification approximated more closely published values for the same pH range. Lakes of about pH 5.5 around Sudbury have manganese concentrations of about 70 μg/l and acidic lakes < pH 5.0 have concentrations of 100 to 350 μg/l (NRCC, 1981).

Change in Predation by Invertebrate Predators

In acidified lakes, invertebrate predators may increase or decrease in abundance. In Scandinavian lakes, reduced predation pressure as fish disappear allows acid-resistant prey species such as Chaoborus flavicans, C. obscuripes, corixids, and dytiscid beetles to become more abundant (Henrikson et al., 1980). These changes can also be seen in non-acid lakes from which the fish are removed (Erickson et al., 1981). As mentioned above, in Lake 223, in contrast, important invertebrate predators, the opossum shrimp, Mysis relicta, and the copepod, Epischura lacustris, declined with acidification.

M. relicta is generally omnivorous although it can be herbivorous (Bowers and Grossnickle, 1978) or carnivorous (Lasenby and Langford, 1972). Guts of Mysis from eutrophic Stony Lake contained 96 percent Daphnia. In oligotrophic Char

Lake, Mysis fed on detritus and diatoms. They readily ate chironomid larvae in the laboratory (Lasenby and Langford, 1972). Its introduction into certain lakes was associated with the decline of cladocerans (Zyblut, 1970; Richards et al., 1975; Goldman et al., 1979). Stomach contents of Mysis collected from Lake Tahoe contained fragments of Epischura, Diaptomus, Bosmina, and Kellicottia (Goldman et al., 1979). In general, Daphnia are preferred over copepods; Epischura, Ceriodaphnia, Bosmina, and cyclopoid copepods are preferred over Diaptomus copepodids and adults; and Diaptomus males are preferred over Diaptomus females (Cooper and Goldman, 1980). Epischura lacustris is a predator on cladocerans, especially B. longirostis (Kerfoot, 1980b; Wong, 1981).

Chaoborus spp are important invertebrate predators generally in ELA lakes. Their abundance or population changes in Lake 223 have been only cursorily examined but they do not appear to be abundant nor to have increased dramatically with acidification. They are rare in daytime zooplankton samples and their abundance in these has not markedly increased with acidification. Nonetheless, most species are benthic by day and pelagic nocturnally and not effectively captured by daytime pelagic sampling. In numerous night dives in Lake 223, Chaoborus were not conspicuous (I. Davies, personal communication). Corixids, predators of cladocerans, have not been studied in Lake 223.

Change in Predation by Minnows or Larger Fish Species

Of the biotic changes with acidification, the decline and disappearance of fish populations have received most attention. The most sensitive fish species are usually lost between pH 6.0 and 5.5 (NRCC, 1981). Number of fish species declines with increasing acidity (Almer et al., 1974; Leivestad et al., 1976; NRCC, 1981).

Prior to acidification, Lake 223 contained lake trout, Salvelinus namaycush; white suckers, Catostomus commersoni; fathead minnows, Pimephales promelas; and small number of slimy sculpins, Cottus cognatus (Beamish et al., 1976). The 1978 year class of fathead minnows failed because of sensitivity to acid. This resulted in extremely low numbers of fatheads in 1979 and their subsequent disappearance by 1980. Pearl dace, Semotilus margarita, consequently increased during 1979 and replaced the fathead minnow by 1980 (K. Mills, personal communication). Fathead minnow and pearl dace overlap a great deal in their feeding. Food is basically periphyton, benthic organisms or detritus with zooplankton as a minor item (R. Tallman, unpublished data).

In 1980, lake trout and white sucker were still abundant in Lake 223 and spawning by both in 1980 was successful. Nevertheless, in 1981, the populations are showing negative effects due to acidification. Compared with preacidification years, the condition factor of the trout and suckers has gradually declined over the last 3 yr and recruitment of suckers in 1981 appears to be unsuccessful (K. Mills, unpublished data). The suckers are benthic feeders. Even though some lake trout stomachs were found to contain zooplankton, the trout are usually piscivorous, preying on the pearl dace, small trout, and slimy sculpins.

Change in Competition

According to Zaret (1980), although competition among species occurs in every community where resources are limited, direct evidence of the determining role of competition in freshwater zooplankton communities has not been demonstrated. Predation and competition are highly interactive with predation determining the "gross morphology" of the prey (body size and shape, behavior) and competition determining which of several possible prey species will become dominants.

The disappearance of D. sicilis, a rare species before acidification, is expected to have little effect on competitive interactions among herbivores. The newly appearing D. catawba x schoedleri presumably alters competiton relationships for D. g. mendotae. The disappearance of M. relicta and E. lacustris would reduce competition with other invertebrates and/or fish for zooplankton prey.

DISCUSSION

Decline in Species Diversity

Number of zooplankton species in Lake 223 declined with acidification as expected from the literature. Decrease in epilimnetic pH from a mean of 6.08 to 5.84 was associated with the disappearance of one species; decrease to pH 5.60 led to loss of two more species. Unexpectedly at pH 5.37, a second daphnid species, Daphnia catawba x schoedleri, appeared.

Order of susceptibility to acid of the three declining species is in agreement with the literature. In Lake 223, D. sicilis and E. lacustris disappeared below pH 6.08 and pH 5.84, respectively. These are slightly lower pH values than those below which they are reported to disappear in other studies;

for example, pH 7.0 for D. sicilis (Carter, 1971) and pH 5.9 to 6.1 for E. lacustris (Carter, 1971; Sprules, 1975; Roff and Kwiatkowski, 1977). Upper metalimnetic water in Lake 223 was 0.2 to 0.5 pH units higher than the epilimnetic water, thus possibly providing a refuge of slightly higher pH until fall turnover. The disappearance of M. relicta at pH between 5.6 and 5.9 is in agreement with Scandinavian results (Nero, 1981).

Changes in Community Structure in Scandinavian and North American Lakes

Daphnids are the first zooplanktonic species to disappear from acidified Scandinavian lakes although the pH at which they decline or disappear is reported to be pH 6.0 in Swedish lakes (Almer et al., 1974) and pH 5.0 in Norwegian lakes (Raddum et al., 1980). Invertebrate predators in these lakes, on the other hand, are more acid-resistant. These predators, such as Chaoborus, corixids, and dytiscid beetles, are normally controlled by fish predation. When fish disappear under acid conditions, the predators expand and exert much more pressure on their zooplanktonic prey. This may change the size structure of the prey, favoring large-bodied herbivores over small-bodied ones (Henrikson et al., 1980; Eriksson et al., 1980). Early signs of acidification in the vertebrate (i.e., fish-dominated) Scandinavian lakes are thus increasing incidence of zooplanktonic phenotypes, providing protection from the intense invertebrate predation (Nilssen, 1980b).

In contrast, in Canadian Shield lakes crustacean predators, rather than herbivores, are the most acid-sensitive planktonic species. Judging from the Lake 223 community structure, early acidification effects on fish have not greatly affected either the zooplankton or the invertebrate predators of zooplankton. Early signs of acidification of North American lakes are thus increasing dominance by herbivores (Nilssen, 1980b). Analysis of community changes with acidification must take into account differing biogeography (Nilssen, 1980b). The discussion of community changes in Lake 223 thus may have greater applicability to other Canadian Shield lakes than to Scandinavian lakes.

Factors Enhancing Herbivore Abundance

With acidification of Lake 223, the zooplankton lost two important predators, E. lacustris and M. relicta, and became dominated by the herbivorous cladocerans and calanoids, especially D. minutus. The decline or loss of M. relicta and E. lacustris in 1979 would be expected to lead to increases in numbers of the preferred prey, particularly Daphnia and Bosmina, respectively.   The data for 1979 indicate no such increase in prey numbers (Tables 1 and 2).   Therefore, it is concluded that release of predation by M. relicta and E. lacustris is not the major factor explaining the herbivore population changes.   The disappearance of M. relicta may be associated with the establishment of D. catawba x schoedleri in 1980, but its success in that year rather than in 1979 correlates better with the abundant algal food supply.   Possibly its appearance is due to reduced competition, not caused by the disappearance of a competing zooplankton species but by the expansion of a required resource, food.

The changes in species and population sizes of fish in Lake 223 appear to have little affected the zooplankton community.   Even though there was a shift in minnow species due to acidification, neither species depends greatly on zooplankton and total feeding pressure on zooplankton by minnows appears to be relatively unchanged.   As of 1980, there had been no abrupt change in level of predation by trout and suckers on zooplankton.

Crustacean abundance is determined by both temperature and food, as was shown by Patalas (1972) in the St. Lawrence Great Lakes.   Numbers of herbivorous crustaceans increased with increasing heat content when food was abundant in Lake Ontario. Temperature controls growth and hatching rates in crustaceans. In Lake 223, mean water column temperature has increased with acidification and phytoplankton of edible size are more abundant, particularly in 1980 (Figure 4).   This suggests that Lake 223 has greater herbivore abundance because of increased transparency and solar heating and associated increased food supply, rather than because of reduction in invertebrate predation.

Factors Causing Species Disappearance

The cause of the disappearance of D. sicilis, E. lacustris, and M. relicta could be decline in food supply, increase in predation or competition, or intolerable changes in the physical or chemical environment.   Food supply for the herbivorous D. sicilis increased with acidification (Figure 4) and there is no reason to expect that competition increased significantly. Nor is there reason to believe predation upon it increased greatly even though it disappeared before the predators E. lacustris and M. relicta.   Its disappearance may be related to toxic effects of the hydrogen ion concentration, or alternatively, the increase in light penetration and average temperature of the water column may not have been tolerable.

D. sicilis is normally an inhabitant of deep, clear glacial lakes (Carter et al., 1980) where it tends to avoid the surface, at least during the day.

Cladoceran food supply of E. lacustris and M. relicta did not decline. In any case, neither species is an obligate predator. Both can switch prey items or consume algae or detritus (Bowers and Grossnickle, 1978; Cooper and Goldman, 1980; Wong, 1981). No change in predation by fish or other predators on these two species was evident. No introduction of competing species was observed. A possible explanation for the disappearance of M. relicta is that the cold, dark, well oxygenated environment required by this species (Smith, 1970) may have been eroded by the increased light penetration and higher average temperature of the water column. Nevertheless, the fact that the population underwent its major decline over the winter when temperature and light levels are low makes this hypothesis less attractive (Nero, 1981).

It is concluded that E. lacustris and M. relicta and probably D. sicilis disappeared because their limits of tolerance to low pH or associated chemical parameters were reached by the time epilimnetic pH was 5.6. The physiological basis of adverse effects of low pH on survival, growth, or reproduction of crustaceans is not well known. Potentially, toxicity may be associated as well with the depletion of bicarbonate ions or the increase in sulfate ions, but the regulation of blood pH is a universal and essential part of physiological homeostasis (Hoar, 1975). Evidence suggests that low pH interferes with ionic balance, pH below 6.0 inhibited adsorption of sodium ions by the crayfish, Astacus pallipes (Shaw, 1960). pH below 5.5 inhibited uptake of sodium ions also by Daphnia magna and Acantholeberis curvirostris. The latter, a cladoceran, is an inhabitant of acid peaty water of low salt content (Potts and Fryer, 1979). pH of 5.5 and below inhibited calcium ion uptake following ecdysis in the crayfish, Orconectes virilis (Malley, 1980). Monocultures of Daphnia pulex experienced reduced population growth rate potential, r, below pH 5.5 due to reduced survivorship and delayed onset of reproductive maturity (Walton et al., in press). The daphnids survived exposure to pH 4.3 for up to 4 days without effect; and if exposed for 21 days, were still able to produce viable offspring albeit at a much lower than normal rate.

For some species, the direct "lethal" effects of low pH (for example, those inhibiting reproduction or causing early death) may occur at only slightly lower pH than that at which the first sublethal physiological effects occur. Alternatively, as illustrated by Walton et al. (in press), the sublethal effects of low pH may extend over a wide pH range for other species.   This suggests that some species could be directly affected by the toxic effects of low pH but be eliminated indirectly because of decreasing ability to compete or withstand predation.   Lastly, other species may suffer no significant adverse physiological effects at low pH but be eliminated because altered competition or predation relationships in the community are unfavorable for them.   Thus, the alternatives of direct or indirect pH effects as used by Nilssen (1980a) may belong to a continuum.

Increase in concentrations of metals with acidification in Lake 223 is another potential source of toxicity.   The effects on ELA zooplankton of cadmium, zinc, and mercury at neutral pH have been examined by adding the metals to plankton in enclosures of various sizes.   Concentrations of cadmium of 1 to 30 µg/l, mercury concentration of 4 µg/l, and zinc at 90 µg/l resulted in lower biomass (density) of planktonic crustaceans (Marshall and Mellinger, 1980; Marshall et al., 1981a; 1981b).   Reductions in species richness and community biomass were also responses of zooplankton in low-pH lakes near Sudbury contaminated with nickel and copper (Yan and Strus, 1980).   The concentrations of cadmium and mercury in Lake 223 were not elevated with acidification.   The concentration of zinc, though increased with acidification, was well below concentrations found previously to reduce biomass of ELA zooplankton, although the amount of complexing with dissolved organic materials in the water may be critical in determining the toxicity of a given zinc concentration (Marshall et al., 1981a).   The planktonic crustaceans in Lake 223 increased rather than decreased in abundance with acidification.   Thus, there is no evidence to suggest that the heavy metals or aluminum in Lake 223 were elevated to levels toxic at the reduced pH present in 1980.

CONCLUSIONS

Three of the 7 factors affecting zooplankton community structure appear to be most influential during early acidification using data from the Lake 223 experiment. Increased herbivore abundance best correlates with increased average temperature of the water column and enhanced food supply, both resulting from increased transparency.   The disappearance of three crustacean species appears to have been caused by intolerable chemical conditions, presumably an increase in the hydrogen ion concentration to levels toxic to them.   The appearance of D. catawba x schoedleri could be related to the enhanced food supply and consequent reduction in competition with closely related D. g. mendotae.

An important difference between the Lake 223 experiment and the case of acidification of lakes by precipitation is that the acid added to Lake 223 did not interact with and leach metals from the watershed. Further years of data on declining pH in Lake 223 are required to evaluate the persistence of the increased primary production and standing algal biomass, lack of decline of algal species diversity, and enhanced crustacean herbivore abundance. Persistent differences at given pHs between the response of Lake 223 and of other Shield lakes exposed to acidic precipitation would imply that effects of acidic precipitation are due to acid plus metals rather than acid alone.

ACKNOWLEDGMENTS

Zooplankton were sampled by field crews under the supervision of E. Matheson, J. Penny, and D. Cruikshank. Zooplankton were identified and counted by I. Delbaere and W. Findlay. G. Mueller assisted in data analysis. Assistance in identification of zooplankton species was given by K. Patalas.

We thank D.W. Schindler, D.M. Rosenberg, and external reviewers for constructive reviews of the manuscript.

LITERATURE CITED

Allen, R.O. and E. Steinnes. 1980. Contribution From Long-Range Atmospheric Transport to the Heavy Metal Pollution of Surface Soil. In: D. Drabløs and A. Tollan, eds., "Ecological Impact of Acid Precipitation," pp. 102-103. SNSF Project, Oslo, Norway.

Almer, B., W. Dickson, C. Ekström, E. Hörnström, and U. Miller. 1974. Effects of Acidification on Swedish Lakes. Ambio 3:30-36.

Anderson, R.S. 1980. Relationships Between Trout and Invertebrate Species as Predators and the Structure of the Crustacean and Rotiferan Plankton in Mountain Lakes. In: W.C. Kerfoot, ed., "Evolution and Ecology of Zooplankton Communities," pp. 635-641. Amer. Soc. Limnol. Oceanogr. Spec. Symp., Vol. 3.

Beamish, R.J., L.M. Blouw, and G.A. McFarlane. 1976. A Fish and Chemical Study of 109 Lakes in the Experimental Lakes Area (ELA), Northwestern Ontario, With Appended Reports on Lake Whitefish Ageing Errors and Northwestern Ontario Baitfish Industry. Fish Mar. Serv. Res. Dev. Tech. Rep. 607. 116 pp.

Bowers, J.A. and N.E. Grossnickle. 1978. The Herbivorous Habits of Mysis relicta in Lake Michigan. Limnol. Oceanogr. 23:767-776.

Brooks, J.L. 1957. The Systematics of North American Daphnia. Memoirs Conn. Acad. Arts Sci. 13:1-180.

Brooks, J.L. 1969. Eutrophication and Changes in the Composition of the Zooplankton. In: "Eutrophication: Causes, Consequences, Correctives," pp. 236-255. National Academy of Sciences, Washington, D.C.

Brooks, J.L. and S.I. Dodson. 1965. Predation, Body Size, and Composition of Plankton. Science 150:28-35.

Carter, J.C.H. 1971. Distribution and Abundance of Planktonic Crustacea in Ponds Near Georgian Bay (Ontario, Canada) in Relation to Hydrography and Water Chemistry. Arch. Hydrobiol. 68:204-231.

Carter, J.C.H., M.J. Dadswell, J.C. Roff, and W.G. Sprules. 1980. Distribution and Zoogeography of Planktonic Crustaceans and Dipterans in Glaciated Eastern North America. Can. J. Zool. 58:1355-1387.

Chang, P.S.S., D.F. Malley, I.L. Delbaere, and G. Mueller. 1981 Species Composition and Seasonal Abundance of Zooplankton in Lake 223, Experimental Lakes Area, Northwestern Ontario: Before and During Acidification, 1974-1979. Can. Data Rep. Fish. Sci. 290:iv. + 42 pp.

Chrisman, T.L., R.L. Schulze, P.L. Brezonik, and S.A. Bloom. 1980. Acid Precipitation: The Biotic Response in Florida Lakes. In: D. Drabløs and A. Tollan, eds., "Ecological Impact of Acid Precipitation," pp. 296-297. SNSF Project, Oslo, Norway.

Cooper, S.D. and C.R. Goldman. 1980. Opossum Shrimp (Mysis relicta) Predation on Zooplankton. Can. J. Fish. Aquat. Sci. 37:909-919.

Cronan, C.S. and C.L. Schofield. 1979. Aluminum Leaching Response to Acid Precipitation: Effects on High-Elevation Watersheds in the Northeast. Science 204:304-306.

DeBruyn, E.R. and J.A. Shearer. 1981 Phytoplankton Primary Production, Chlorophyll and Suspended Carbon in the Experimental Lakes Area - 1980 Data. Can. Data Rep. Fish. Aquat. Sci. 260:iv. +52 pp.

Dickson, W. 1978. Some Effects of the Acidification of Swedish Lakes. <u>Verh. Int. Verein. Theor. Angew Limnol. 20</u>:851-856.

Dickson, W. 1980. Properties of Acidified Waters. <u>In</u>: D. Drabløs and A. Tollan, eds., "Ecological Impact of Acid Precipitation," pp. 75-83. SNSF Project, Oslo, Norway.

Dillon, P.J., D.S. Jeffries, W. Snyder, R. Reid, N.D. Yan, D. Evans, J. Moss, and W.A. Scheider. 1978. Acidic Precipitation in South-Central Ontario: Recent Observations. <u>J. Fish. Res. Bd. Can. 35</u>:809-815.

Drabløs, D. and A. Tollan (eds.). 1980. Ecological Impact of Acid Precipitation: Proceedings of an International Conference, Sandefjord, Oslo, Norway, March 11-14, 1980. SNSF Project: Acid Precipitation - Effects on Forest and Fish. (Address of SNSF Project: P.O. Box 61, 1432 Ås-NLH, Norway.)

Eriksson, M.O.G., L.Henrikson, K. Larsson, B-I. Nilsson, G. Nyman, H.G. Oscarson, and A.E. Stenson. 1980. Predator-prey Relations Important for the Biotic Changes in Acidified Lakes. <u>Ambio 9</u>:248-249.

Findlay, D.L. and G. Saesura. 1980. Effects on Phytoplankton Biomass, Succession and Composition in Lake 223 as a Result of Lowering pH levels from 7.0 to 5.6. Data from 1974 to 1979. <u>Can. MS Rep. Fish Aquat. Sci. 1585</u>. 16 pp.

Franzin, W.G., G.A. McFarlane, and A. Lutz. 1979. Atmospheric Fallout in the Vicinity of a Base Metal Smelter at Flin Flon, Manitoba, Canada. <u>Environ. Sci. Technol. 13</u>:1513-1522.

Goldman, C.R., M.D. Morgan, S.T. Threlkeld, and N. Angeli. 1979. A Population Dynamics Analysis of the Cladoceran Disappearance from Lake Tahoe, California-Nevada. <u>Limnol. Oceanogr. 24</u>:289-297.

Hall, D.J., W.E. Cooper, and E.E. Werner. 1970. An Experimental Approach to the Production Dynamics and Structure of Freshwater Animal Communities. Limnol. Oceanogr. 15:839-928.

Henriksen, A. 1980. Acidification of Freshwaters - A Large Scale Titration. <u>In</u>: D. Drabløs and A. Tollan, eds., "Ecological Impact of Acid Precipitation," pp. 68-74. SNSF Project, Oslo, Norway.

Henrikson, L., H.G. Oscarson, and J.A.E. Stenson. 1980. Does the Change of Predator System Contribute to the Biotic Development in Acidified Lakes? <u>In</u>: D. Drabløs and A. Tollan, eds., "Ecological Impact of Acid Precipitation," p. 316. SNSF Project, Oslo, Norway.

Hillbricht-Ilkowska, A. and T. Weglenska. 1970. Some Relations Between Production and Zooplankton Structure of Two Lakes of a Varying Trophy. Pol. Arch. Hydrobiol. 17:233-240.

Hoar, W.S. 1975. General and Comparative Physiology, 2nd ed. Prentice-Hall, Englewood Cliffs, New Jersey. 848 pp.

Hrbáček, J. 1969. Relations Between Some Environmental Parameters and the Fish Yield as a Basis for a Predictive Model. Verh. Int. Verein. Theor. Angew Limnol. 14:1069-1081.

Hrbáček, J., M. Dvořáková, V. Kořínek, and V . Procházková. 1961. Demonstration of the Effect of the Fishstock on the Species Composition of Zooplankton and the Intensity of Metabolism of the Whole Plankton Association. Verh. Int. Verein. Theor. Angew Limnol. 14:192-195.

Hutchinson, G.E. 1957. The Thermal Properties of Lakes. In: "A Treatise on Limnology," Vol. I: Geography, Physics, and Chemistry, pp. 426-540. John Wiley & Sons, Inc., New York, London.

Kerfoot, W.C. (ed.). 1980a. Evolution and Ecology of Zooplankton Communities. Amer. Soc. Limnol. Oceanog. Spec. Symp., Vol. 3, University Press of New England, Hanover, New Hampshire and London, England. 793 pp.

Kerfoot, W.C. 1980b. Perspectives on Cyclomorphosis: Separation of Phenotypes and Genotypes. In: W.C. Kerfoot, ed., "Evolution and Ecology of Zooplankton Communities," pp. 470-496. Amer. Soc. Limnol. Oceanogr. Spec. Symp., Vol. 3.

Kerfoot, W.C. and W.R. DeMott. 1980. Foundations for Evaluating Community Interactions: The Use of Enclosures to Investigate Coexistence of Daphnia and Bosmina. In: W.C. Kerfoot, ed., "Evolution and Ecology of Zooplankton Communities," pp. 725-741. Amer. Soc. Limnol. Oceanogr. Spec. Symp., Vol. 3

Kwiatkowski, R.E. and J.C. Roff. 1976. Effects of Acidity on the Phytoplankton and Primary Productivity of Selected Northern Ontario Lakes. Can. J. Bot. 54:2546-2561.

Lasenby, D.C. and R.R. Langford. 1972. Growth, Life History, and Respiration of Mysis relicta in an Arctic and Temperate Lake. J. Fish. Res. Bd. Can. 29:1701-1708.

Leivestad, H., G. Hendrey, I.P. Muniz, and E. Snekvik. 1976. Effects of Acid Precipitation on Freshwater Organisms. In: F.H. Braekke, ed., "Impact of Acid Precipitation on Forest and Freshwater Ecosystems in Norway," pp. 87-111. SNSF-Project, FR 6/76 NISK, 1432 Ås-NLH, Norway.

Lewis, W.M., Jr. 1979. Zooplankton Community Analysis. Springer-Verlag, New York, Heidelberg, Berlin. 163 pp.

Lynch, M. 1979. Predation, Competition, and Zooplankton Community Structure: An Experimental Study. Limnol. Oceanogr. 24:253-272.

Lynch, M. 1980. Aphanizomenon Blooms: Alternate Control and Cultivation by Daphnia pulex. In: W.C. Kerfoot, ed., "Evolution and Ecology of Zooplankton Communities," pp. 229-304. Amer. Soc. Limnol. Oceanogr. Spec. Symp., Vol. 3.

Malley, D.F. 1980. Decreased Survival and Calcium Uptake by the Crayfish Orconectes virilis in Low pH. Can. J. Fish. Aquat. Sci. 37:364-372.

Malley, D.F. and P.S.S. Chang. 1981. Response of Zooplankton in Precambrian Shield Lakes to Whole-lake Chemical Modifications Causing pH Change. In: "Restoration of Lakes and Inland Waters," pp. 108-114. EPA 440/5-81-010, U.S. Environmental Protection Agency.

Malley, D.F., S.G. Lawrence, I.L. Delbaere, P.S.S. Chang, and D.W. Schindler. Response of Zooplankton to the Experimental Acidification of Lake 223, Experimental Lakes Area, Northwestern Ontario, 1974-1980. Can. J. Fish. Aquat. Sci. (in manuscript).

McFarlane, G.A., W.G. Franzin, and A. Lutz. 1979. Chemical Analysis of Flin Flon Area Lake Waters and Precipitation: 1973-1977. Can. J. Fish. Mar. Serv. MS Rep. 1486. 42 pp.

McNaught, D.C. 1975. A Hypothesis to Explain the Succession from Calanoids to Cladocerans During Eutrophication. Verh. Int. Verein. Theor. Angew Limnol. 19:724-731.

Marshall, J.S. and D.L. Mellinger. 1980. Dynamics of Cadmium-Stressed Plankton Populations. Can. J. Fish. Aquat. Sci. 37:403-414.

Marshall, J.S., D.M. Nelson, D.L. Mellinger, C. Lei, S.G. Lawrence, and D.F. Malley. 1981a. Lake Water Chemistry and Zinc Toxicity to Plankton Communities--A Preliminary Report. Argonne National Laboratory, Radiological and Environmental Research Division Annual Report, Ecology. Jan.-Dec., 1980, ANL-80-115, Part III:64-67.

Marshall, J.S., J.I. Parker, D.L. Mellinger, and S.G. Lawrence. 1981b. An In Situ Study of Cadmium and Mercury Stress in the Plankton Community of Lake 382, Experimental Lakes Area, Northwestern Ontario. Can. J. Fish. Aquat. Sci. 38:1209-1214.

NRCC. 1981. Acidification in the Canadian Aquatic Environment. NRCC No. 18475, National Research Council of Canada, Ottawa. 369 pp.

Nero, R.W. 1981. The Decline of Mysis relicta Loven in Response to Experimental Acidification of a Whole Lake. M.Sc. Thesis, University of Manitoba, Winnipeg, Manitoba.

Nilssen, J.P. 1980a. Acidification of Freshwater and Limnetic Organisms: Complex Biotic Interactions or Statistics? In: D. Drablós and A. Tollan, eds., "Ecological Impact of Acid Precipitation," p. 344. SNSF Project, Oslo, Norway.

Nilssen, J.P. 1980b. Early Warning Signals of Acidification. In: D. Drablós and A. Tollan, eds., "Ecological Impact of Acid Precipitation," p. 344. SNSF Project, Oslo, Norway.

Patalas, K. 1971. Crustacean Plankton Communities in Forty-Five Lakes in Experimental Lakes Area, Northwestern Ontario. J. Fish. Res. Bd. Can. 28:231-244.

Patalas, K. 1972. Crustacean Plankton and the Eutrophication of St. Lawrence Great Lakes. J. Fish. Res. Bd. Can. 29:1451-1462.

Porter, K.G. 1977. The Plant-Animal Interface in Freshwater Ecosystems. Amer. Sci. 65:159-170.

Porter, K.G. and J.D. Orcutt, Jr. 1980. Nutritional Adequacy, Manageability, and Toxicity as Factors That Determine the Food Quality of Green and Blue-Green Algae for Daphnia. In: W.C. Kerfoot, ed., "Evolution and Ecology of Zooplankton Communities," pp. 268-281. Amer. Soc. Limnol. Oceanogr. Spec. Symp., Vol. 3.

Potts, W.T.W. and G. Fryer. 1979. The Effects of pH and Salt Content on Sodium Balance in Daphnia magna and Acantholeberis curvirostris (Crustacea:Cladocera). J. Comp. Physiol. B129:289-294.

Raddum, G.G., A. Hobaek, E.R. Lømsland, and T. Johnson. 1980. Phytoplankton and Zooplankton in Acidified Lakes in South Norway. In: D. Drabløs and A. Tollan, eds., "Ecological Impact of Acid Precipitation," pp. 332-333. SNSF Project, Oslo, Norway.

Richards, R.C., C.R. Goldman, T.C. Frantz, and R. Wickwire. 1975. Where Have All the Daphnia Gone? The Decline of a Major Cladoceran in Lake Tahoe, California-Nevada. Verh. Int. Verein. Theor. Angew Limnol. 19:835-842.

Roff, J.C. and R.E. Kwiatkowski. 1977. Zooplankton and Zoobenthos Communities of Selected Northern Ontario Lakes of Different Acidities. Can. J. Zool. 55:899-911.

Schindler, D.W. 1980. Experimental Acidification of a Whole Lake. A Test of the Oligotrophication Hypothesis. In: D. Drabløs and A. Tollan, eds., "Ecological Impact of Acid Precipitation," pp. 370-374. SNSF Project, Oslo, Norway.

Schindler, D.W., R.H. Hesslein, R. Wagemann, and W.S. Broecker. 1980a. Effects of Acidification on Mobilization of Heavy Metals and Radionuclides From the Sediments of a Freshwater Lake. Can. J. Fish. Aquat. Sci. 37:373-377.

Schindler, D.W. and M.A. Turner. Physical, Chemical, and Biological Responses of Lakes to Experimental Acidification. Water Air Soil Pollut.:(in press).

Schindler, D.W., R. Wagemann, R.B. Cook, T. Ruszczynski, and J. Prokopowich. 1980b. Experimental Acidification of Lake 223, Experimental Lakes Area: Background Data and the First Three Years of Acidification. Can. J. Fish. Aquat. Sci. 37:342-354.

Shaw, J. 1960. The Absorption of Sodium Ions by the Crayfish Astacus pallipes Lereboullet. III. The Effect of Other Cations in the External Solution. J. Exp. Biol. 37:548-556.

Smith, W.E. 1970. Tolerance of Mysis relicta to Thermal Shock and Light. Trans. Amer. Fish. Soc. 99:418-422.

Sprules, W.G. 1975. Midsummer Crustacean Zooplankton Communities in Acid-Stressed Lakes. J. Fish. Res. Bd. Can. 32:389-395.

Sprules, W.G.   1980.   Zoogeographic Patterns in the Size Structure of Zooplankton Communities, With Possible Applications to Lake Ecosystem Modeling and Management.   In: W.C. Kerfoot, ed., "Evolution and Ecology of Zooplankton Communities," pp. 642-656. Amer. Soc. Limnol. Oceanogr. Spec. Symp., Vol. 3.

Stockner, J.G. 1977. Lake Fertilization as a Means of Enhancing Sockeye Salmon Populations: The State of the Art in the Pacific Northwest. Fish. Mar. Serv. Tech. Rep. 740.   14 pp.

Threlkeld, S.T., J.T. Rybock, M.D. Morgan, C.L. Folt, and C.R. Goldman.   1980.   The Effects of an Introduced Invertebrate Predator and Food Resource Variation on Zooplankton Dynamics in an Ultraoligotrophic Lake.   In: W.C. Kerfoot, ed., "Evolution and Ecology of Zooplankton Communities," pp. 555-568. Amer. Soc. Limnol. Oceanogr. Spec. Symp., Vol. 3.

Walton, W.E., S. Compton, J.D. Allan, and R.E. Daniels.   The Effect of Acid Stress on Survivorship and Reproduction of *Daphnia pulex* (Crustacea:Cladocera). *Can. J. Zool.*:(in press).

Wong, C.K.   1981.   Predatory Feeding Behavior of *Epischura lacustris* (Copepoda, Calanoida) and Prey Defense. *Can. J. Fish. Aquat. Sci.* 38:275-279.

Wright, R.F., N. Conroy, W.T. Dickson, R. Harriman, A. Henriksen, and C.L. Schofield. 1980. Acidified Lake Districts of the World: A Comparison of Water Chemistry of Lakes in Southern Norway, Southern Sweden, Southwestern Scotland, the Adirondack Mountains of New York, and Southeastern Ontario. In: D. Drabløs and A. Tollan, eds., "Ecological Impact of Acid Precipitation," pp. 377-379. SNSF Project, Oslo, Norway.

Wright, R.F., T. Dale, E.T. Gjessing, G.R. Hendrey, A. Henriksen, M. Johannessen, and I.P. Muniz. 1976. Impact of Acid Precipitation on Freshwater Ecosystems in Norway. *Water Air Soil Pollut.* 6:483-499.

Yan, N.D. and P. Stokes.   1978.   Phytoplankton of an Acidic Lake, and Its Response to Experimental Alterations of pH. *Environ. Conserv.* 5:93-100.

Yan, N.D. and R. Strus.   1980.   Crustacean Zooplankton Communities of Acidic, Metal-Contaminated Lakes Near Sudbury, Ontario. *Can. J. Fish. Aquat. Sci.* 37:2282-2293.

Zaret, T.M. 1980. Predation and Freshwater Communities. Yale University Press, New Haven and London. 187 pp.

Zyblut, E.R.   1970.   Long-Term Changes in the Limnology and Macrozooplankton of a Large British Columbia Lake.   J. Fish. Res. Bd. Can. 27:1239-1250.

CHAPTER 15

EFFECTS OF ACIDIC PRECIPITATION ON BENTHOS

Robert Singer
Department of Biology
Colgate University
Hamilton, New York 13346

IMPORTANCE OF THE BENTHIC COMMUNITY

The community of organisms, the benthos, which inhabit aquatic sediments interact with biological and chemical components of the water column by processing detritus, recycling inorganic nutrients, mixing sediments, and serving as a principal food source for fish, waterfowl, and riparian mammals. Studies of the effects of acidic precipitation on this community have begun only in the past few years (Almer et al., 1974).

Benthic macroinvertebrates rapidly process debris and render it palatable to microorganisms. Macroinvertebrate "shredders" rip and chew leaves, vastly increasing the surface area of leaves, and partially digesting material as it passes through their guts. Traaen (1976, 1977) showed that leaf packs in lakes were processed much more slowly at lower pH (5.0) than at higher pH (6.0) values, but he also cautioned (1977) that many other factors besides acidity can affect leaf processing. Friberg et al. (1980) reported on increased accumulation of detritus and a reduction in numbers of scrapers in an acidic stream (pH 4.3-5.9) as compared to a neutral stream (pH 6.5-7.3). Hall et al. (1980) and Hall and Likens (1980a, 1980b) artificially acidified a stream in Hubbard Brook, New Hampshire, and also showed that scrapers were largely lost; additionally, they reported that collectors were inhibited.

After the macroinvertebrates have broken up the detritus, fungi, bacteria, and protozoa complete the digestion and restore the material as inorganic nutrients. The pH of the water column controls the solubility equilibria of these constituents and largely determines whether they will be available for recycling by plants; but the rate of decay depends on the efficiency of this microbial community, which is also pH-dependent (Gahnstrom et al., 1980; Laake, 1976). This topic will be discussed at length in the next section.

Macroinvertebrates act to mix sediments by their burrowing movements. The top few centimeters of sediments generally demonstrate large gradients of pH, Eh, dissolved oxygen, and other constituents (Hutchinson, 1957). Losses or alterations of plant and animal communities have profound effects on the chemistry of this top layer of sediments (Mortimer's, 1941, 1942, "oxidized microzone"), yet little work has centered on this habitat in acidified lakes. Mitchell et al. (1981) found that the presence of burrowing mayflies (Hexagenia) affected sulfur dynamics in sediment cores taken from acidic lakes. This whole topic of sediment/water column biological and chemical interactions is difficult to study because the events proceed across strong chemical gradients over short distances (Mitchell et al., 1981) which are easily perturbed by experimental procedures, including even in situ measurements.

Benthic animals are at the base of most food chains which lead to game fish. The elimination of the amphipod Gammarus lacustris and most molluscs below pH 6.0 might reduce trout production in acidified streams by 10 to 30 percent (J. Økland and K.A. Økland, 1980), but this prediction has not been verified. Rosseland et al. (1980) reported that trout in acidified waters shifted their diet from acid-sensitive invertebrates like mayflies and bivalves, to acid-tolerant forms such as corixids and beetles. Although loss of fish populations due to alterations of the benthic community has not been studied, sub-lethal stress on fish due to nutrient changes should be considered.

Finally, changes in the benthic plant community affect macroinvertebrate distribution. The littoral habitat is an important area of benthos, and alterations in plant community structures are likely to affect all other trophic levels. These interactions remain to be investigated.

MICROBIAL COMMUNITY

Studies of the effects of acidification on protozoans have not been conducted. Other members of this community include bacteria and fungi. In general, acidification of lakes causes bacterial decomposers to be replaced by fungi (Hendrey and Barvenik, 1978; Hendrey et al., 1976). It has been proposed (Grahn, 1976; 1977; Hultberg and Grahn, 1976) that the shift to fungi accounts for the observed (Leivestad et al., 1976) accumulation of detritus in acidic lakes. Liming of lakes to increase the pH brings a rapid restoration of normal microbial activity (Gahnstrom et al., 1980; Scheider et al., 1975, 1976).

Hall et al. (1980) observed a growth of basidiomycete fungus on birch leaves in an artificially acidified portion of a stream which was lacking in the non-acidified reference section. Hultberg and Grahn (1976) and Grahn et al. (1974) described an accumulation of a "fungal mat" on the bottom of many acidified Scandinavian lakes, but it is understood now that this coarse particulate material is a mixture of detritus, some fungi, and a blue-green alga, possibly Phormidium sp. (Hendrey, personal communication).

There is some controversy regarding the effects of microbial metabolism brought about by acidification (Baath et al., 1979). The accumulation of detritus indicated a lowering of decomposition rates by bacteria (Leivestad, 1976). The reduction of oxygen utilization by acidified cores (Harvey et al., 1976) supported this view. Furthermore, liming increased oxygen consumption of previously acidic cores (Gahnstrom et al., 1980) and at pH levels below 5, oxygen consumption, ammonia oxidation, peptone decomposition, and total bacterial numbers all declined (Bick and Drews, 1973). However, Schindler (1980) reported no change in decomposition rates in an artificially acidified lake and Traaen (1978) observed no clear changes in the planktonic bacterial populations from 7 lakes of pH < 5 as compared to 7 lakes of pH > 5. Traaen argued that acidic inputs should affect the plankton populations prior to benthic algae. His results showed that organic inputs and temporal and spatial (depth) patchiness were more important factors than pH. Gahnstrom et al. (1980) reported that inhibition of oxygen uptake by sediments increased in acidic lakes as compared to reference lakes only in the littoral sediments. They argued that the inhibition of microbial activity in the littoral zone might be due to the inflow of acidic runoff, which is restricted to the epilimnion (Hendrey et al., 1980) during snowmelt and autumn rains.

Microbial effects on other systems probably involve alterations of sulfur and nitrogen dynamics. Methylation of mercury (Tomlinson, 1978) and other heavy metals may have profound effects on higher trophic levels (Galloway and Likens, 1979).

EFFECTS ON MACROPHYTES

Acidification brings about a simplification and reduction of the macrophytic community by an elimination of many angiosperms and the development of vast beds of the moss, Sphagnum

(Grahn, 1976, 1977; Grahn et al., 1974). With a drop in pH
from 6.0 to about 4.5, Sphagnum replaced the typical Isoetes-
Lobelia community over 50 percent of the substrate in 6 Swedish
lakes (Grahn et al., 1974). In addition, Grahn (1976, 1977)
and Hultberg and Grahn (1976) discussed the possible depletion
of nutrients by adsorption onto Sphagnum.

The dominance of submersed mats of Sphagnum in acidified
lakes is highly significant because:

(1)   Sphagnum actively transports calcium ions from the
      water by exchanging 2 hydrogen ions for each calcium
      ion, thus adding to the acidity and further reducing
      the buffering capacity, and
(2)   Sphagnum is a poor substrate for most epiphytes.

Two notable exceptions are the green alga Mougeotia and the red
alga Batrachospermum, both of which form long filaments on
Sphagnum (Grahn et al., 1974). The increased light penetration
brought about by acid-induced oligotrophication probably
enhances the growth of submersed Sphagnum in depths as great as
6 m.

Sphagnum does not just out-compete the typical softwater
flora like Lobelia. It has been shown that primary productivity
was inhibited in L. dortmanna when plants were incubated at low
pH (Leivestad et al., 1976). These authors also speculated
that one cause of the upset of the microbial community discussed
above may be the loss of oxygen transport to the sediments by
the lacunal system of angiosperms when the moss, Sphagnum,
becomes dominant.

Nilssen (1980) extensively studied 15 lakes of various
acidities in a single watershed in Norway and reported that the
most acidic lakes (pH 4.2-5.2) had replaced the typical Lobelia
flora with a floating growth form of Juncus bulbosus. Once
again, the dominant epiphyte was Mougeotia. These data are in
variance with several reports from North America (Hendrey and
Barenik, 1978).

A macrophyte survey was performed at an acidic (pH 4.8-
5.0) Adirondack lake and compared with a similar survey which
was done in 1932 (Hendrey and Vertucci, 1980). In the recent
survey, the littoral zone was reported to be covered with a mat
of Sphagnum pylaesii. This moss was absent in 1932. At that
time, Eleocharis acicularis, Nuphar luteum, and Lobelia dortmanna were much more common. Eriocaulon septangulare and
Utricularia sp., however, had increased in abundance since the
earlier survey.

Hellquist (1980) recently reported the distribution of 31 species and 38 varieties of the common pondweed Potamogeton with respect to alkalinities at 321 sites in New England. Within this one genus, there is considerable specialization and adaptation in response to alkalinity. Hellquist recognized 6 statistically significant groups of Potamogeton, from P. confervoides, which coexists in peaty ponds with Sphagnum with almost undetectable alkalinities, to P. pectinatus, which is found only at hardwater sites with bicarbonate alkalinities greater than 110 mg/l. The physiological mechanisms by which species of this one genus have adapted to such different chemical environments are totally unknown, and deserve further investigation. The hazards of generalizing to a whole genus or even higher taxon from presence-absence data based on one species are obvious.

Gorham and Gordon (1963) speculated that it was the release of metal ions, not the hydrogen ion concentration, which was toxic to macrophytes. In an analysis of waters around Sudbury, Ontario, they found that macrophytes were impacted by metal ions even in waters which were well buffered.

Moore and Clarkson (1967) studied the macrophytes from the acidic Monongahela River system, West Virginia. Although these waters were acidic because of mine drainage, careful statistical analysis permitted the isolation of pH as a factor. The total lack of other studies on the response of lotic macrophyte communities to acid precipitation justifies the inclusion of these data here. They found the following taxa in order of abundance: Eleocharis acicularis, Sagittaria graminea, S. latifolia, Potamogeton epihydrus, P. nodosus, Sparganium americanum, and Myriophyllum heterophyllum.

Different groups of plants may be vulnerable by various mechanisms during their life cycle. The motile spermatozoids of terrestrial ferns, for instance, were shown to be more sensitive to acidification than sporophytic and gametophytic tissue (Evans and Conway, 1980). This phenomenon remains to be investigated in aquatic ferns.

In a recent survey of 9 remote, small (10-30 ha), Adirondack lakes of different acidities, it was observed that species richness decreased at low pH (C.W. Boylen, D.A. Roberts, R. Singer, in preparation). The acidic lakes (pH < 5.0) had a simple macrophyte community consisting of Utricularia sp., U. resupinata, Eriocaulon septangulare, Myriophyllum tenellum, Eleocharis sp., Isoetes sp., Sparganium, Lobelia dortmanna, Juncus pelocarpus, Nuphur sp., Nymphaea sp., Brasenia schreberi, and in most cases, extensive beds of Sphagnum sp. The 4 lakes with pH >6.0 had far more extensive lists of species.

In the acidic lakes, an algal mat grew over the submersed plants, except for <u>Utricularia</u> sp., which floated above the mat, and the <u>Sphagnum</u> sp. beds, which seemed to inhibit the algal growth. Another unpublished survey confirmed many of the these findings (M. Best, personal communication).

PERIPHYTON

The community of algae which lives upon macrophytes and directly on sediments makes important contributions to primary production and to nutrient cycling, particularly in lotic systems. The species composition of this community reflects changes in the chemistry of both the water column and the sediments. These algae also serve as an important source for the grazing macroinvertebrates which are the principal source of food for fish.

The pH tolerance of many species of algae, particularly diatoms, has been tabulated elsewhere (e.g., Lowe, 1974). Acid lakes develop communities dominated by species which have been determined elsewhere to be acidophilous, but dramatic decreases in species diversity below pH 5.5 have been observed (Almer <u>et al</u>., 1974). Acid rain, with its load of nitrates and sulfates, might tend to increase the growth rate of algae. Berge (1976) compared the diatom assemblages in sediments from 7 Norwegian sites with the communities from the same sites as reported in 1949 and found no quantitative change in the diatoms in the 26-yr period; however, there was a marked shift towards species which required or preferred low pH. During an even longer period (ca. 1920-1978), Dam <u>et al</u>. (1980) reported a more obvious shift towards acidophilic diatoms in sediments from acidic Swedish lakes. Three hundred years of diatom deposition in sediments were used to calculate pH values in two Norwegian lakes (Davis and Berge, 1980), and the results fit well with inferred pH from other methods. These studies of diatoms from sediment cores represent a mixture of both planktonic and benthic algae.

Del Prete and Schofield (1981) used sediment cores to study the succession of diatom species in three Adirondack lakes. They observed an increase in dominance by acidobiontic and acidophilous species in the most acid-impacted lakes. A trend towards oligotrophication was also reported. <u>Tabellaria fenestrata</u> and <u>Cyclotella stelligera</u> increased in numbers most directly with increasing acidity, although some of the results were equivocal.

Coesel <u>et al</u>. (1978) have compared the desmid populations from a group of lakes in the Netherlands with community composition reported in studies done in 1916 to 1925, 1950 to 1955, and with their own survey in 1977. Many of the species from the rich flora in the earliest survey were lost due to cultural eutrophication. In the most recent survey, those ponds which were not impacted by nutrient additions were affected by acidic precipitation, as reflected by the paucity of desmid species. These ponds appeared to have undergone oligotrophication. The eutrophied ponds remained well-buffered and unchanged.

Müller (1980) studied the succession of periphyton in artificially acidified chambers held in situ in Lake 223 of the Experimental Lakes area in northwestern Ontario (Schindler et al., 1980). At the control pH of 6.25, there was a succession from dominance by diatoms in the spring to dominance by Chlorophyta in mid-July. At pH < 6, Chlorophyta dominated the periphyton throughout the sampling period. The abundance of Cyanophyta was reduced and almost eliminated under the most acidic conditions. This inhibition of blue-green periphyton was verified in a study of an extremely acidic river (pH < 3.0) in Japan (Kamijo et al., 1974). Müller observed no trend with respect to changes in biomass, but there was a sharp decrease in Hill's species diversity in the acidified (pH 4.0) samples. Primary production (^{14}C) showed no clear trend. The dominance of the periphyton by Chlorophyta in the acidified samples was due almost entirely to the growth of Mougeotia sp., which represented 96 percent of the biomass and cell numbers at pH 4.0 by June. This taxon was responsible for less than 4 percent of the biomass and cell numbers in the natural lake water. Interestingly, in May the blue-green alga, Anabaena sp., rose in numbers from 3.4 percent of the biomass in the lake water (pH 6.2) to 4.3 percent at pH 4.0, but Anabaena sp. was almost absent in June. Due to the small size of this alga, it accounted for 25 and 41 percent of the total cell numbers in these two samples in spite of the low biomass which was present.

Higher standing crops of periphyton but lower rates of carbon fixation per unit chlorophyll occurred in periphyton growing in artificial stream channels at reduced pH. The total rate of ^{14}C-uptake was similar over a wide range of the hydrogen ion concentration. Increased standing crop was attributed to a combination of three mechanisms:

(1) enhanced growth by acidophilic taxa,
(2) reduction in grazing by the reduced macroinvertebrate population, and
(3) the inhibition of microbial decomposition (Hendrey, 1976).

In an artificially acidified section of a softwater stream in New Hampshire, Hall et al. (1980) reported an increase in periphyton numbers and substrate chlorophyll a concentration. They did not perform a taxonomic analysis of the periphyton community.

MICROINVERTEBRATES

The responses of several minor groups of invertebrates to acidification have been studied. The Nematoda and Gastrotricha are both common, but poorly studied inhabitants of interstitial water in sediments (meiofauna). The ubiquitous meiobenthic gastrotrich, Lepidodermella squammata, was almost totally eliminated under laboratory conditions below pH 6.4 (Faucon and Hummon, 1976). Unfortunately, the pH gradient was achieved by mixing unpolluted creek water with water from a stream receiving acidic strip mine drainage, so it is not easy to generalize on streams receiving acidic precipitation. Hummon and Hummon (1979) added calcium carbonate to the acidic mine drainage and showed that at the same pH, water with more carbonate ameliorated the deleterious effects of acid stress. The extreme sensitivity of these animals to some component of the acid water, possibly low carbonate or high concentrations of metal ions, bears further investigation. Roundworms (Nematoda) normally have a ubiquitous distribution (Ferris et al., 1976); but in an extensive survey of Norwegian lakes, the sub-littoral sediments of acidic lakes had a scarcity of roundworms compared to shallow sediments from the same lakes (Raddum, 1976). No other mention is made of the Nematoda in the literature pertaining to acidic deposition to aquatic systems.

The freshwater Porifera are epifaunal and directly subject to water column alterations but have not been studied with respect to their responses to acid precipitation. Jewell (1939), however, studied the distribution of Spongillidae from 63 lakes, bogs, and rivers in Wisconsin with various levels of hardness and pH. She found that most of the species did have limited ranges of calcium ion concentrations in which they flourished. The distributions correlated better with calcium ion concentration than magnesium ion concentration. Six common species were exposed to chemically modified water and growth was observed. The lowest pH in this experiment was 5.9, but there were indications that the most important parameter was the availability of calcium bicarbonate.

Aquatic mites (Acarina) are not generally collected in surveys of benthic fauna, but Raddum (1976) noted that mites occurred in great abundance in the shallow water of an acid-impacted lake. At a depth of 0.5 m, mites were third in abundance after nematodes and chironomids. At depths > 2.0 m, there were almost no mites collected. The shallow mites probably receive their nutrition from the shore or the surface of the water rather than substrate. In constrast, Wiederholm and Eriksson (1977) observed mites in deep water (> 10 m) in an acid lake in Sweden, and Collins et al. (1981) reported no differences between the distribution of mites in acidic and control lakes. Clearly, much work needs to be performed on the distribution of this group.

CRUSTACEA

Benthic crustaceans include a diversity of forms. The distribution and chemistry of habitats for the isopod Asellus aquaticus and the amphipod Gammarus lacustris were summarized from 275 sites for A. aquaticus and 1097 sites for G.lacustris in Norway (K.A. Økland, 1979a, 1980a). A. aquaticus populations were reduced at sites below pH 5.2 and absent below a pH of 4.8. Competition between G. lacustris and A. aquaticus at pH ca 7.0 favored Gammarus. Asellus out-competed Gammarus only at those sites that were stressed with either acid or organic enrichment. A. aquaticus was widely distributed in acid-stressed lakes at pH 5 (K.A. Økland, 1980b), but G. lacustris was inhibited at pH 6.0 (K.A. Økland, 1980c).

Laboratory work with Gammarus pulex (Costa, 1967) demon- strated that this amphipod swam through pH gradients between pH 6.4 to 9.6. Within 12 to 15 minutes after the pH was lowered to 6.2 in one part of the tank, the amphipods began to stay near the alkaline side. Immature Gammarus performed this avoidance behavior faster than adults.

Sutcliffe and Carrick (1973) verified that G. pulex is not normally found below pH 6.0, but they pointed out that it was found in France at pH 4.5 to 6.0. They suggested that the avoidance response (Costa, 1967) might explain its limitation to near neutral water, instead of direct mortality due to low pH. Laboratory studies (Borgström and Hendrey, 1976) suggest, however, that direct mortality is important at pH ≤ 5.0. K.A. Økland (1980a) ascribed these differences to the variable sensitivity of different populations.

Steigen and Raddum (1981) noted that A. aquaticus responded to acidification by leaving the water, so they confined some of the animals in wire-enclosed tubes. The confined individuals resorted to cannibalism, but the increased energetic demands caused by hydrogen ion stress resulted in losses of total caloric values; the unconfined specimens left the water, but returned to feed, sometimes cannibalistically, and the survivors gained in caloric content. This behavioral response may be the mechanism by which Asellus can tolerate higher acidities than Gammarus.

Eggs of the tadpole shrimp, Lepidurus arcticus (Eubranchiopoda; Notostraca) took longer to hatch and the larvae matured more slowly than normal at pH < 5.5 than at pH values > 5.5 (Borgstrom and Hendrey, 1976). At pH< 4.5, eggs of L. arcticus never hatched and larvae died in two days. A survey from Sweden (Borgstrom et al., 1976) reported the L. arcticus was not found below pH 6.1. An increased abundance of benthic cladocerans has been reported (Collins et al., 1981) from 2 of 3 acidic lakes studied in Ontario.

There is some variation in sensitivity to pH between species of crayfish. Malley (1980) indicated that Orconectes virilis, in softwater of ca 22 μmhos/cm conductivity and calcium ion concentration of 2.8 mg/l, was stressed by pH<5.5. However, Cambarus sp. was reported (Warner, 1971) in a stream receiving acidic mine drainage at pH 4.6, calcium ion concentration of 12 mg/l, and conductivity of 96 μmhos/cm. Cambarus bartoni was found (Collins et al., 1981) in 3 acidic lakes (pH 4.6-4.9, ca 3 mg/l Ca^{2+}) and Orconectes propinquis was collected at one of 3 acidic lakes.

This apparent discrepancy in pH tolerances of various crayfish may not be due entirely to interspecific or inter-population differences. The crayfish, Orconectes virilis, has difficulty recalcifying its exoskeleton after molting at pH < 5.5. Uptake of $^{45}Ca^{2+}$ by crayfish stopped at pH 4.0 and was inhibited at pH 5.7 (Malley, 1980). Hence, the tolerance of Cambarus to pH 4.6 from an acidic mine drainage stream may be due to the higher calcium ion concentration in the stream as compared to habitats affected by acidic precipitation. The ameliorative effect of cations is suggested by the inability of the crayfish, Astacus pallipes, to transport $^{22}Na^+$ below pH 5.5 (Shaw, 1960). Stress is a function of both low pH levels and low calcium levels.

INSECTA

Studies of benthic insects undergoing acid stress include surveys, mostly from Europe and Canada, and some experimental manipulations. Survey work involves presence-absence data from which tolerances have been assumed. The general conclusion drawn from many surveys of lakes and streams (Friberg et al., 1980; Overrein et al., 1980; Raddum, 1979; Wiederholm and Eriksson, 1977; Conroy et al., 1975; Hendrey and Wright, 1976; Leivestad et al., 1976; Wright et al., 1976, 1976; Sutcliffe and Carrick, 1973) is that species richness, diversity, and biomass are reduced with increasing acidity. Since predation by fish is eliminated in some watersheds, and food should be abundant due to the accumulation of detritus (Grahn et al., 1974), one might suppose that insect biomass would increase. However, acid places stresses on the insects which are as severe as predation (Henrikson et al., 1980) and the lack of bacterial decomposition of detritus (Traaen, 1976, 1977) may render the detritus unpalatable to insects (Hendrey, 1976; Hendrey et al., 1976).

The sensitivity of benthic insects to pH stress varies considerably between different taxa and between different life cycle stages (Raddum and Steigen, 1981; Gaufin, 1973). Female mayfly adults (<u>Baetis</u>) did not lay eggs on otherwise suitable substrates in water pH < 6.0, although three different species were found within 200 to 300 m in neutral brooks with similar substrates (Sutcliffe and Carrick, 1973). The adult presumably can detect levels of acid by the dipping motion of her abdomen into the water as she flies. Besides <u>Baetis</u>, the common mayflies, <u>Ephemerella ignita</u> and <u>Heptagenia lateralis</u>, were absent only from the acidic region of the River Duddon, England (Sutcliffe and Carrick, 1973). Mayflies were found to be sensitive to pH stress from a Swedish survey, too (Nilssen, 1980). A plot of number of mayfly species <u>vs</u> pH of 35 lakes and 24 rivers indicted that the number of mayfly species decreased logarithmically with decreasing pH. Species were lost in two groups; one group did not appear below pH 6.5 and another decline occurred below pH 4.5 (Borgstrom et al., 1976; Leivestad et al., 1976). In another survey (Fiance, 1978), the distributional pattern of the mayfly, <u>Ephemerella funeralis</u>, was studied in the Hubbard Brook, New Hampshire, watershed during a two-year period. Nymphs were absent from waters of pH <5.5. The 2-yr life cycle of this mayfly makes it particularly sensitive to irregular episodic stresses.

In an experimentally acidified section of stream (pH 4.0), mayfly (<u>Epeorus</u>) emergence was inhibited but drift of nymphs increased (Pratt and Hall, 1981; Hall et al., 1980; Hall and Likens, 1980a, 1980b). These responses suggest that mayflies make both behavioral and physiological responses to acidity. Laboratory bioassays verified (Bell, 1971; Bell and Nebeker, 1969) that mayflies were the most sensitive insect order to acid (Harriman and Morrison, 1980)(Table 1). Experimental exposure of transplanted insects in cages to acidified river water also showed (Raddum, 1979) that mayflies could not survive and try to leave in the drift.

In contrast, dragonflies and damselflies (Table 1) are much more resistant to pH stress (Borgstrom et al., 1976; Bell, 1971; Bell and Nebeker, 1969). The dragonfly nymph, <u>Libellula</u>

Table 1

Results of Laboratory Studies on pH Tolerance of
Selected Insect Nymphs - The pH Which Was Lethal to 50
Percent of the Organisms (TL_{50}) was Determined for
Exposures of 96 Hr (Bell and Nebeker, 1969), 30
Days, and For Emergence of Adults (Bell, 1971)

	pH		
Insect	96-hr TL_{50}	30-day TL_{50}	50 Percent Successful Emergence
Trichoptera			
Brachycentrus americanus	1.50	2.45	4.0
Hydropsyche betteni	3.15	3.38	4.7
Plecoptera			
Taeniopteryx maura	3.25	3.71	4.0
Acroneuria lycorias	3.32	3.85	5.0
Isogenus frontalis	3.68	4.50	6.6
Pteronarcys dorsata	4.25	5.00	5.8
Ephemeroptera			
Stenonema rubrum	3.32	--	--
Ephemerella subvaria	4.65	5.38	5.9
Odonata			
Boyeria vinosa	3.25	4.42	5.2
Ophiogomphus rupinsulensis	3.50	4.30	5.2

pulchella, tolerated pH 1.0 for several hours (Stickney, 1922).
Dragonfly nymphs (Anisoptera) may be able to endure episodic
acid stress by closing their anus, through which they respire,
but this behavior has not been investigated.

Tolerance to acidification within the Plecoptera
(stoneflies) is variable according to surveys (Leivestad et
al., 1976; Sutcliffe and Carrick, 1973), field manipulations
(Hall and Likens, 1980a, 1980b; Raddum, 1979), and laboratory
studies (Bell, 1971; Bell and Nebeker, 1969). Lethal
sensitivity of this group varies between pH 4.5 to 5.5, but
their distribution generally follows that of mayflies, except

for some tolerant forms like Taeniopteryx, Nemoura, Nemurella, and Protonemura (Raddum, 1979).

Caddisflies (Trichoptera) have been found at ca pH 4.5 in field surveys (Leivestad et al., 1976; Raddum, 1976; Sutcliffe and Carrick, 1973) but not at pH 4.0 (Hall and Likens, 1980a, 1980b; Raddum, 1979). Raddum (1979) observed that the caddisflies Rhyacophila nubila, Hydropsyche sp., Polycentropus flavomaculatus, and Plectrocnemia conspersa all survived pH 4.0 in the laboratory, but only P. conspersa did well in situ at pH 4.8. Raddum explained the loss of Rhyacophila and Hydropsyche in the field by alterations in their food supply. P. flavomaculatus became cannibalistic at pH 4.0, which may account for its absence in the stream but its survival in isolation in vitro.

Most other insects are largely unaffected, or slightly favored, in acidic lakes and streams. The alderfly, Sialis (Megaloptera), increased its emergence rates in an artificially acidified stream (Hall and Likens, 1980a, 1980b) and was found commmonly in shallow water in an acidic (pH 3.9-4.6) Swedish lake (Wiederholm and Eriksson, 1977) and in a highly variable (pH 6.2-4.2) Norwegian lake (Hagen and Langeland, 1973).

Several true flies (Diptera) increase in relative abundance (Collins et al., 1981; Raddum and Saether, 1981; Raddum, 1979; Wiederholm and Eriksson, 1977; Hagen and Langeland, 1973). The most successful dipterans are the midges (Chironomidae), the predacious phantom midge (Chaoborus, Chaoboridae), and in streams, the black fly (Simulidae). Often the principal insects in acidic lakes are the midges (Chronomidae) Procladius sp., Limnochironomous sp., Sergentia coracina, Stichtochironomus sp., and phantom midges (Chaoborus)(Leivestad et al., 1976) which comprised 56 and 41 percent of the benthos of a Swedish acid lake (pH 3.9-4.6)(Wiederholm and Eriksson, 1977). Chironomids appear to be preadapted for acidification, as the same species are found in clean water acid lakes as were found in humic acid lakes (Raddum and Saether, 1981). Uutala (1981) reported that the chironomid fauna of two acidic Adirondack lakes was reduced in biomass as compared to nearby control lakes. The different life cycle stages have variable responses to pH stress, but the molting period is the most sensitive (Bell, 1970). Other insects which are abundant in acidic waters are the true bugs (Hemiptera) and beetles (Coleoptera) of the families Dytiscidae and Gyrinidae (Nilssen, 1980; Raddum et al., 1979; Raddum, 1976).

Important generalizations are better made by analyzing the data after grouping the taxa by functional guilds (Merritt and Cummins, 1978) rather than by phylogenetic associations.

Collins et al. (1980) compared 3 acidic softwater lakes (pH 4.6-4.9) with 11 neutral softwater lakes in central Ontario and reported no significant decreases in populations of animals living in sediments (infauna). Observations of epifauna by scuba divers concurred with the general observation that acidic lakes have depauperate mollusc and insect populations.

It is hardly surprising that infaunal communities, which are protected by the buffering capacity of the substrate, are less stressed than epifauna. Yet few studies have organized data in such a manner as to verify that epifaunal insects are indeed the targets of acid stress. Also, a perusal of the data presented above suggests that it is those epifaunal forms with filamentous gills which are most sensitive. Air-breathing beetles and bugs do well, as do infaunal forms with filamentous gills such as the burrowing mayfly, Hexagenia. Actions of Hexagenia nymphs increased the Eh, ammonia, inorganic sulfur, and sulfate levels and decreased the pH as compared to control microcosms lacking nymphs or with dead nymphs (Mitchell et al., 1981).

Organizing species lists into guilds based on eating methods showed that total invertebrate biomass in an acidic (pH 4.3-5.9) Swedish stream was ca 2.6 times less than that of a neutral stream (pH 6.5-7.3) 6 km away in southern Sweden (Friberg et al., 1980), and that shredders increased in relative abundance at the expense of scrapers in the acidic water. These data differ from those reported by Hall and Likens (1980a, 1980b) from an artifically acidified stream in New Hampshire, where shredders and predators were not affected. The tolerance of predators, mostly predacious diving beetles (Dytiscidae), whirligig beetles (Gyrinidae), water striders (Gerridae), and water boatmen (Corixidae), is corroborated from the surveys (Raddum et al., 1979; Leivestad et al., 1976).

It is likely that other factors besides hydrogen ion concentration are the actual causal mechanisms which induce stress (Overrein et al., 1980). Malley's (1980) work suggests that reduced calcium deposition may be inhibiting for insects as well as crustaceans. Havas (1981) suggested that sodium ion transport may be affected. Effects of increased aluminum concentrations have not been studied on invertebrates as they have with fish (Baker and Schofield, 1980). Other metals, such as mercury (Tomlinson, 1978), may also be important factors. Nutrient depletion, inefficient microbial digestion, substrate alteration, dissolved oxygen stress, and changes in other populations (e.g., fish predation) all may act on insect populations. Different taxa respond in various ways. Some may make behavioral adaptations, others, like the water boatmen (Corixidae), can alter rates of sodium ion pumping (Vangenechten and Vanderborght, 1981; Vangenechten et al., 1979). For reasons which are not clear, there is a shift towards species within a higher taxon which are larger in size (Raddum, 1980). This may be due to reduced predation pressure on larger insects in the absence of fish or because larger species have less surface/volume and can cope better with chemical and osmotic stress. The increase in the abundance of insect predators may be due to the opening of this niche due to the loss of fish (Henrikson et al., 1980) or due to the larger size of most predators, as discussed above.

Alterations in insect populations are likely to affect fish populations (J. Økland and K.A. Økland, 1980). Rosseland et al. (1980) reported that corixids composed 15 percent of the gut content volume of trout from neutral waters but 44 percent in a declining population, but no causal relationship between shifts in diet and population decline can be made at this time.

MOLLUSCA

The impact of acidic precipitation on molluscan populations is dramatic. The calcareous shell of these animals is highly soluble at pH < 7.0 and requires the animals to precipitate fresh calcium carbonate faster than it can dissolve. The only thorough survey of clams and snails in acid-impacted waters was done in Norway (K.A. Økland and Kuiper, 1980; K.A. Økland, 1980b, 1979b, 1979c, 1971; J. Økland and K.A. Økland, 1980, 1978; J. Økland, 1980, 1979, 1976, 1969a, 1969b). Surveys of about 1500 localities, mostly lakes in Norway, were performed between 1953 and 1973. Fingernail clams (Sphaeriidae) and snails (Gastropoda) were recorded. Sphaeriidae are infaunal and no surveys of the more epifaunal unionid mussels have been conducted. Norway has 17 species of Pisidium and 3 of Sphaerium. None of these clams occurred below pH 5.0. The 6 most common sphaeriids were eliminated below pH 6.0. These common species were found in lakes with low alkalinities but with pH values ca 6.0, so their absence from these poorly buffered lakes serves as an indication of acidification, not just calcium carbonate stress (K.A. Økland and Kuiper, 1980).

Freshwater gastropods were reported to be stressed much like the clams from the Norwegian survey. There were 27 species of snails reported in Norway. Of these, only 5 were found below pH 6.0 (J. Økland, 1980). Snails could tolerate higher hydrogen ion concentrations if the total hardness were higher, indicating that pH may stress snails by reducing the calcium carbonate availability (J. Økland, 1979). These authors (J. Økland and K.A. Økland, 1980) estimated that the crustacean Gammarus lacustris and the molluscs accounted for 45 percent of the caloric input of trout and predicted that trout production should be reduced 10 to 30 percent below pH 6.0 due to the loss of food resources.

Some additional distributional data, which corroborates the conclusions of the Økland's cited above, has been recorded from Sweden (Wiederholm and Eriksson, 1977), Norway (Nilssen, 1980; Hagen and Langeland, 1973), and from a river in England (Sutcliffe and Carrick, 1973). These last authors emphasized the absence of the freshwater limpet Ancylus fluviatilis as an indicator of pH levels below 5.7. They also concluded that pH served to limit the distribution of molluscs by reducing the availability of calcium carbonate as measured by hardness.

A physiological response of molluscs to pH stress was studied by Singer (1981). In Anodonta grandis (Unionidae) from 6 lakes in New York and Ontario with various levels of pH and hardness, marked differences in shell morphometry and ultrastructure were observed. The clams from alkaline lakes had thick shells with fine layers of organic conchiolin interspersed. The clams from softwater neutral lakes had thinner shells, with relatively thick prismatic layers, and the clams from a slightly acidic lake (pH 6.6) had thin shells with heavy plates of organic material substituting for the normal calcium carbonate matrix. The use of unionid shells from museum collections as indicators of pre-acidification water quality was suggested.

ANNELIDS

Aquatic worms have been used extensively as indicators of organic (Brinkhurst, 1974; Goodnight, 1973) and toxic pollution (Hart and Fuller, 1974). With an increase in organic detritus and a decrease in oxygen concentrations, the benthic community is typically dominated by Tubifex spp. and Limnodrilus hoffmeisteri (Howmiller, 1977; Brinkhurst, 1965).

Considering their tolerance to other sorts of abuse, and the abundance of detritus, it is surprising that oligochaetes are reduced in biomass in acidic lakes. Raddum (1980, 1976) found few oligochaetes in water deeper than 20 m in 19 acidic lakes and normal fauna in 16 other neutral Norwegian oligotrophic lakes, yet the acidic lakes had more oligochaetes than the non-acidic lakes at a depth of 0.5 m. The difference in numbers at greater depths was more pronounced in the spring and autumn. Non-acidified lakes had 3 to 4 times the total number of oligochaetes per m^2 and the abundance correlated with excess sulfate ion. Orciari and Hummon (1975) found an acidic mine drainage lake (pH 4.0) in Ohio to have much lower levels of species diversity than a non-impacted neutral lake, with dominance in the acidic lake by L. hoffmeisteri. Although various depths are sampled in Ohio, the authors did not report depth-dependent distributions. Raddum (1980) attributed the reduction in numbers of oligochaetes in acidic lakes to pollutants associated with acid precipitation (e.g., heavy metals and aluminum). These worms, however, are routinely collected in vast numbers directly below sewage and industrial effluents with far greater concentrations of pollutants (Chapman et al., 1980; Hart and Fuller, 1974). An alternative explanation for their reduction in numbers might be the unpalatability of their detrital food due to the slower decomposition rates in acidic lakes (Traaen, 1977). Orciari and Hummon (1975) found an acidic mine drainage lake (pH 4.0) in Ohio to have much lower values of species diversity than a non-impacted neutral lake, with dominance in the acidic lake by L. hoffmeisteri. Although various depths were sampled in Ohio, the authors did not report depth-dependent distributions.

In one study which mentioned the distribution of leeches (Hirudinea) in acidic lakes (Nilssen, 1980), it was reported that these worms disappeared below pH 5.5. Leeches characteristic of eutrophic waters (Hirudo medicinalis, Glossiphonia heteroclita) were absent from even mildly impacted lakes. Raddum (1980) reported that Hirudinea were restricted to waters above pH 5.5, largely because of the loss of prey items below this pH, yet many leeches are detritivores and scavengers, not obligate carnivores (Pennak, 1978).

VERTEBRATES

Other aquatic or semi-aquatic animals which may be susceptible to artificial acidification include salamanders, frogs, waterfowl, and mammals. Direct toxicity is only a serious factor for the amphibians, which are most sensitive during their larval stages (Pough, 1976; Pough and Wilson, 1977). The salamanders, Ambystoma maculatum and A. jeffersonianum, breed in shallow hilltop ponds which rapidly reflect changes in hydrogen ion concentration brought about by acidic precipitation. Egg mortality in the spotted salamander, A. maculatum, increased to > 60 percent in water less than pH 6.0 from the normal mortality of < 1 percent at pH 7.0; in contrast, the Jefferson salamander, A. jeffersonianum, was most successful at pH 5.0 to 6.0. Adult A. maculatum preferred neutral substrates, too (Mushinsky and Brodie, 1975). These efts may also be affected by food availability in acid lakes, since they feed frequently on sticklebacks, Eucalia (Bishop, 1941).

Variation in pH preference within a genus occurs in newts from Britain as well (Cooke and Frazer, 1976). Smooth newts (Triturus vulgaris) were rarely encountered in water with pH < 6.0, but the Palmate newt (T. helveticus) was routinely captured in bogs at pH ca 4 and once as low as 3.9. The variable responses of these two species were most strongly correlated with potassium and calcium levels.

The contribution of salamanders to the energy flow of a forest-aquatic ecosystem is considerable. In one study (Burton and Likens, 1975a), 20 percent of the energy available to birds and mammals passed through salamanders, and these amphibians represented twice as much standing crop of biomass as birds and an amount equal to that of small mammals (Burton and Likens, 1975b). Most (94%) of the salamanders were terrestrial, but all salamanders are aquatic as efts.

Adult newts (Notophthalmus viridescens) have been reported as deep as 13 m in a neutral (pH 7.4) Adirondack lake (George et al., 1977) but were also observed in acidic (pH 4.8) Woods Lake at 6 m (Singer, Boylen, and Roberts, unpublished results). Other adult amphibians in lakes and streams include the mudpuppy (Necturus maculosus) and frogs. Bullfrogs, Rana catesbeiana, are very sensitive to acidic bog water (Saber and Dunson, 1978), but in neutral waters populations are controlled by parasitism and predation (Cecil and Just, 1979). Although pH < 4.75 delayed development and increased embryonic mortality in leopard frogs, Rana pipiens (Noble, 1979, cited in Glass and Brydges, 1981), this species was anecdotally observed in Adirondack acid lakes (Singer, Boylen, and Roberts, unpublished results) at pH 4.8. Gosner and Black (1957) observed that the effects of low-pH water on embryos of several species of frogs were similar to effects brought about by high salt concentrations and speculated that acidity interferes with the osmotic mechansims responsible for maintaining the perivitelline space. Subsequent to a shrinking of the perivitelline volume, mechanical damage to the egg membrane ensued, which brought about mortality. This is probably only one of several mechanisms by which pH acts to stress anurans. These authors also observed interspecific differences in tolerance, but the source of their acid-tolerant species was bog water from the New Jersey Pine Barrens, which may not be directly applicable to populations receiving atmospheric inputs.

Waterfowl and mammals which feed on fish are likely to avoid lakes devoid of prey. Indeed, species richness is positively correlated with pH (Nilsson and Nilsson, 1978), but other factors like watershed area and nutrient loading may provide better causal explanations. These authors surveyed 11 Swedish lakes which ranged between pH 5.6 to 6.6. In North America, populations of the Common Loon, Gavia immer, have declined around Adirondack lakes (McIntye, 1979; Trivelpiece et al., 1979), but no causal relationship between acidification and declining bird populations was implied. Impairment of the deposition of eggshell by birds in Sweden feeding on aquatic invertebrates laden with heavy metals mobilized by acidic deposition has also been reported (Nyholm and Myrberg, 1977). In contrast, competition for aquatic insects by fish and goldeneye ducks (<u>Bucephala</u> <u>clangula</u>) was severe enough that the ducks favored acidic lakes which lacked fish (Ericksson, 1979).

Mammals which feed on aquatic plants and animals, such as muskrats, minks, otters, shrews, racoons, and even bear and deer, will be affected variously, depending on their ability to choose alternative food sources. Beavers, which live in water but feed on land, have not been affected noticeably by acidic precipitation. Responses by mammals to alterations in their food supplies and habitats brought about by acidic deposition have not been studied with vigor. Alterations of some biotic components of lakes have effects on many other components (Eriksson <u>et</u> <u>al</u>., 1980), and it is reasonable that those changes may extend to terrestrial fauna, too.

CONCLUSIONS AND ASSESSMENT

Although the effects of acidic precipitation on benthos are still largely conjectural, a few generalizations can be made:

(1) The biomass and activity of both microbial and invertebrate decomposers are reduced. This is in contrast to planktonic communities.

(2) Invertebrate communities shift in abundance toward air-breathing carnivores.

(3) Epifaunal forms are more strongly affected than infaunal forms.

It is difficult to make such generalizations about larger taxonomic units, but the expected effects on the biota of acidifying a softwater, neutral lake or stream to ca pH 4.5 are summarized in Table 2. Most taxa show considerable intergeneric differences in tolerances.

pH-induced stress may act in several ways on benthic organisms:

(1) Nutrient limitation can occur by the flocculation of phosphate and other anions.

(2) Alternatively, nutrient additions by changing loading and equilibria for sulfate and nitrate ions may take place.

(3) Oxygen depletion in sediments may occur if gas exchange via the lacunal system of plants is lost as angiosperms are replaced.

Table 2

Summary of the Effects of Acidification to ca
pH 4.5 by Acid Precipitation on Benthic Biota -
See Text for Literature Citations

Biota	Effects
Bacteria	Inhibition of growth rate.
Fungi	Increase in abundance and importance.
Periphyton	Species shift to acidobiontic/acidophilic taxa and increase in standing crop.
Crustacea	Decapods - results equivocal. <u>Gammarus</u> lost. <u>Asellus</u> survives.
Insects	Diversity and biomass decreased. Ephemeroptera and most Plecoptera killed. Some Plecoptera, Trichoptera, and Odonata resistant. <u>Simulium</u> sp., many Chironomidae, Coleoptera, Hemiptera, tolerant. Within each order, genera with larger individuals are favored.
Oligochaeta and Hirudinea	Probably reduced.
Mollusca	Eliminated.

(4) In contrast, lowered decomposition rates may raise the oxygen concentration.
(5) The increased solubility of metals may provoke toxic reactions.
(6) Calcium deposition and transport is largely disrupted.
(7) Sodium pumping may be altered with high levels of hydrogen ions.
(8) Reductions in fish predation and changes in plankton and macrophyte communities affect the benthos.

Those animals that can acclimate to low pH do so, undoubtedly, by a variety of physiological and behavioral mechanisms, but little is known about their responses. Major shifts in the benthic community are likely to affect other communities. pH is, of course, only one of many variables which influences the distribution of organisms, and its effects on communities cannot be studied without an understanding of the full suite of physical, chemical, and biological components of the niche.

Fish shift their food to available prey, but the nutritional effects of switching from a diet of largely amphipods, mayflies, and stoneflies, to one of water boatmen, beetles, and water striders are not known. Effects on different age classes of fish are likely to vary. Changes in the rates of detrital processing and decomposition rates affect primary productivity and hence the whole ecosystem, but it has not been determined if the changes in the water column are the causes or the consequences of changes in the benthos. It has even been suggested (Eriksson et al., 1980) that many of the changes in non-fish components of the ecosystem of acidified lakes may be secondary effects of the elimination of fish predation and grazing. These authors studied the benthos and plankton of a neutral lake to which fish were removed by a poison specific to vertebrates and found that typical "acid lake" fauna became dominant.

RECOMMENDATIONS FOR FUTURE RESEARCH

During the U.S. Environmental Protection Agency/North American Benthological Society's "Special Symposium on the Effects of Acid Rain on Benthos," held at Colgate University in August, 1980, questionnaires were handed out to all those present soliciting comments on future research needs. Similar forms were distributed to the Organization for Economic Development (OECD) Conference on Lake Management held in Portland, Maine, in September, 1980. Results of these surveys showed a remarkable uniformity in the identification of research needs. In order of importance, they were listed as:

(1) survey and monitoring,
(2) field and laboratory manipulations to determine responses to controlled levels of acid-related parameters, and
(3) physiological studies to determine mechanisms of the responses.

Appreciation is expressed to the approximately 100 scientists who participated in this survey.

Even the most rudimentary surveys of benthic organisms will be of great value in this country. Large projects, as exemplified by The Norwegian Interdisciplinary Research Programme (SNSF project)(Wright et al., 1975, 1976; Overrein et al., 1980) could be of critical importance. Regional surveys must be performed in order to predict the extent and severity of future impacts.   Currently, we even lack knowledge about which groups of organisms to watch for signs of acid impacts. A summary of what is known about the effects of acidification on various taxa is presented in Table 2.   The transient nature of precipitation events, and the varying acidities of the throughfall, make monitoring all potentially impacted watersheds impractical.   Monitoring the benthos on a periodic basis, and educating the public to look for problems, in the same way fishermen can alert wildlife experts of a fish kill, is a cost-effective way of identifying problem lakes and streams.   Policy makers can then choose to either initiate restoration procedures or expect to lose the fishery at that site.   The mobilization of heavy metals from sediments, with their potential for direct human health effects, makes routine monitoring and baseline inventories all the more important. Baseline data will be particularly useful during restoration attempts.

The benthic community, which has almost been completely overlooked outside of Scandinavia, is of importance because of its role in energy budgets, its relative immunity to normal seasonal fluctuations, and its sensitivity to change.   Yet, groups which have been established as sensitive indicators of stress such as the Protozoa, Oligochaeta, and Mollusca, have not been studied in acid-impacted waters in the United States. A recommended methodology for using insect drift to monitor acidification was offered (Pratt and Hall, 1981) and Mitchell et al. (1981) recommended using the mayfly, Hexagenia, as an indicator. Since the nature of the precipitation is chemically different in North America and many of the species are not cosmopolitan, the lack of Nearctic data is particularly acute (Likens et al., 1980).

Besides surveys of flora and fauna in impacted and potentially impacted areas, there is an urgent need for experimental studies.   Laboratory bioassays such as those performed by Bell (1970; 1971) and Bell and Nebeker (1969) are invaluable in distinguishing theoretical lower limits of tolerance and distinguishing between the effects of interacting variables like hardness, pH, nutrient availability, and temperature.   Field manipulations are in their infancy.   The artificial acidification of a lake in Canada (Schindler et al., 1980) and a river in New Hampshire (Hall et al., 1980; Hall and Likens, 1980a; Pratt and Hall, 1981) are notable examples of the types of research which should be supported.

Finally, an understanding of the physiological mechanisms by which acid inputs exert effects on organisms should be studied. Work such as that by Vangenechten and Vanderborght (1980) on the physiological responses of Hemiptera is an example of an attempt at determining how acidic precipitation brings about changes in species composition. Acid inputs have different effects on organisms at different stages in their life cycles and times of the year. For most groups, age and seasonal-specific mortality have not been considered. As we begin to understand that acidification acts in consort, and perhaps synergistically with other stresses, an understanding of these processes at a mechanistic level becomes imperative.

ACKNOWLEDGMENTS

Thanks are expressed to all those scientists who responded to the questionnaires regarding future research needs and to those who generously sent reprints, preprints, and manuscripts. Deborah Rhyde, Martha Cordova, and Frank Dugan of the Library staff of Colgate University helped gather together many, sometimes obscure, references. Translation of articles written in Norwegian was done with the aid of Penny Kirkwood. Lorraine Jones, Robin Swindell, and Ann Werges aided by checking and transcribing citations. Lucinda Johnson-Singer and Diane Malley reviewed the manuscript, and it was typed by Karen Armour. This research was supported by a grant from the U.S. Environmental Protection Agency/North Carolina State Acid Precipitation Program.

LITERATURE CITED

Almer, B., W. Dickson, W. Ekstrom, E. Hornstrom, and U. Miller. 1974. Effects of Acidification on Swedish Lakes. Ambio 3:30-36.

Baath, E., B. Lundgren, and B. Soderstrom. 1979. Effects of Artificial Acid Rain on Microbial Activity and Biomass. Bull. Environ. Contam. Toxicol. 23:737-740.

Baker, J.P. and C.L. Schofield. 1980. Aluminum Toxicity to Fish as Related to Acid Precipitation and Adirondack Surface Water Quality. In: D. Drabløs and A. Tollan, eds., "Proc. Int. Conf. Ecol. Impact Acid Precip.," pp. 292-293. SNSF Project, Norway.

Bell, H.L. 1970. Effects of pH on the Life Cycle of the Midge Tanytarsus dissimilis. Can. Entomol. 102:636-639.

Bell, H.L. 1971. Effect of Low pH on the Survival and Emergence of Aquatic Insects. Water Res. 5:313-319.

Bell, H.L. and A.V. Nebeker. 1969. Preliminary Studies on the Tolerance of Aquatic Insects to Low pH. J. Kans. Entomol. Soc. 42(2):230-237.

Berge, F. 1976. Kiselalger og pH i noen elver og innsjøer i Agder og Telemark. En sammenlikning mellom årene 1949 og 1975. Internal Report 18/76, SNSF Project, Oslo-Ås, Norway.

Bick, H. and E.F. Drews. 1973. Selbstreinigung und Ciliatenbesiedlung in saurem Milieu (Modellversuche). Hydrobiologia 42:393-402.

Bishop, S. 1941. The Salamanders of New York. New York State Museum, Bulletin 324:1-365.

Borgstrom, R., J. Brittain, and A. Lillehammer. 1976. Everte-brater og surt vann. Oversikt over innsamlings-lokaliteter. Internal Report 21/76, SNSF Project, Oslo-Ås, Norway.

Borgstrom, R. and G.R. Hendrey. 1976. pH Tolerance of the First Larval Stages of Lepidurus arcticus (Pallas) and adult Gammarus lacustris G. O. Sars. Internal Report 22/76, SNSF Project, Oslo-Ås, Norway. 37 pp.

Brinkhurst, R.O. 1965. The Biology of the Tubificidae with Special Reference to Pollution. In: "Proc. 3rd Seminar on Water Quality Criteria," pp. 57-66, Cincinnati, Ohio.

Brinkhurst, R.O. 1974. The Benthos of Lakes. St. Martin's Press, New York, New York.

Burton, T.M. and G.E. Likens. 1975a. Energy Flow and Nutrient Cycling in Salamander Populations in the Hubbard Brook Experimental Forest, New Hampshire. Ecology 56:1068-1080.

Burton, T.M. and G.E. Likens. 1975b. Salamander Populations and Biomass in the Hubbard Brook Experimental Forest, New Hampshire. Copeia 1975(3):541-546.

Cecil, S.G. and J.J. Just. 1979. Survival Rate, Population Density and Development of a Naturally Occurring Anuran Larvae (Rana catesbeiana). Copeia 1979(3):447-453.

Chapman, P.M., L.M. Churchland, P.A. Thomson, and E. Michnowsky. 1980. Heavy Metal Studies with Oligochaetes. In: R.O. Brinkhurst and D.G. Cook, eds., "Aquatic Oligochaete Biology," pp. 477-502. Plenum Press, New York, New York.

Coesel, P.F.M., R. Kwakkestein, and A. Verschoor. 1978. Oligotrophication and Eutrophication Tendencies in Some Dutch Moorland Pools as Reflected in Their Desmid Flora. Hydrobiologia 61:21-31.

Collins, N.C., A.P. Zimmerman, and R. Knoechel. 1981. Comparisons of Benthic Infauna and Epifauna Biomasses in Acidified and Nonacidified Ontario Lakes. In: R. Singer, ed., "Effects of Acidic Precipitation on Benthos," pp. 35-48. Proceedings of a Regional Symposium on Benthic Biology, North American Benthological Society, Hamilton, New York.

Conroy, N., K. Hawley, W. Keller, and C. LaFrance. 1975. Influences of the Atmosphere on Lakes in the Sudbury Area. J. Great Lakes Res. 2:146-165.

Cooke, A.S. and J.F.D. Frazer. 1976. Characteristics of Newt Breeding Sites. J. Zool. Lond. 178:223-236.

Costa, H.H. 1967. Responses of Gammarus pulex (L.) to Modified Environment. II. Reactions to Abnormal Hydrogen Ion Concentrations. Crustaceana 13:1-10.

Dam, H. van, G.Suurmond, and C. ter Braak. 1980. Impact of Acid Precipitation on Diatoms and Chemistry of Dutch Moorland Pools. In: D. Drabløs and A. Tollan, eds., "Proc. Int. Conf. Ecol. Impact Acid Precip.," pp. 298-299. SNSF Project, Norway.

Davis, R.B. and F. Berge. 1980. Atmospheric Deposition in Norway During the Last 300 Years as Recorded in SNSF Lake Sediments, USA and Norway. II. Diatom Stratigraphy and Inferred pH. In: D. Drabløs and A. Tollan, eds., "Proc. Int. Conf. Ecol. Impact Acid Precip.," pp. 270-271. SNSF Project, Norway.

Del Prete, A. and C. Schofield. 1981. The Utility of Diatom Analyses of Lake Sediments for Evaluating Acid Precipitation Effects on Dilute Lakes. Arch. Hydrobiol. 91(3):332-340.

Eriksson, M.O.G. 1979. Competition Between Freshwater Fish and Goldeneyes, Bucephalia clangula (L.) for Common Prey. Oecologia (Berl.) 41:99-107.

Eriksson, M.O.G., L. Henrikson, B-I. Nilsson, G. Nyman, H.G. Oscarson, A.E. Stenson, and K. Larsson. 1980. Predatory-prey Relations Important for the Biotic Changes in Acidified Lakes. Ambio 9:248-249.

Evans, L.S. and C.A. Conway. 1980. Effects of Acidic Solutions on Sexual Reproducton of Pteridium aquilinum. Am. J. Bot. 67:866-875.

Faucon, A.S. and W.D. Hummon. 1976. Effects of Mine Acid on the Longevity and Reproductive Rate of the Gastrotricha Lepidodermella squammata (Dujardin). Hydrobiologia 50:205-209.

Ferris, V.R., L.M. Ferris, and J.P. Tjepkema. 1976. Genera of Freshwater Nematodes (Nematoda) of Eastern North America. U.S. Environmental Protection Agency, Cincinnati, Ohio.

Fiance, S.B. 1978. Effects of pH on the Biology and Distribution of Ephemerella funeralis (Ephemeroptera). Oikos 31:332-339.

Friberg, F., C. Otto, and B.S. Svensson. 1980. Effects of Acidification on the Dynamics of Allochthonous Leaf Material and Benthic Invertebrate Communities in Running Waters. In: D. Drabløs and A. Tollan, eds., "Proc. Int. Conf. Ecol. Impact Acid Precip.," pp. 304-305. SNSF Project, Norway.

Gahnstrom, G., G. Andersson, and S. Fleischer. 1980. Decomposition and Exchange Processes in Acidified Lake Sediment. In: D. Drabløs and A. Tollan, eds., "Proc. Int. Conf. Ecol. Impact Acid Precip.," pp. 306-307. SNSF Project, Norway.

Galloway, J.N. and G.E. Likens. 1979. Atmospheric Enhancement of Metal Deposition in Adirondack Lake Sediments. Limnol. Oceanogr. 24(3):427-433.

Gaufin, A.R. 1973. Water Quality Requirements of Aquatic Insects. EPA-660/3-73-004, U.S. Environmental Protection Agency, Cincinnati, Ohio.

George, C.J., C.W. Boylen, and R.B. Sheldon. 1977. The Presence of the Red-Spotted Newt, Notophthalmus viridescens Rafinesque (Amphibia, Urodela, Salamandridae), in Waters Exceeding 12 Meters in Lake George, New York. J. Herpetol. 11(1):87-90.

Glass, G.E. and T.G. Brydges (eds.). 1981. United States-Canada Memorandum of Intent on Transboundary Air Pollution. Aquatic Impact Assessment (Final Draft Copy). U.S. Environmental Protection Agency, Environmental Research Laboratory, Duluth, Minnesota. 150 pp.

Goodnight, C.J. 1973. The Use of Aquatic Macroinvertebrates as Indicators of Stream Pollution. Trans Amer. Microsc. Soc. 92:1-12.

Gosner, K.L. and I.H. Black. 1957. The Effects of Acidity on the Development and Hatching of New Jersey Frogs. Ecology 38(2):256-262.

Grahn, O. 1976. Macrophyte Succession in Swedish Lakes Caused by Deposition of Airborne Acid Substances. In: A.S. Dochinger and T.A. Seliga, eds., "Proceedings of the International Symposium on Acid Precipitation and the Forest Ecosystem (1st), Columbus, Ohio," pp. 519-530. USDA Forest Service General Technical Report NE-23, Upper Darby, Pennsylvania.

Grahn, O. 1977. Macrophyte Succession in Swedish Lakes Caused by Deposition of Airborne Acid Substances. Water Air Soil Pollut. 7:295-306.

Grahn, O., H. Hultberg, and L. Landner. 1974. Oligotrophication: A Self-Accelerating Process in Lakes Subjected to Excessive Supply of Acid Substances. Ambio 3:93-94.

Hagen, A. and A. Langeland. 1973. Polluted Snow in Southern Norway and the Effect of the Meltwater on Freshwater and Aquatic Organisms. Environ. Pollut. 5:45-57.

Hall, R.J. and G.E. Likens. 1980a. Ecological Effects of Experimental Acidification on a Stream Ecosystem. In: D. DrablØs and A. Tollan, eds., "Proc. Int. Conf. Ecol. Impact Acid Precip.," pp. 375-376. SNSF Project, Norway.

Hall, R.J. and G.E. Likens. 1980b. Ecological Effects of Whole-Stream Acidification. In: D.S. Shriner, C.R. Richmond, and S.E. Lindberg, eds., "Atmospheric Sulfur Deposition, Environmental Impact and Health Effects," pp. 443-451. Ann Arbor Science Publishers, Inc., Ann Arbor, Michigan.

Hall, R.J., G.E. Likens, S.B. Fiance, and G.R. Hendrey. 1980. Experimental Acidification of a Stream in the Hubbard Brook Experimental Forest, New Hampshire. Ecology 61(4):976-989.

Harriman, R. and B. Morrison. 1980. Ecology of Acid Streams Draining Forested and Non-Forested Catchments in Scotland. In: D. Drablos and A. Tollan, eds., "Proc. Int. Conf. Ecol. Impact Acid Precip.," pp. 312-313. SNSF Project, Norway.

Hart, C.W., Jr., and S.L.H. Fuller. 1974. Pollution Ecology of Freshwater Invertebrates. Academic Press, New York, New York.

Havas, M. 1981. Physiological Response of Aquatic Animals to Low pH. In: R. Singer, ed., "Effects of Acidic Precipitation on Benthos," pp. 49-65. Proc. of a Regional Symposium on Benthic Biology, North American Benthological Survey, Hamilton, New York.

Hellquist, C.B.   1980.   Correlation of Alkalinity and the Distribution of Potamogeton in New England. Rhodora 82:331-344.

Hendrey, G.R. 1976. Effects of pH on the Growth of Periphytic Algae in Artificial Stream Channels. Internal Report 25/76, SNSF Project, Oslo-Ås, Norway.

Hendrey, G.R., K. Baalsrud, T.S. Traaen, M. Laake, and G.G. Raddum. 1976. Acid Precipitation: Some Hydrobiological Changes. Ambio 5:224-227.

Hendrey, G.R. and F.W. Barvenik. 1978. Impacts of Acid Precipitation on Decomposition and Plant Communities in Lakes. In: H.H. Izard and J.S. Jacobson, eds., "Scientific Papers from the Public Meeting on Acid Precipitation, Science and Technology Staff, New York State Assembly," pp. 92-103. Center for Environmental Research, Cornell University, Ithaca, New York.

Hendrey, G.R., J.N. Galloway, and C.L. Schofield. 1980. Temporal and Spatial Trends in the Chemistry of Acidified Lakes Under Ice Cover. In: D. Drabløs and A. Tollan, eds., "Proc. Int. Conf. Ecol. Impact Acid Precip.," pp. 266-267. SNSF Project, Norway.

Hendrey, G.R. and F.A. Vertucci. 1980. Benthic Plant Communities in Acidic Lake Colden, New York: Sphagnum and the Algal Mat. In: D. Drabløs and A. Tollan, eds., "Ecological Impact of Acid Precipitation," pp. 314-315. SNSF Project, Sandefjord, Norway.

Hendrey, G.R. and R.F. Wright. 1976. Acid Precipitation in Norway: Effects on Aquatic Fauna. J. Great Lakes Res. 2:192-207.

Henrikson, L., H.G. Oscarson, and J.A.E. Stenson. 1980. Does the Change of Predator System Contribute to the Biotic Development in Acidified Lakes? In: D. Drabløs and A. Tollan, eds., "Proc. Int. Conf. Ecol. Impact Acid Precip.," pp. 316-317. SNSF Project, Norway.

Howmiller, R. 1977. On the Abundance of Tubificidae (Annelida: Oligochaeta) in the Profundal Benthos of Some Wisconsin Lakes. Amer. Midl. Nat. 97:211-216.

Hultberg, H. and O. Grahn. 1976. Effects of Acid Precipitation on Macrophytes in Oligotrophic Swedish Lakes. J. Great Lakes Res. 2(Suppl. 1):208-217.

Hutchinson, G.E.   1957.   A Treatise on Limnology.   Vol. I.
Geography, Physics, and Chemistry.   John Wiley and Sons, New
York.

Hummon, M.R. and W.D. Hummon.   1979.   Reduction in Fitness of
the Gastrotrich Lepidodermella squammata by Dilute Acid Mine
Water and Amelioration of the Effect by Carbonates.   Int. J.
Invert. Reprod. 1:297-306.

Jewell, M.E.   1939.   An Ecological Study of the Freshwater
Sponges of Wisconsin.   II.   The Influence of Calcium.   Ecology
20:11-28.

Kamijo, H., T. Watanabe, and K. Mashiko.   1974.   The Attached
Algal Flora of the Nagase-Gawa, a Strong Acid Water River and
Its Tributaries, Fukushima Prefecture.   Japanese J. Ecol.
24(2):147-152.

Laake, M.   1976.   Effekter av lav pH på produkcjon, nedbrytning
og stoffkretsløp i littoralsonen.   Internal Report 29/76, SNSF
Project, Oslo-Ås, Norway.

Leivestad, H., G. Hendrey, I.P. Muniz, and E. Snekvik.   1976.
Effects of Acid Precipitation on Freshwater Organisms.   In:
F.H. Braekke, ed., "Impact of Acid Precipitation on Forest and
Freshwater Ecosystems in Norway," pp. 87-111.   Research Report
6/76, SNSF Project, Oslo-Ås, Norway.

Likens, G.E., F.H. Bormann, and J.S. Eaton.   1980.   Variations
in Precipitation and Streamwater Chemistry at the Hubbard Brook
Experimental Forest During 1964 to 1977.   In: T.C. Hutchinson
and M. Havas, eds., "Effects of Acid Precipitation on
Terrestrial Ecosystems," pp. 443-464.   Plenum Press, New York.

Lowe, R.L.   1974.   Environmental Requirements and Pollution
Tolerance of Freshwater Diatoms.   EPA-670/4-74-004, U.S.
Environmental Protection Agency, Cincinnati, Ohio.

Malley, D.F.   1980.   Decreased Survival and Calcium Uptake by
the Crayfish Orconectes virilis in Low pH.   Can. J. Fish Aquatic
Soi. 37:364-372.

McIntyre, J.W.   1979.   Status of Common Loons in New York from
an Historical Perspective.   In: S.A. Sutcliffe, ed., "The
Common Loon," pp. 117-121.   Proc. Second N.A. Conf. Common Loon
Research and Management, National Audubon Society, New York.
162 pp.

Merritt, R.W. and K.W. Cummins.   1978.   An Introduction to the
Aquatic Insects of North Ameica.   Kendall/Hunt Publishing Co.,
Dubuque, Iowa.

Mitchell, M.J., D.H. Landers, and D.F. Brodowski. 1981. Sulfur Constituents of Sediments and Their Relationship to Lake Acidification. Water Air Soil Pollut. 16:177-186.

Mitchell, M.J., G.B. Lawrence, D.H. Landers, and K.A. Stucker. 1981. Role of Benthic Invertebrates in Affecting Sulfur and Nitrogen Dynamics of Lake Sediments. In: R. Singer, ed., "Effects of Acidic Precipitation on Benthos," pp. 67-76. Proceedings of a Regional Symposium on Benthic Biology, North American Benthological Society, Hamilton, New York.

Moore, J.A. and R.B. Clarkson. 1967. Physical and Chemical Factors Affecting Vascular Aquatic Plants in Some Acid Stream Drainage Areas of the Monongahela River. Proc. West Virginia Acad. Sci. 42:83-89.

Mortimer, C.H. 1941. The Exchange of Dissolved Substances Between Mud and Water in Lakes (Parts I and II). J. Ecol. 29:280-329.

Mortimer, C.H. 1942. The Exchange of Dissolved Substances Between Mud and Water in Lakes (Parts III and IV). J. Ecol. 30:147-201.

Müller, P. 1980. Effects of Artificial Acidification on the Growth of Periphyton. Can. J. Fish Aquatic Sci. 37:355-363.

Mushinsky, G.R. and E.D. Brodie, Jr. 1975. Selection of Substrate pH by Salamanders. Amer. Midl. Nat. 93(2):440-443.

Nilssen, P.J. 1980. Acidification of a Small Watershed in Southern Norway and Some Characteristics of Acidic Aquatic Environments. Int. Rev. Gesamten Hydrobiol. 65:177-207.

Nilsson, S.G. and I.N. Nilsson. 1978. Breeding Bird Community Densities and Species Richness in Lakes. Oikos 31:214-221.

Nyholm, N.E.I. and H.E. Myhrberg. 1977. Severe Eggshell Defects and Impaired Reproductive Capacity in Small Passerines in Swedish Lapland. Oikos 29:336-341.

Økland, J. 1969a. Om forsuring av vassdrag og betydningen av surhetsgraden (pH) for fishkens naeringsdyr i ferskvann. Fauna (Oslo) 22:140-147.

Økland, J. 1969b. Distribution and Ecology of the Fresh-Water Snails (Gastropoda) of Norway. Malacologia 9(1):143-151.

Økland, J. 1976. Utbredelsen av noen ferskvannsmuslinger i Norge, og litt om European Invertebrate Survey. Fauna (Oslo) 29:29-40.

Økland, J. 1979. Kalkinnhold, surhetsgrad (pH) og snegler i norske innsjøer. Fauna (Oslo) 32:96-111.

Økland, J. 1980. Environment and Snails (Gastropoda): Studies of 1000 Lakes in Norway. In: D. Drabløs and A. Tollan, eds., "Ecological Impact of Acid Precipitation," pp. 322-323. SNSF Project, Sandefjord, Norway.

Økland, J. and K.A. Økland. 1978. Use of Fresh-Water Littoral Fauna for Environmental Monitoring: Aspects Related to Studies of 1000 Lakes in Norway. In: "The Use of Ecological Variables in Environmental Monitoring," Proceedings from a conference arranged by the Swedish Society Oikos and the Nordic Council for Ecology. SNSF Contribution FA 37/78. Swedish University of Agricultural Sciences, Uppsala, Sweden.

Økland, J. and K.A. Økland. 1980. pH Level and Food Organisms for Fish: Studies of 1000 Lakes in Norway. In: D. Drabløs and A. Tollan, eds., "Ecological Impact of Acid Precipitation," pp. 326-327. SNSF Project, Sandefjord, Norway.

Økland, K.A. 1971. On the Ecology of Sphaeriidae in a High Mountain Area in South Norway. Norway J. Zool. 19:133-143.

Økland, K.A. 1979a. Localities with Asellus aquaticus (L.) and Gammarus lacustris G.O. Sars in Norway, and a Revised System of Faunistic Regions. Technical Note 49/79, SNSF Project, Oslo-Ås, Norway.

Økland, K.A. 1979b. A Project for Studying Ecological Requirements and for Mapping Distribution Patterns of Small Bivalvia (Sphaeriidae) and Some Crustacea (Asellus, Gammarus) in Fresh Water in Norway. C. R. Soc. Biogeog. (Mem. 3e Ser., No. 1) 1979:37-43.

Økland, K.A. 1979c. Sphaeriidae of Norway: A Project for Studying Ecological Requirements and Geographical Distribution. Malacologia 18:223-226.

Økland, K.A. 1980a. Ecology and Distribution of Asellus aquaticus (L.) in Norway, Including Relation to Acidification in Lakes. Internal Report 52/80, SNSF Project, Oslo-Ås, Norway.

Økland, K.A. 1980b. Mussels and Crustaceans: Studies of 1000 Lakes in Norway. In: D. Drabløs and A. Tollan, eds., "Ecological Impact of Acid Precipitation," pp. 324-325. SNSF Project, Sandefjord, Norway.

Økland, K.A. 1980c. Økologi og utbredelse til Gammarus lacustris G. O. Sars i Norge, med vekt pa forsuringsproblemer. Internal Report 67-80, SNSF Project, Oslo-Ås, Norway.

Økland, K.A. and J.G.J. Kuiper. 1980. Distribution of Small Mussels (Sphaeriidae) in Norway, With Notes on Their Ecology. Haliotis 10(2):109.

Orciari, R.D. and W.D. Hummon. 1975. A Comparison of Benthic Oligochaete Populations in Acid and Neutral Lentic Environments in Southeastern Ohio. Ohio J. Sci. 75:44-49.

Overrein, L.N., H.M. Seip, and A. Tollan. 1980. Acid Precipitation: Effects on Forest and Fish. Final Report SNSF Project 1972-1980, Oslo-Ås, Norway.

Pennak, R.W. 1978. Fresh-Water Invertebrates of the United States. John Wiley and Sons, New York, NY.

Pough, F.H. 1976. Acid Precipitation and Embryonic Mortality of Spotted Salamanders, Ambystoma maculatum. Science 192:68-70.

Pough, F.H. and R.E. Wilson. 1977. Acid Precipitation and Reproductive Success of Ambystoma Salamanders. Water Air Soil Pollut. 7:307-316.

Pratt, J.M. and R.J. Hall. 1981. Acute Effects of Stream Acidification on the Diversity of Macroinvertebrate Drift. In: R. Singer, ed., "Effects of Acidic Precipitation on Benthos," pp. 77-95. Proceedings of a Regional Symposium on Benthic Biology, North American Benthological Society, Hamilton, New York.

Raddum, G.G. 1976. Preliminaere data om bunnfaunaen i øvre botnat jønn. Technical Note 27/76, SNSF Project, Oslo-Ås, Norway.

Raddum, G.G. 1979. Virkninger av lav pH på insektlarver. Internal Report 45/79, SNSF Project, Oslo-Ås, Norway.

Raddum, G.G. 1980. Comparison of Benthic Invertebrates in Lakes with Different Acidity. In: D. Drabløs and A. Tollan, eds., "Ecological Impact of Acid Precipitation," pp. 330-331. SNSF Project, Sandefjord, Norway.

Raddum, G.G., J. Gastry, B.O. Rosseland, and I. Seveldrud. 1979. Vannteger i Sør-Norge og deres betydning some fiskefode i vann med ulik pH. Internal Report 50/79, SNSF Project, Oslo-Ås, Norway.

Raddum, G.G. and O.E. Saether. 1981. Chironomid Communities in Norwegian Lakes with Different Degrees of Acidification. Verh. Int. Verein. Limnol. 21:367-373.

Raddum, G.G. and A.L. Steigen. 1981. Reduced Survival and Calorific Content of Stoneflies and Caddis Flies in Acidic Water. In: R. Singer, ed., "Effects of Acid Precipitation on Benthos," pp. 97-101. Proceedings of a Regional Symposium on Benthic Biology, North American Benthological Society, Hamilton, New York (in press).

Rosseland, B.O., I. Sevaldrud, D. Svalastog, and I.P. Muñiz. 1980. Studies on Freshwater Fish Populations: Effects of Acidification on Reproduction, Population Structure, Growth and Food Selection. In: D. Drabløs and A. Tollan, eds., "Proc. Int. Conf. Ecol. Impact Acid Precip.," pp. 336-337. SNSF Project, Norway.

Saber, P.A. and W.A. Dunson. 1978. Toxicity of Bog Water to Embryonic and Larval Anuran Amphibians. J. Exp. Zool. 204:33-42.

Scheider, W.A., J. Adamski, and M. Paylor. 1975. Reclamation of Acidific Lakes Near Sudbury, Ontario. Ontario Ministry of the Environment, Rexdale, Ontario, Canada.

Scheider, W.A., J. Jones, and B. Cave. 1976. A Preliminary Report on the Neutralization of Nelson Lake Near Sudbury, Ontario. Ontario Ministry of the Environment, Rexdale, Ontario, Canada.

Schindler, D.W. 1980. Experimental Acidification of a Whole Lake: A Test of the Oligotrophication Hypothesis. In: D. Drabløs and A. Tollan, eds., "Proc. Int. Conf. Ecol. Impact Acid Precip.," pp. 370-374. SNSF Project, Norway.

Schindler, D.W., R. Wagemann, R. Cook, T. Ruszczynski, and J. Prokopowich. 1980. Experimental Acidification of the Lake 223 Experimental Lakes Area: Background Data and the First Three Years of Acidification. Can. J. Fish Aquatic Sci. 37:342-354.

Shaw, J. 1960. The Absorption of Sodium Ions by the Crayfish Astacus pallipes Lereboulett. III. The Effect of Other Cations in the External Solution. J. Exp. Biol. 37:548-572.

Singer, R. 1981. Notes on the Use of Shells of Anodonta grandis Say (Bivalvia; Unionidae) as a Paleoecological Indicator of Trophic Status and pH. In: R. Singer, ed., "Effects of Acidic Precipitation on Benthos," pp. 103-111. Proceedings of a Regional Symposium on Benthic Biology, North American Benthological Society, Hamilton, New York (in press).

Steigen, A.L. and G.G. Raddum. 1981. Effects of Acidified Water on Behavior and Energy Content in the Water Louse Asellus aquaticus (L.). In: R. Singer, ed., "Effects of Acidic Precipitation on Benthos," pp. 113-118. Proceedings of a Regional Symposium on Benthic Biology, North American Benthological Society, Hamilton, New York (in press).

Stickney, F. 1922. The Relation of the Nymphs of a Dragonfly (Libellula pulchella Drury) to Acid and Temperature. Ecology 3(3):250-254.

Sutcliffe, D.W. and T.R. Carrick. 1973. Studies on the Mountain Streams in the English Lake District. I. pH, Calcium, and the Distribution of Invertebrates in the River Duddon. Freshwater Biol. 3:437-462.

Tomlinson, G.H. 1978. Acidic Precipitation and Mercury in Canadian Lakes and Fish. In: H.H. Izard and J.S. Jacobson, eds., "Scientific Papers from the Public Meeting on Acid Precipitation, Science and Technology Staff, New York State Assembly," pp. 104-118. Center for Environmental Research, Cornell University, Ithaca, New York.

Traaen, T.S. 1976. Nedbrytning av organisk materiale. Forsøk med "litterbags". Technical Note 19/76, SNSF Project, Oslo-Ås, Norway.

Traaen, T.S. 1977. Nebrytning av organisk materiale. Forsøk med "litterbags", resultater etter 2 ars feltforsøk. Technical Report, SNSF Project, Oslo-Ås, Norway.

Traaen, T.S. 1978. Bakterieplankton i innsjøer. Technical Note 41/78, SNSF Project, Oslo-Ås, Norway.

Trivelpiece, W., S. Brown, A. Hicks, R. Fekete, and N.J. Volkman. 1979. An Analysis of the Distribution and Reproductive Success of the Common Loon in the Adirondack Park, New York. In: S.A. Sutcliffe, ed., "The Common Loon," pp. 45-55. Proc. Second N.A. Conf. on Common Loon Research Management, National Audubon Society, New York. 162 pp.

Uutala, A.J. 1981. Composition and Secondary Production of the Chironomid (Diptera) Communities in Two Lakes in the Adirondack Mountain Region, New York. In: R. Singer, ed., "Effects of Acidic Precipitation on Benthos," pp. 139-154. Proceedings of a Regional Symposium on Benthic Biology, North American Benthological Society, Hamilton, New York (in press).

Vangenechten, J.H.D., S. Van Puymbroeck, and O.L.J. Vanderborght. 1979. Effect of pH on the Uptake of Sodium in the Waterbugs *Corixa dentipes* (Thoms.) and *Corixa punctata* (Illig.)(Hemiptera, Heteroptera). Comp. Biochem. Physiol. 64A:509-521.

Vangenechten, J.H.D. and O.L.J. Vanderborght. 1980. Effect of pH on Sodium and Chloride Balance in an Inhabitant of Acid Freshwaters: The Waterbug *Corixa punctata* (Illig.)(Insecta, Hemiptera). In: D. Drabløs and A. Tollan, eds., "Ecological Impact of Acid Precipitation," pp. 342-343. SNSF Project, Sandefjord, Norway.

Warner, R.W. 1971. Distribution of Biota in a Stream Polluted by Acid Mine-Drainage. Ohio J. Sci. 71:202-215.

Wiederholm, T. and L. Eriksson. 1977. Benthos of an Acid Lake. Oikos 29:261-267.

Wright, R.F., T. Dale, E.T. Gjessing, G.R. Hendrey, A. Henriksen, M. Johannessen, and I.P. Muniz. 1975. Impact of Acid Precipitation on Freshwater Ecosystems in Norway. Research Report FR 3/75, SNSF Project, Oslo-Ås, Norway.

Wright, R.F., T. Dale, E.T. Gjessing, G.R. Hendrey, A. Henriksen, M. Johannessen, and I.P. Muniz. 1976. Impact of Acid Precipitation on Freshwater Ecosystems in Norway. Water Air Soil Pollut. 6:483-499.

CHAPTER 16

THE EFFECTS OF ACID PRECIPITATION ON WATER QUALITY
IN ROOF CATCHMENT-CISTERN WATER SUPPLIES

William E. Sharpe and Edward S. Young
School of Forest Resources and The Institute
for Research on Land and Water Resources
The Pennsylvania State University
University Park, Pennsylvania 16802

INTRODUCTION

Roof catchment-cistern water supplies are common in regions of the country where groundwater supplies are either unavailable or unusable. Cistern systems consist of a roof, usually the house roof, which serves as an impervious catchment for precipitation, and a below-ground cistern to store the collected water. The stored water is pumped from the cistern to points of use within the house. Very little is known about the prevalence of this type of water supply in the United States. A recent paper by Kincaid (1979) cites Ohio Department of Health records that report a total of 67,000 cistern systems in the state of Ohio alone.

Roof catchment-cistern systems are fairly common in the coal mining regions of Pennsylvania where groundwater has been polluted by mining and public water supplies are unavailable. These systems are also common where groundwater development has been unsuccessful and surface water sources are either polluted or nonexistent.

Published data (Lazrus et al., 1970; Hutchinson, 1973; Cogbill and Likens, 1974) indicated that precipitation falling on the northeastern United States could not meet the drinking water standards established by the U.S. Environmental Protection Agency. This fact resulted in the initiation of a study to determine the impact of acid precipitation on the quality of water being supplied by roof-catchment cistern water systems. Inputs of heavy metals by both direct input in precipitation and as corrosion products resulting from the contact of acid precipitation with household plumbing were the principal areas of concern.

365

PROCEDURE

The main study area was located in Clarion County, Pennsylvania, and involved 35 roof catchment-cistern systems. An additional 5 systems were located in widely scattered locations in Indiana County, Pennsylvania. Study participants were solicited via a news release published in local newspapers. All of the systems studied were in actual use and were sampled without interfering with normal operations. In some systems, bottled drinking water was purchased or hauled from local springs. Hauled water was used to provide for needs during dry periods.

Field data collection was begun in the winter of 1979 in the Clarion County study area and is continuing. A bulk precipitation collection site was established just outside Clarion, Pennsylvania, and samples were collected in two different types of polyethylene containers. Snow was collected in 46 x 30.5 x 46 cm Nalgene tubs located on wooden platforms 2 m above the ground surface. Rain was collected with 25.5 cm diameter polyethylene funnels connected to 2 liter polyethylene bottles. Condensation and evaporation of the collected samples were minimized by barriers designed into the collection apparatus. Weekly bulk precipitation samples were collected from January through March and from June through August in 1979 and 1980. The station at Clarion was used from January through March, 1979. For the 1979 summer sampling period, two additional stations were established in northern Clarion County, near the towns of Leeper and Snydersburg. In 1980, the Snydersburg station was continued, the Clarion station was discontinued, and the Leeper station was moved to a site in southern Clarion County, near the town of Mechanicsville. Also, a third station was established in Indiana County, a few miles south of Indiana, Pennsylvania.

Bulk precipitation samples were analyzed for the metals lead, cadmium, zinc, and copper. A Langelier saturation index value was determined for each sample from the parameters pH, total alkalinity, specific conductance, and calcium. The Langelier saturation index is a measure of instability of water with regard to calcium carbonate deposition and solution. Positive Langelier saturation index values indicate that calcium carbonate will be deposited and negative values indicate that calcium carbonate will be dissolved; consequently, negative values indicate corrosion. The quality of the bulk precipitation samples obtained was assumed to be representative of the quality of precipitation falling on the roof catchments of the homes in the study.

Cistern water supplies were sampled at all participating homes in 1979 and 1980. Each system was sampled twice in 1979 and three times in 1980. The 1979 samples were analyzed for lead and cadmium. The 1980 analyses also included zinc, copper, specific conductance, calcium, pH, and total alkalinity.

The cistern water samples were collected at the end of each precipitation collection period so that the water in each system would be representative of the precipitation over the previous 10 weeks. Thirty-two systems were sampled in 1979. During 1980, an additional 8 systems were added bringing the total sampled that year to 40.

The cistern sampling scheme varied somewhat from 1979 to 1980. In 1979, samples were collected at a cold-water tap in the house (usually the kitchen) and from the cistern at a point just below the surface of the stored water. The tap samples were taken at random times throughout the day and no attempt was made to control sample timing based on length of residence time for the water in the plumbing system. Samples were collected on March 12-13 and July 30-31, 1979.

In 1980, samples were collected from three additional locations within each cistern water system and the sampling scheme was standardized. A sample was collected by cistern owners from the kitchen cold-water tap prior to any other water use that day. An additional tapwater sample was collected from this same tap at some time later in the day, after allowing the cold-water faucet to remain full open for 30 sec. Samples were also taken from the surface and 15 cm above the bottom of the cistern. Cistern bottom sediment/water samples were also obtained from the bottom 0.5 cm of water in the cistern. These samples were a mixture of sediment and water. Results of the 1980 tapwater analyses are as yet incomplete; consequently they are omitted from further discussion. Samples were collected on March 26-28, July 8-10, and August 12-14, 1980.

Following sample collection, each metals sample was acidified with concentrated nitric acid to pH < 2.0. Next, a 25 ml aliquot of each sample was digested at $95^{\circ}C$ for 3 hr (U.S. EPA, 1979). After digestion, distilled water was added to each sample until its volume was again 25 ml. Metals concentrations were determined with a Perkin-Elmer Model 703 atomic absorption spectrophotometer with a heated graphite atomizer Model 2200. Metals concentrations so determined are reported as "total recoverable metals" concentrations (U.S. EPA, 1979).

RESULTS AND DISCUSSION

Six weekly snow and 22 weekly rain samples were collected in 1979. Results of the analysis of these samples are presented in Table 1. Mean snow pH values were significantly higher than rain pH values. The mean corrosivity of rain as indicated by the Langelier saturation index was slightly higher than that of snow. The average acidity of winter bulk rain was almost three times that of bulk snow and the maximum acidity was more than three times that of bulk snow. The average bulk snow lead concentration was 134 μg/l and the average cadmium concentration ws 17.2 μg/l. Both of these values exceed the recommended U.S. Environmental Protection Agency drinking water maxima for lead and cadmium set at 50 μg/l and 10 μg/l, respectively. All pH values recorded in Table 1 fall well below the U.S. Environmental Protection Agency's recommended drinking water minimum of 6.5. The condition "non-corrosive" is also recommended in current secondary drinking water standards. The negative Langelier saturation index values listed in Table 1 indicate that bulk snow and rain are highly corrosive.

The 1980 bulk precipitation results vary somewhat from the results obtained in 1979 (Table 2). It should be remembered that a single bulk precipitation site, located in Clarion Borough, was used in winter 1979, while three rural sites were used in 1980. The data presented in Table 2 are average values for the three collection sites.

Although the specific values for 1980 presented in Table 2 indicate the variability of bulk precipitation data, they corroborate several 1979 findings. The Langelier saturation index values from both years indicate that both rain and snow were very corrosive. The mean pH of summer rain was slightly lower in 1980 than in 1979. Both 1979 and 1980 values are well under the suggested drinking water minimum for pH. Lead concentrations were considerably lower in 1980, possibly owing in part to the rural locations of the collection sites. However, the 1980 maximum lead concentrations for snow and mixed precipitation of 81 and 147 μg/l, respectively, exceed the recommended safe drinking water limit of 50 μg/l. Also, the mean lead concentration for snow samples collected in 1980 of 49 μg/l nearly equals the drinking water limit for lead. Although 1980 rain samples generally contained lower concentrations of lead than 1979 rain samples, the maximum concentration in 1980 summer rain of 42 μg/l is close to the drinking water limit. Cadmium concentrations in 1980 precipitation were also somewhat lower than in 1979 and were well under the drinking water limit in all cases. Copper concentrations were appreciable but well under the suggested

Table 1

Results of Analyses of 1979 Bulk Precipitation

Collection Period	Parameter	pH	Acidity (mg CaCO$_3$/l)	Langelier Saturation Index at 20°C	Pb (µg/l)	Cd (µg/l)
Winter	Snow[a] – max.	4.11	10.2	-7.3	352	36.0
	min.[b]	4.04	7.8	-6.7	<20	<1
	mean	4.08	9.0	-7.2	134	17.2
	Rain – max.	4.73	34.6	-7.7	56	3.3
	min.[b]	3.47	14.8	-7.1	<20	<1
	mean[b]	3.57	26.2	-7.6	21	2.1
Summer	Rain – max.	4.31	26.1	-7.7	68	10.5
	min.[b]	3.71	7.0	-7.1	<20	<1
	mean[b]	3.91[c]	12.6	-7.5	25[c]	1.8[c]

[a]Includes two samples that were a mixture of rain and snow.

[b]Values less than detection limit were not used in computing means.

[c]Summer rain mean was significantly different from snow mean at 0.05 α level.

Table 2

Results of Analyses of 1980 Bulk Precipitation

Collection Period	Parameter	pH	Acidity (mg CaCO$_3$/l)	Langelier Saturation Index at 20°C	Zinc (µg/l)	Lead (µg/l)	Cadmium (µg/l)	Copper (µg/l)
Winter	Snow – max.	4.23	43.0	-8.1	190	81	3.0	160
	min.[a]	3.23	10.6	-6.7	<30	25	<1	100
	mean	3.82	24.1	-7.4	80	49	1.5	120
	Mixed – max.	4.20	23.5	-7.8	300	147	3.5	180
	min.[a]	3.65	6.9	-7.2	<30	13	<1	70
	mean	3.93	12.3	-7.5	120	41	1.7	110
Summer	Rain – max.	4.19	62.5	-8.0	70	42	2.0	80
	min.[a]	3.40	9.5	-7.1	<30	<20[b]	<1	30[b]
	mean	3.88	19.1	-7.5	50[b]	12[b]	1.2	53[b]

[a] Values less than detection limit were not used in computing means.
[b] Summer rain mean was significantly different from snow mean at 0.05 α level.

drinking water maximum of 1 mg/l. All zinc concentrations were also well under the drinking water maximum of 5 mg/l.

The mean metals concentrations of 1980 snow samples for the 3 sampling sites are portrayed graphically in Figure 1. Zinc and copper were always present in higher concentrations than lead or cadmium. This result is consistent with findings reported by other researchers (NADP, 1980). Lead levels were relatively high in all of the samples that were collected.

To help keep the metals concentrations in perspective, it would be well to note that the mean values presented are only for samples that exceeded the detection limits of the measurement apparatus. Lead concentrations in all winter 1980 precipitation samples exceeded the detection limit. The same is true of copper concentrations. However, cadmium concentrations exceeded the detection limit in 9 of 10 samples (90%) for snow, 5 of 15 samples (33%) for mixed snow and rain, and in 6 of 26 samples (24%) for summer rain.

The data presented in Tables 1 and 2 indicate that the bulk precipitation falling on roof catchments in the study areas always failed to meet secondary drinking water quality standards for pH and corrosivity. Twelve of 83 precipitation samples (14%) failed to meet the mandatory drinking water standard for lead.

Water samples were taken from 2 points in the cistern systems of 32 homes in 1979. At no point in any of the systems investigated in 1979 was cadmium found to exceed the drinking water limit of 10 µg/l. Results of the 1979 analysis of lead in cistern water supplies are presented in Table 3. Of the 32 cistern water supplies sampled, only the 8 that had lead concentrations in excess of detection limits at either the tap or in the cistern appear in Table 3. Of these, 5 had tap samples which exceeded the drinking water limit for lead on one or both of the two sampling dates.

Although 1979 precipitation values for lead were often in excess of the drinking water limit, no cistern water samples exceeded 26 µg/l and only 2 had lead concentrations above the detection limit. A determination of tapwater Langelier saturation index values for those systems with high lead concentrations revealed highly corrosive water. However, the water was considerably less corrosive than bulk precipitation.

Although each cistern system was unique, most cisterns were constructed of cinder block coated with a waterproof sealant. Many of the systems had sand and gravel filters to

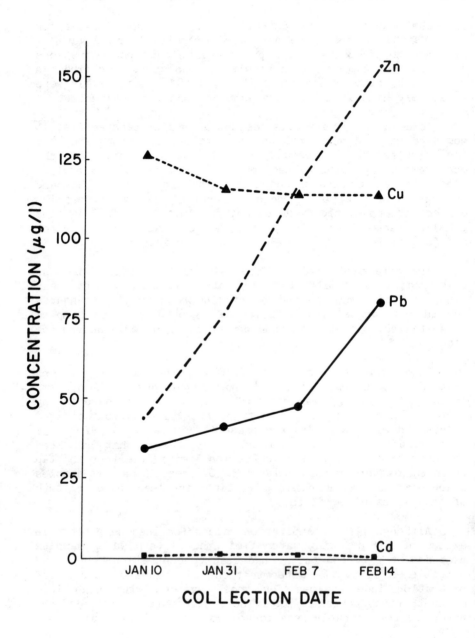

*Figure 1.   Mean (values above detection limits) concentra-
tions of lead, cadmium, copper, and zinc in 1980
snow samples.*

Table 3

Concentrations of Lead Found in Household Cistern and Tap
Water in 1979 in Homes With Appreciable Lead

House Number	March 12-14, 1979		July 31-August 1, 1979	
	Cistern	Tap	Cistern	Tap
	------------------------μg/l-------------------------			
1	ns[a]	77	ns	<20[b]
2	26	23	20	52
3	<20	28	<20	26
4	<20	65	<20	64
5	<20	294	<20	92
6	20	<20	23	<20
7	<20	80	<20	25
8	<20	36	<20	<20
Mean	23	86	22	52
Minimum	<20	<20	<20	<20
Maximum	26	294	23	92

[a]No sample.
[b]The detection limit for lead was 20 μg/l.

remove particulates from the incoming precipitation. It was
thought that the corrosive water entering the cisterns dissolved
calcium carbonate from the concrete blocks and/or the sealant,
with a resultant decrease in the relative corrosivity of the
water. In those systems with carbonate gravel filters, calcium
carbonate may have also been added as the precipitation passed
through the filter. It was further hypothesized that the
increased calcium carbonate content resulted in precipitation
of dissolved lead from incoming precipitation and that
particulate lead was also settling out of the cistern water
during the long period of quiescent storage. The corrosive
cistern water, upon entering the plumbing system, dissolved lead
from solder joints, thus accounting for the high tapwater lead
concentrations.

To test these hypotheses, a new sampling scheme was devised
and implemented in 1980. A sampling apparatus was developed
to allow sampling of the water and sediment within 1 cm of the

bottom of the cistern.   Cistern samples were also collected just beneath the surface and at a depth 15 cm above the floor of the cistern.

The data presented in Table 4 show high concentrations of lead and cadmium in the cistern sediment/water samples.   It should be remembered that sediment/water samples were unfiltered samples of water containing sediments from the bottom of the cistern.   Maximum concentrations for lead and cadmium greatly exceed drinking water limits.   Of even greater concern is the frequency of occurrence of these high concentrations. Lead concentrations in cistern sediment/water samples exceeded drinking water limits in 55, 42, and 59 percent of the systems for the three 1980 sampling dates.

The frequency of occurrence of lead and cadmium values in excess of drinking water limits is shown graphically in Figure 2.   Lead concentrations in precipitation, tapwater, and cistern water occasionally exceeded drinking water limits.   However, cistern sediment/water samples showed a very high number of homes with lead concentrations in excess of established limits. Sediment/water cadmium levels also exceeded drinking water limits in a relatively large number of homes.

Considerable differences between mean bulk precipitation and mean cistern sediment/water values for lead and cadmium were found.   This fact offers support for the hypothesis that the lead and cadmium contained in bulk precipitation settled out of the quiescent water stored in the cisterns.   The relatively high concentrations of lead and cadmium found indi-cate that these metals accumulate over time in the cistern; consequently, average bulk precipitation values which fall within drinking water limits offer no assurance that drinking water limits will not be exceeded in cistern water systems.

The analysis of 1980 cistern surface samples for lead, cadmium, and copper yielded results very similar to those obtained in 1979.   Results from 1980 cistern water samples are presented in Tables 5 and 6.   Maximum values for cadmium and copper were below drinking water limits for all cistern samples in 1980.   For most systems, cistern-water cadmium was below the detection limit of 1 μg/l for all sampling dates.

Lead maxima were over the drinking water limit for three cistern water samples.   The greatest number of systems with lead and cadmium concentrations in excess of the detection limits occurred in the March, 1980, sampling.   In both 1979 and 1980, the highest concentrations of lead and cadmium occurred in the bulk precipitation during the months January through March, preceding the cistern sampling.   This suggests that the

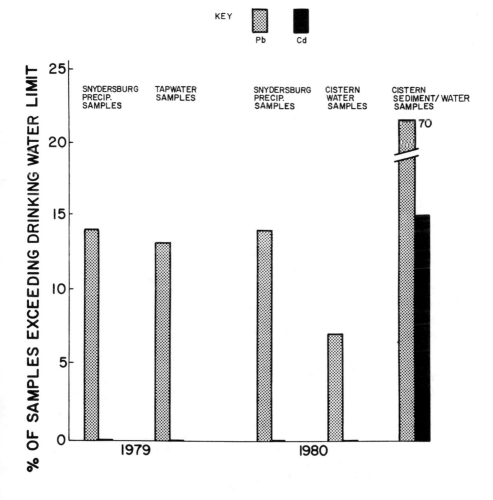

Figure 2. Frequency of occurrence of lead and cadmium concentrations in excess of mandatory drinking water standards.

Table 4

Results of 1980 Cistern Sediment/ Water
Analysis for Cadmium, Copper, and Lead

Sampling Date	n^a	Metal	Concentration (µg/l)			Samples Exceeding Drinking Water Limits	
			Max	Min^b	Mean	No.	%
March 26-28[c]	18	Cu	400	50	130	0	0
	24	Cd	33	1	4.5	4	11
	31	Pb	2610	10	283	21	55
July 8-10[d]	36	Cu	310	30	110	0	0
	19	Cd	30	1	4.1	1	3
	29	Pb	657	5	126	15	42
August 12-14[e]	21	Cu	310	30	90	0	0
	19	Cd	25	1	5.3	3	9
	22	Pb	1430	20	287	20	59

[a]Number of samples with concentration $\geq$ detection limit.
[b]Values given are detection limits in all cases.
[c]Samples obtained from 38 systems.
[d]Samples obtained from 36 systems.
[e]Samples obtained from 34 systems.

concentrations of lead and cadmium in cistern water were related to their concentrations in the bulk precipitation prior to cistern sampling.

Mean heavy metal concentrations and Langelier saturation index values for 1980 summer bulk precipitation and August cistern samples are summarized by a series of bar graphs in Figure 3. Mean lead and cadmium concentrations were a great deal higher in the sediment/water samples than at any other point in the cistern and many times higher than the mean rain concentrations. They were also higher in cistern water than they were in bulk rainwater, indicating a possible accumulating effect. Copper values were also highest in the cistern sediment/water samples. Mean Langelier saturation index values clearly show the neutralizing effects of cistern storage with a negative value less than half that of rainwater. Since the

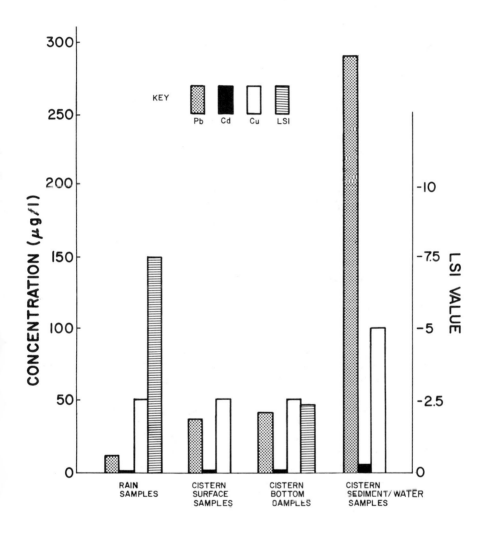

Figure 3.   Mean (values above detection limits) heavy metals concentrations and mean Langelier saturation index values for summer, 1980.

Table 5

Results from 1980 Cistern Surface Samples
for Copper, Cadmium, and Lead

Sampling Date	n	Metal	Max[a]	Mean[b]	n $\geq$ dl[c]
			------µg/l-----		
March 26-28	40	Cu	90	70	6
	40	Cd	2.6	1.4	5
	40	Pb	57	19	19
July 8-10	38	Cu	90	60	35
	38	Cd	3.0	2.0	2
	37	Pb	18	9.0	9
August 12-14	37	Cu	60	40	11
	37	Cd	3.0	1.7	6
	37	Pb	72	37	5

[a]All minimums were less than detection limits.
[b]Values less than detection limits were not used in computing
means.
[c]dl--Detection Limit

concentrations of lead, cadmium, and copper were much greater
in cistern sediment/water samples, it seems apparent that all
of these metals tended to accumulate in the sediments.

Data on the pH, alkalinity, specific conductance, and
calcium concentration of the cistern water have not yet been
summarized for 1980. However, it is evident that the majority
of the cisterns sampled show a marked increase in pH,
alkalinity, and calcium concentrations over levels found in the
bulk precipitation. The increases were less dramatic in
cisterns that were most effectively sealed. Four cisterns with
vinyl liners and no gross-particulate filters had a mean
Langelier saturation index value of -6.9. Although saturation
index values for water stored in most cisterns tended to
improve, in most cases they remained negative, indicating that
serious potential corrosion problems still existed. A report
on the corrosion of household plumbing by this water is planned
when analysis of pertinent data is completed.

Table 6

Results from 1980 Cistern Samples Obtained 15 cm Above
Cistern Floor for Copper, Cadmium, and Lead

Sampling Date	n	Metal	Max[a]	Mean[b]	n ≥ dl[c]
			------μg/l-----		
March 26-28	40	Cu	120	80	5
	40	Cd	2.0	1.3	8
	40	Pb	28	16	10
July 8-10	38	Cu	110	60	37
	38	Cd	3.0	1.5	4
	38	Pb	26	10	8
August 12-14	38	Cu	60	40	10
	38	Cd	3.0	1.7	3
	38	Pb	52	42	3

[a]All minimums were less than detection limits.
[b]Values less than detection limits were not used in computing
means.
[c]dl--Detection Limit

SUMMARY AND CONCLUSIONS

The bulk precipitation collected at several rural
locations in western Pennsylvania always failed to meet the
U.S. Environmental Protection Agency's drinking water limits
for pH and corrosivity. In addition, it failed to meet drinking
water limits for lead 14 percent of the time. The bulk
precipitation collected was representative of that being
collected by roof catchment-cisterns located in the local area
surrounding the collection sites. The quality of water in the
cistern water supplies studied appeared to reflect the quality
of bulk precipitation collected.

The concentrations of lead and cadmium in cistern water
were relatively low and usually below drinking water limits and
maximum values obtained for bulk precipitation. This was true
because the cistern water in most systems acquired calcium
carbonate by virtue of its contact with the calcium carbonate
material (i.e., cinder block and cement sealer) composing the
walls of the cistern. Cisterns with vinyl liners had more

corrosive water. Because of the addition of calcium carbonate, soluble metals were removed by precipitation. However, the settling of particulates was probably even more important in the removal of lead and cadmium from cistern water.

Much of the lead and cadmium that entered the system in the bulk precipitation appeared to be deposited in the sediment on the bottom of the cistern. Measurements of lead concentrations in cistern sediment/water samples supported this thinking in that the mean lead concentrations in cistern sediment/water samples were almost double the bulk precipitation maxima for 1980. Mean sediment/water cadmium concentrations also exceeded the maximum 1980 bulk precipitation cadmium concentration. Analysis of tapwater in 1979 revealed lead concentrations in excess of the 50 µg/l drinking water limit in 13 percent of the homes studied that year. In addition, 7 percent of the homes had cistern water in excess of the drinking water limit for lead in 1980. Cistern sediment/water lead concentrations exceeded the drinking water limit in 70 percent of the homes studied on at least one occasion. Cistern sediment/water cadmium concentrations exceeded the drinking water limit in 15 percent of the homes on one or more occasions.

It appears from the data presented that acid precipitation and its burden of heavy metals have a profound impact on roof catchment-cistern water supplies. As acid water enters the system, it reacts chemically with the materials with which it comes in contact producing significant concomitant changes in its chemical nature. Calcium carbonate is dissolved from the masonry cistern structure and heavy metals are deposited on the bottom of the cistern. The less acidic, but still highly corrosive water, is pumped into the domestic plumbing system where it corrodes metal pipes and soldered pipe joints, picking up additional heavy metals. Since there are no safeguards in most roof catchment-cistern systems to prevent this lead- and cadmium-contaminated cistern water and sediment from being ingested by system users, the changes caused by the acidic character of precipitation pose a serious health threat to the users of such systems.

ACKNOWLEDGMENTS

Funding for this study was provided by The Pennsylvania State University, College of Agriculture, under Title V of the Rural Development Act.

LITERATURE CITED

Cogbill, C.V. and G.E. Likens. 1974. Acid Precipitation in the Northeastern U.S. Water Resources Res. 10(6):1133-1137.

Hutchinson, T.C. 1973. Lead Pollution, the Automobile, and Health. Institute of Environmental Sciences and Engineering, University of Toronto, Toronto, Ontario. Publ. EF-5, pp. 1-10.

Kincaid, T.C. 1979. Alternative Individual Water Sources. In: "Quality Water for Home and Farm," pp. 73-78. Proc. Third Domestic Water Quality Symposium, ASAE.

Lazrus, A.L., E. Lorange, and J.P. Lodge, Jr. 1970. Lead and Other Metal Ions in U.S. Precipitation. Environ. Sci. Tech. 4(1):55-58.

NADP. 1980. Report of a Workshop on Toxic Substances in Atmospheric Deposition: A Review and Assessment, J.N. Galloway, S.J. Eisenreich, and B.C. Scott (eds.), National Atmospheric Deposition Program, Jekyll Island, Georgia, November, 1979.

U.S. EPA. 1979. Methods for Chemical Analysis of Water and Wastes. EPA-600/4-79-020, U.S. Environmental Protection Agency, Cincinnati, Ohio.

PART 4

EFFECTS OF ACID PRECIPITATION
ON TERRESTRIAL SYSTEMS

CHAPTER 17

THE INTERACTION OF WET AND DRY DEPOSITION
WITH THE FOREST CANOPY

Steven E. Lindberg and David S. Shriner
Environmental Sciences Division
Oak Ridge National Laboratory
Oak Ridge, Tennessee 37830

William A. Hoffman, Jr.
Department of Chemistry
Denison University
Granville, Ohio 43023

INTRODUCTION

Vegetation is a particularly important sink for atmospheric emissions; because of their reactivity and large surface areas, foliar canopies are effective receptors of airborne material delivered by both wet and dry deposition processes. Dry deposition cannot be neglected in studies of the role of the atmosphere in geochemical cycling, particularly when a constituent of dry deposition is delivered to the canopy in an available (water-soluble) form. Whereas elements transported to a canopy in an insoluble form may be of limited consequence, a particle-associated element of high solubility can be mobilized when exposed to moisture on the foliage. This can have important nutritional or toxic consequences for the receptor plant community.

The purposes of this chapter are to present a brief overview of current knowledge regarding the effects of wet- and dry-deposition on the terrestrial environment, and then to describe results and implications of some recent research on the interactions of wet and dry deposited acidity, sulfate, and trace metals with the forest canopy at the Walker Branch Experimental Watershed in eastern Tennessee.

Before considering the potential interactions between wet and dry deposition of acidic substances in the forest canopy, it is appropriate to review the effects of wet deposition _per se_ on vegetation surfaces. Several recent reviews are available

385

(Jacobson, 1980; Evans and Hendrey, 1979; Shriner, 1981), and the reader is encouraged to refer to them for more detailed discussion.

Rain, fog, dew, and other forms of wet deposition play an important role for vegetation and soils as sources of nutrient inputs and as removal or leaching mechanisms for mineral nutrients, amino acids, carbohydrates, and growth regulators (Tukey, 1975). Recent monitoring data demonstrate the role precipitation plays in scavenging pollutants from the atmosphere (Dana, 1980), and suggest changes in the character of wet deposition over the last 50 to 80 yr (Shriner and Henderson, 1978). Considerable speculation has been generated on the potential effects that increases in concentrations of rain-scavenged pollutants might have on biotic receptors (Cowling, 1980).

The case study described in the following section documents some interactions in the forest canopy between wet- and dry-deposited pollutants. Interpretation of the implications of the case study requires several assumptions based on the conclusions of earlier research on effects of wet deposition:

(1)   Both direct and indirect effects on vegetation may occur as the result of wet deposition (Shriner, 1978a).

(2)   Direct effects are directly measurable consequences of exposure   i.e., altered physiological response (Ferenbaugh, 1976); visible injury (Evans and Curry, 1979); or growth, yield, quality reduction or increase (Jacobson, 1980).

(3)   Increased inputs of hydrogen, sulfate, and nitrate ions may have both positive and negative effects on soil-plant systems. These effects may occur simultaneously (Jacobson et al., 1980; Irving, 1979).

(4)   The major components of wet deposition which are of documented concern in vegetation effects are dissolved gases (SO_2, NO_2), sulfate, nitrate, and hydrogen ions (Jacobson et al., 1980; Jacobson and Van Leuken, 1977); the effects of heavy metals are not well documented.

(5)   The occurrence of direct effects on vegetation depend upon: (a) precedent and antecedent conditions e.g., soil nutrient status and plant nutrient requirements (Noggle, 1980; Turner and Lambert, 1980), plant sensitivity and growth stage (Jacobson and Van Leuken, 1977; Shriner, 1978b; Evans and Curry,

1979) ; and (b) the total loading or deposition of critical ions (H^+, $SO_4^=$, NO_3^-)(Jacobson and Van Leuken, 1977; Wood and Bormann, 1974; Irving, 1979).

(6)   Substantial differences exist between greenhouse and field response of vegetation to wet deposition (Shriner, 1981).

(7)   The following leaf surface characteristics may play a critical role in wet deposition effects (Evans et al., 1977):   (a) presence of sensitive structures; (b) wettability (Shriner, 1981); (c) leaf morphology (e.g., shape, patterns of venation, etc.)(Evans et al., 1977); and (d) tissue age (Evans and Bozzone, 1977; Shriner, 1978b; Evans et al., 1977).

(8)   Apart from possible direct effects resulting in visible tissue damage, the effects of acid precipitation alone can be considered as primarily a nutrition problem. In circumstances where nutrient elements become growth-limiting or more available as a result of acid rain, reduced or increased forest productivity may result (Abrahamsen, 1980).

(9)   Wet deposition may play a role in the individual plant's relative fitness or ability to survive or otherwise compete at the community or ecosystem level. Such effects are, for the most part, extremely difficult to characterize.

DEPOSITION/CANOPY INTERACTIONS: A CASE STUDY

There have been few studies of atmospheric deposition to the forest canopy which utilize separate collection of wet and dry deposition on an event basis.   During 1977-1978, we investigated the rates and processes of atmospheric deposition of sulfate, several trace metals, and precipitation acidity to the forest canopy in Walker Branch Watershed (WBW, see Figure 1).   Walker Branch is a 98 ha deciduous forest catchment in a rural area of eastern Tennessee (35° 58'N, 84° 17'W) and is typical of the oak-hickory forests of the eastern United States. The facilities and related research projects are described in detail in Harris (1977) and Lindberg et al. (1979).   The watershed is a suitable field laboratory for study of the deposition and accumulation of coal-combustion-derived emissions because it is situated within 20 km of three coal-fired power plants (total electric generation capacity about 2200 MW) and within 350 km of 22 coal-fired power plants (about 2×10^4 MW).   As indicated in Figure 1, several major emission sources are frequently upwind of the watershed.   Our methods

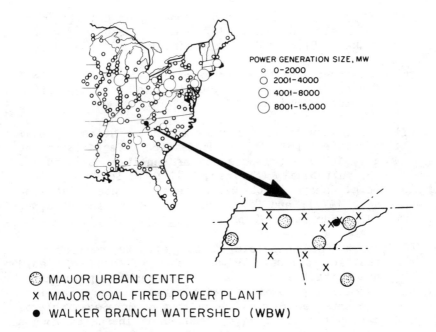

POWER GENERATION SIZE, MW
- o 0-2000
- ○ 2001-4000
- ◯ 4001-8000
- ◯ 8001-15,000

◉ MAJOR URBAN CENTER
X MAJOR COAL FIRED POWER PLANT
● WALKER BRANCH WATERSHED (WBW)

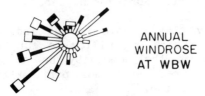

ANNUAL
WINDROSE
AT WBW

Figure 1.   *Geographical location of Walker Branch Watershed
in relation to continental, regional, and local
sources of atmospheric emissions from fossil fuel
utilization. Shown are the spatial distribution
of centers of coal and oil-fired steam and electric generation capacity in the eastern United
States and the location of specific regional coal-
fired power plants and major urban centers.*

of collection and analysis of wet and dry deposition involved
modified   HASL-type wetfall-only collectors for rain and
throughfall and a combination of vegetation and flat-plate-
collector sampling in the upper canopy for dry-deposited
particles. A 46-m meteorological tower at WBW (Figure 2) was
used for the above-canopy precipitation, dry deposition, and
aerosol sampling as well as for collection of upper-canopy
vegetation. These methods and our laboratory extraction and
analytical techniques are described in detail elsewhere
(Lindberg et al., 1977; Lindberg et al., 1979; Hoffman et al.,
1980a; Lindberg and Harriss, 1981).

Figure 2. Aerial photograph of the 46 m meteorological tower at Walker Branch Watershed used for above-canopy sampling of deposition and for access to upper-canopy sampling sites.

Precipitation Acidity

Our recent work on rain chemistry (Hoffman et al., 1980a) has explored the role of strong and weak acids in rain above and below the chestnut oak canopy in Walker Branch. The contribution to the free acidity as measured by conventional pH electrodes is from all protons in solution, regardless of source. Strong acids completely dissociate in solution, while weak acids, both organic and inorganic, partially dissociate. The degree of weak acid dissociation increases with increasing pH such that at the pH of most acid precipitation samples (<5.6) they contribute negligibly to the free acidity and primarily to the total acidity (Galloway et al., 1976). For 29 storm samples in WBW during the 1977-1978 growing seasons, we found weak acids to account for approximately 30 percent of the total acidity in incident rain ($\bar{X} \pm SE = 32 \pm 3\%$). Total acidity averaged 137 ($\pm$ 8, SE) µeq/l while strong acidity averaged 93 $\pm$ 11 µeq/l. This is similar to the early data published on weak acids in rain by Galloway et al. (1976) who reported a contribution of 32 percent weak acids to the total acidity in one storm sampled during July in the northeast.

In throughfall collected beneath the chestnut-oak canopy for the period described above, we found weak acids to contribute approximately twice as much to the total acidity (55 $\pm$ 4%) as was the case for incoming rain. The contribution of weak acids to throughfall chemistry has not been previously quantified. In general, the contribution of the weak acids to the free acidity was negligible in both rain and throughfall. The canopy had little influence on total acidity of throughfall (135 $\pm$ 8 µeq/l); however, strong acidity was decreased significantly (61 $\pm$ 9 µeq/l). On a concentration basis, the average net loss of strong acidity from the incoming rain to the canopy was approximately 30 µeq/l.

The role of rain acidity in both cation and weak acid leaching in throughfall should be considered. Although the composition of leaf cuticle is not well known, it is thought to be a polyester of $C_{16}-C_{18}$ acids (Albersheim, 1965). The action of the cuticular material as an exchange site following interception of rain could be expected to result in some degree of structural alteration (Garrells and Christ, 1965) such that some components of the negative framework would dissolve into the surrounding solution. Thus, the interaction between hydrogen ions in rain and the canopy may be responsible for both cation displacement and weak acid leaching as discussed in more detail by Lindberg et al. (1979) and Hoffman et al. (1980).

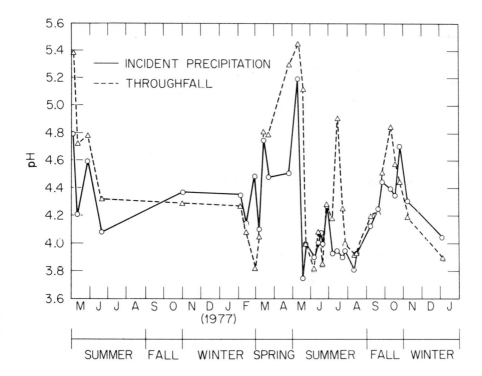

*Figure 3.   Seasonal variation of observed pH of rain and
            throughfall in individual events sampled
            simultaneously above and below the canopy at
            Walker Branch Watershed over about a 2-yr
            period.*

With this background, it is useful to consider the event-
by-event variations in rain and throughfall acidity during a
2-yr period (Figure 3). During the summer of 1976, the influence
of the forest canopy was apparent, with consistently higher pH
values in throughfall than in rain above the canopy. Following
leaf fall, and up to canopy budbreak, the throughfall pH values
were similar to but consistently lower than the incident rain
pH values. This was attributed to leaching of primarily weak
organic acids and, in some cases, strong acids from bark and
twigs (Hoffman et al., 1980a, 1980b). Following budbreak and
canopy refoliation, the pH again reversed such that the
throughfall pH exceeded that of incoming rain, indicating
hydrogen ion exchange at the newly formed leaf surface.
However, this trend was not maintained throughout the growing
season as it was during the previous year. For five successive
storms in May to June, the throughfall pH values were less than

or comparable to those in rain above the canopy, with one exception. This suggests that the fully developed canopy had temporarily lost the ability to remove hydrogen ions from incoming precipitation. This phenomenon did not coincide with a concomitant decrease in net leaching of other cations, however (Lindberg et al., 1979). The indication is that leaching during this period must have involved some mechanism other than the cation exchange process described above, or that the hydrogen ions, once at exchange sites, were not retained by the leaves.

We found that this phenomenon of higher throughfall acidity relative to incident rain during full canopy development is caused by increases in both strong and weak acids in throughfall (Hoffman et al., 1980a). This indicates increased weak acid leaching from the leaves and solubilization of particles on the leaf surface having a strong acid reaction. Particles with a strong acid nature are abundant both in ambient air and on various surfaces at WBW (Lindberg et al., 1979). Interestingly, the May to June "pH reversal" period was preceded by 15 days of no precipitation (except for a brief 10 min shower of 0.1 cm on the 5th day following the start of this period) during which time several local air stagnation bulletins were issued. The period was characterized by calm, hazy conditions and frequent inversion/fumigation events, during which time the ozone, sulfur dioxide, total suspended particle, or nitrogen oxide concentrations in ambient air exceeded air quality standards (Knoxville - Knox County Air Quality Board, personal communication).

The interaction of the strong oxidant ozone at and internal to the leaf surface is known to attack double bonds and affect membrane permeability (Mudd and Kozlowski, 1975). The esterified $C_{16}-C_{18}$ acids, which comprise the cuticle, include free carboxyl groups which would be likely sites of cation exchange. If these groups are points of reaction with ozone, as would be predicted by a strict chemical consideration, their ability to behave as hydrogen ion exchange sites would be significantly impaired. This may partially explain the subsequent pH reversal. The storm events occurring during this period (every event was sampled) were generally small in volume (<2.5 cm). However, the first major storm following this period (> 11 cm) resulted in return of the throughfall/rain pH relationship to that expected for the developed canopy (throughfall pH $\geq$ rain pH). It may be speculated that this large volume of rain efficiently cleansed the leaves and in some way restored the exchange nature of the surfaces, since the subsequent storms generally resulted in considerably higher throughfall pHs relative to incoming rain. This "normal" relationship was maintained throughout the remainder of the growing season, although there were two more events where the pH of rain was essentially unchanged following canopy interception (Figure 2). During the period following leaf fall (approximately October 20), throughfall pH was again consistently less than the pH of the incident precipitation as during the previous year. Detailed information on the strong and weak acid content, sulfate and nitrate concentratons, and levels and identities of several organic compounds in rain and throughfall collected during these periods has been published (Hoffman et al., 1980a, 1980b).

Continuing examination of the chemistry resulting from the rain-leaf interaction is both prudent and necessary to achieve a fuller understanding of internal and external mechanisms affecting leaf processes. The phenomenon of strong acid scavenging from rain by the forest canopy is a well-known but poorly understood phenomenon resulting in the release of plant-related weak acids and nutrients (Eaton et al., 1973; Likens et al., 1977; Hoffman et al., 1980a). The resulting pH of throughfall is often considerably above the values characteristic of "acid rain." Future research in forests where this occurs (e.g., eastern deciduous forests) should concentrate on canopy effects in addition to soil effects, particularly on the role of hydrogen uptake and subsequent organic carbon and nutrient loss from foliage.

Dry-Deposited Particles

Our data indicate that atmospheric deposition plays a significant role in the cycling of metals and sulfate through the canopy and to the forest floor in Walker Branch Watershed and that dry deposition cannot be neglected in the forest environment (Lindberg et al., 1979). The relative proportion of total atmospheric deposition attributed to dry processes has been largely ignored in field studies until quite recently. Table 1 (right-hand section) indicates the relative contribution of dry deposition to total deposition measured at WBW over several time scales. The atmospheric input during the short-term experiments (periods W2, W3, and W6) was generally dominated by dry deposition; only during period W3 was the wetfall process of greater relative importance and then only for cadmium and sulfate-sulfur. Over the two longer time scales of 207 to 365 days, dry deposition constituted a significant fraction of the total atmospheric input of cadmium and zinc (~20%), sulfate (~35%), and lead (~55%).

Particle dry deposition is generally considered to be a chronic, cumulative exposure of vegetation to atmospheric constituents. However, rainfall is episodic, and a sudden inundation of the leaf surface can result in an unusually harsh

Table 1

Comparison of Wet, Dry, and Total Deposition Measured at Walker Branch Watershed – Wet and Dry Deposition Rates to Individual Upper Canopy Surfaces Are Compared When Normalized to a Unit Time Basis in the Center Section of the Table – The Right-Hand Portion of the Table Indicates the Relative Contribution of Dry Deposition to the Total Deposited Quantity of an Element to the Entire Canopy Over Several Time Periods – Dry Deposition is Estimated From Particle Accumulation on Inert, Flat Plates Exposed During These Periods as Described in Detail in Lindberg et al. (1979)

Period	Normalized Deposition Rates ($\mu g/m^2/h$)				Relative Contribution of Dry to Total Deposition (dry/total) · 100%			
	Zn	Cd	Pb	$SO_4^=$-S	Zn	Cd	Pb	$SO_4^=$-S
W2[a]								
Wet	23	2.7	24	42,000	88	82	100	70
Dry	0.10	0.01	0.62	50				
W3[b]								
Wet	11	2.7	45	17,000	84	18	86	47
Dry	0.03	0.0003	0.12	10				
W6[c]								
Wet	37	3.1	25	23,000	82	67	100	80
Dry	0.03	0.0001	0.24	17				
W2, 3, 6								
Mean wet/dry ratio[d]	610	4,100	170	1,300				
Growing season[e]	--	--	--	--	20	26	26	34
Annual	--	--	--	--	18	17	53	33

[a]May 16 to May 20, 1977; wet event duration = 0.007 d, total period duration = 4.2 d.

[b]May 30 to June 6, 1977; wet event duration = 0.02 d, total period duration = 7.0 d.

[c]July 12 to July 18, 1977; wet event duration = 0.007 d, total period duration = 6.0 d.

[d]April 1 to October 24, 1977; total period duration = 207 d.

[e]Calendar year 1977 (data from Lindberg et al., 1979).

exposure of the vegetation to potentially toxic material. Wet
deposition rates of zinc, cadmium, lead, and sulfate measured
during events of short duration and low rainfall volume, and
hence generally high concentrations, were considerably higher
than any of the measured dry deposition rates when expressed on
a comparative unit time basis (center column of Table 1). The
episodic wet deposition rates were calculated as the areal
wetfall input divided by the duration of the precipitation
event. Although the duration of the rain events during W2, W3,
and W6 was short (10 to 30 min), the wet deposition rates during
the rain events were approximately two to three orders of
magnitude greater than the dry deposition rates measured for
each period between rain events (when expressed on a per hour
basis).

For the initial bioreceptor, this episodic flux of
potentially toxic material can play an important role in
physiological effects. Precipitation events of short duration,
low volume, and high elemental concentrations were common during
the growing season at WBW. In many cases, these events followed
relatively long (5 to 10 days) dry periods characterized by
elevated air concentrations and dry deposition rates. When the
subsequent precipitation event was very small (~0.5 to 1.5 mm),
much of the initial precipitation remained on the leaf surface,
not being washed off or diluted by subsequent rainfall. The
potential for physiological effects is enhanced when this
solution contacts previously dry-deposited material on the leaf
surface. The high concentrations which develop under these
conditions are further enhanced during the evaporation of
droplets on the leaf surface.

The event which occurred during period W2 provides an
example of this phenomenon. Using the dry deposition rate of
the water-soluble fraction of elements measured for surfaces
in the upper canopy, it is possible to estimate the approximate
surface-area concentrations of dry-deposited material on a leaf
prior to the precipitation event. These values are summarized
in Table 2, including calculations of the total quantity of
wet-deposited material, dissolved concentrations in water
droplets on the leaf surface, and total deposition to the leaf,
expressed relative to the leaf internal content. Precipitation
of 1.3 mm falling on a 50 cm^2 leaf would deposit about 6 ml of
water. As this solution begins to evaporate, the leaf is
exposed, although briefly, to extremely high concentrations
resulting from the interaction between surface moisture and dry
deposition. Evaporation of the surface moisture would result
in concentrations of dissolved constituents several hundred to
several thousand times higher than typical rain concentrations
(compare the concentrations in Table 2 with average rain

concentrations at WBW of Cd = 0.42 μg/l, Zn = 5.6 μg/l, and SO_4^{-2}-S = 1200 μg/l; Lindberg, in press).

Quantification of these interactions is particularly important in relating precipitation chemistry to vegetation effects observed in the field; the need for both event sampling and wet/dry segregation is obvious. Figure 4 illustrates the importance of sampling strategy in determining actual field

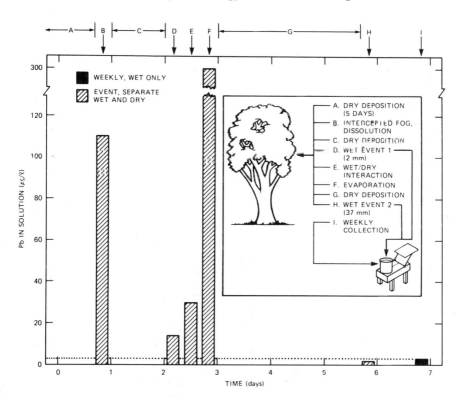

Figure 4. *The importance of sampling strategy in quantifying pollutant flux to and concentrations on canopy surfaces. The ordinate indicates the range in concentrations of lead measured at a leaf surface due to several wet/dry events over the course of a week as determined by separate collection of wet and dry deposition on an event basis (hatched bars), and by nonevent compositing of wet deposition only over a 7 day period using methods currently employed by several deposition networks [including the National Atmospheric Deposition Program (solid bar)].*

Table 2

Potential Concentrations of Several Elements in Solution on a Typical Upper-Canopy Chestnut Oak Leaf Surface (50 cm^2) Following a Brief Summer Shower Which Was Preceded by a 10-Day Dry Period

Parameter	Units	Water-Soluble Constituents		
		Cd	Zn	$SO_4^=$-S
Surface area concentration of previously dry-deposited, water-soluble material[a]	ng/cm^2	0.1	1.2	1200
Mass of dry-deposited, water-soluble material	ng	5	60	70x10^3
Concentration of soluble fraction dissolved by precipitation	µg/l	1	9	10x10^3
Concentration of elements in incident precipitation	µg/l	0.35	2.9	5x10^3
Estimated total concentration in solution on the leaf surface	µg/l	1.3	12	15x10^3
Potential concentration following evaporation	µg/l	130	1200	1.5x10^6
Total mass of elements in solution on the leaf surface	µg	0.007	0.08	100
Total quantity of elements bound within the leaf	µg	0.065	7.0	2000
Total quantity of available fraction of elements delivered to the leaf surface during the growing season	µg	1.1	30	1700
Ratio of soluble element deposition during W2 to total leaf content[a]	--	0.11	0.01	0.05
Ratio of available element deposition during the growing season to total leaf content	--	17	4.3	0.8

[a]Wet and dry deposition were separately sampled at Walker Branch Watershed from 5/16 to 5/20/77 (period W2). Although the rain event, which occurred on 5/18, was preceded by a 10-day dry period, dry-deposited element concentrations were determined only for the ~2.4-day sampling period preceding the

rain. Dry-deposited element concentrations are thus conservative estimates of total dry material on the leaf surfaces prior to the rain (data from Lindberg et al., 1979).

exposure levels during a hypothetical 1-wk period. Shown are
the concentrations of lead in solution on a leaf surface in the
upper canopy during and following a series of events (A through
I), including two rainstorms (D and H) of considerably different
rainfall amount (2 and 37 mm), several periods of dry deposition
(A, C, and G), one period of intercepted fog (B), and one example
of dissolution of dry deposition by rain (E-F). Interactions
of moisture with previously deposited particles on the leaf
surface can result in extremely high concentrations of soluble
material (~100 to 300 µg/l) which exceed concentrations in
wetfall alone by factors of approximately 10 to 100. These
exposure levels could not be quantified without analysis of dry
deposition rates. Another complicating factor involves the
choice of collection period for wetfall. If precipitation had
been sampled on a weekly basis (I), instead of by event during
this same period, the occurrence of the larger storm (H)
following the 2 mm event (D) would have further diluted the
apparent concentrations to which the vegetation was exposed
because of the inverse relationship between precipitation
concentrations and rain volume (Lindberg, in press). The effect
of weekly wetfall sampling in this situation would be to
obliterate the details of the series of events which may or may
not have resulted in an observed effect on the vegetation.
These calculations assume a uniform surface concentration of
dry-deposited material, which we have shown is a gross
simplification (Lindberg et al., 1979). Scanning electron
photomicrographs of leaf surfaces indicate consider-able
heterogeneity in deposited particle distribution (Shriner and
Henderson, 1978; Lindberg et al., 1979).

The physiological effects of surface-deposited metals on
vegetation, either in particulate or dissolved form, require
considerable study, as they are very poorly understood. There
are several conflicting reports on toxicity in the literature
(as reviewed by Krause and Kaiser, 1977; Zimdahl, 1976).
However, the estimated pH of the resulting solution on the leaf
surface (~2, not indicated in Table 2) can cause adverse effects
in several plant species (Shriner, 1981). Our calculation of
acidity assumes no neutralizing capacity of the surface-
deposited particles or of the leaf itself, a situation which
is unusual but which has been documented (Lindberg et al., 1979;
Hoffman et al., 1980a).

Table 2 also summarizes estimates of the magnitude of the
deposition inputs during one event and over the growing season
relative to the total leaf internal content of cadmium, zinc,
and sulfate. The quantity of lead estimated to be in solution
on the leaf surface following the single event (not shown) was
nearly comparable to the total lead content of the leaf
(Lindberg et al., 1979). Lesser but still significant
quantities of soluble cadmium and sulfate were deposited during

this event. During the full growing season, the leaf surface was exposed to one to two orders of magnitude more dissolved cadmium and lead, approximately 4 times more zinc, and a nearly equal amount of sulfate relative to the leaf content. While the effects of such exposures are unknown, the importance of atmospheric deposition in the cycling of these elements in the landscape is obvious and warrants continued research.

POTENTIAL EFFECTS OF POLLUTANT INTERACTIONS

Wet deposition of acidic substances occurs over wide-spread areas of the eastern United States and northern Europe as a result of long-range transport of the precursors and secondary reaction products involved. As a major consequence, vegetation in these regions is commonly exposed not only to those gaseous phytotoxicants and fine particulates common to urban and some rural areas but also to acid rain. In terms of effects of pollutant interactions, two major phenomena may be classed as processes; one which takes place at the leaf surface, and one which uses the leaf surface as a "reaction surface." The leaf surface itself may be either an active or a passive participant in these processes.

Little information is available which will permit a conclusive evaluation of the potential for interactive effects of wet- and dry-deposited pollutants as far as plant response is concerned. With dry deposition of gases, preliminary work by Shriner (1978b), Irving (1979), and Jacobson et al. (1980) suggests that interactions may occur. In controlled exposure greenhouse experiments, Shriner (1978b) reported no significant interaction between multiple rain exposures (pH 4.0) and 4 sulfur dioxide exposures (3 mg/l peak, 1 hr, see method described by McLaughlin et al., 1979) as far as growth of bush beans was concerned. Irving (1979), working with field-grown soybeans, found that simulated acid precipitation (pH 3.1) apparently diminished a photosynthetic decrease she observed that resulted from 17 exposures to 0.19 mg/l of sulfur dioxide during the growing season. Shriner (1978b) also exposed plants to 0.15 mg/l ozone (four 3-hr exposures) in between 4 weekly exposures to rainfall of pH 4.0 and observed a significant growth reduction at the time of harvest. Jacobson et al. (1980), using open-top exposure chambers with field-grown soybeans, compared growth and yield among three pH levels of simulated rain (pH 2.8, 3.4, and 4.0) and two levels of ozone (<0.03 and ≤ 0.12 mg/l). Results demonstrated that ozone depressed both growth and yield of soybeans with all three rain treatments, but that the depression was greatest with the most acidic rain. The above investigations clearly point to the need for research in this area, because ozone levels equal to or greater than those used in the studies are relatively common in most of the areas of the eastern United States where acid deposition is also a problem (Jacobson et al., 1980).   Mechanisms of these interactions are currently unknown.

During significant portions of the growing season, the foliage of vegetation will be wetted by rain, fog, or dew formation in mid-latitude temperate climates.   During these periods, one would expect leaf surfaces to become more efficient sinks for the dry deposition of gases, and perhaps more efficient in particle retention due to surface-tension effects. Unsworth and Fowler (1976), in describing sulfur dioxide fluxes to a wheat canopy, defined surface resistance of the crop canopy.   By their definition, if that crop surface resistance term should be zero, then the crop is a perfect sink for sulfur dioxide. During the course of their experimentation, they found that when the leaves of the crop were wet with dew, the surface resistance was very small (near or equal to zero).   As sulfur dioxide dissolved in the film of moisture on leaf surfaces, they found that the pH of the dew decreased, suggesting the potential for direct acid-type effects on wet vegetation surfaces until the pH-dependent sulfur dioxide solubility limits the process.

The leaf surface is involved in dynamic interaction with the atmosphere.   Vaughan (1976) lists a number of extrinsic factors operating at the leaf surface which can significantly modify the subsequent sorption and translocation processes associated with particle deposition.   Those factors include humidity, precipitation, light cycle, mass loading of the leaf, and size of the aerosol particle.   Vaughan (1976) stated that mass loading in excess of about 1 $\mu g/cm^2$ of leaf surface area significantly inhibits translocation of photosynthate from the leaf to the root by impairing gas exchange.   Data of Lindberg et al. (1979) suggest that if the 1 μg figure is accurate, quite large areas of deciduous forest canopies may be operating under conditions of impaired gas exchange during at least portions of the growing season.

Total mass loading of the leaf surface is influenced by many factors.   Our research (Lindberg et al., 1979) indicates that the pattern of residual particles on leaf surfaces may be strongly influenced by the drainage patterns of wet deposition passing from the leaves as throughfall, with collection occurring along major veins and leaf margins.   An additional significant factor in particle retention during short periods in the growing season is the presence of excretions of biological origin, such as aphid "honey dew," which results in the leaves becoming sticky and efficient collection surfaces.

Limited data are available from which to draw conclusions regarding the full role of interactions between wet- and dry-deposited pollutants on the leaf surface. However, Hosker and Lindberg (1981) recently reviewed a number of factors which must be considered in the context of the potential fate and effects of these pollutants in soil/plant systems. The interaction of particles with vegetation is similar in some respects to that with gases. Exceptions arise due to the following processes:

(1) aerodynamic effects,
(2) diffusion of small particles,
(3) gravitational sedimentation of larger particles (>1 μm), and
(4) particle bounce leading to resuspension from the receptor surface.

Physical entrapment by structural features of the leaf surface (leaf hairs, irregularities associated with vein patterns, hyphae of leaf surface microflora, etc.) may also be an important factor, but it is virtually uncharacterized.

Chamberlain (1975) discussed the role that wetted leaf surfaces play in dampening the particle bounce mentioned above, suggesting increased particle retention by dry deposition to wet surfaces. However, impaction of water droplets during wet deposition may effectively dislodge and resuspend dry deposited particles, resulting in secondary movement of the particles through the canopy and ultimately to the soil surface either in throughfall or stemflow.

Hosker and Lindberg (1981) also listed: (a) physical properties of particle size, density, and shape; (b) chemical composition, especially solubility in water or degree of hydration; and (c) electric charge, as properties likely to affect the interaction of particles with the leaf surface. Of these, the first two would appear to be of special significance in the potential interaction of liquids and particles on the leaf surface. Carlson et al. (1976), investigating wind reentrainment versus rainwash of lead aerosols deposited on plant leaves, found that wind reentrainment was a minor factor in comparison to rainwash. Simulated rainfall removed up to 95 percent of the topically applied lead. Of the lead removed in rainwash, the mechanisms of dissolution or suspension and of subsequent runoff appeared to predominate over that of splash-off from leaf surfaces.

An additional factor influencing particle retention on leaf surfaces is that of cuticle weathering. Rainwash of leaf surfaces results in weathering of epicuticular waxes which, in turn, results in increased wettability of the leaf surface (Martin and Juniper, 1970). This increased wettability will result in increased retention time by the leaf of both moisture and particles. Hosker and Lindberg (1981) suggest that this retention time is probably "the most critical parameter" influencing effects, including those resulting from particles interacting subsequently with wet- or dry-deposited pollutants.

Particles deposited on the leaf surface include soluble compounds which, when mobilized, may penetrate into the leaf via natural flaws in the cuticle, as well as across cell membranes of trichomes and other specialized cells devoid of epicuticular waxes. Direct cuticular penetration is possible, with rates dependent on compound concentration, cuticle thickness, and the affinity of cuticular constituents for the solutes involved (Hosker and Lindberg, 1981). It is interesting to speculate that the cuticle could represent a significant sink for nonpolar compounds deposited on the leaf surface.

Particles on a leaf surface, in essence, increase the reactive areas of that leaf surface, potentially increasing the sorptive capacity of the surface for gases as well as serving as a source of soluble constituents to be leached by dew, fog, and rain. The roles that strong and weak acids in precipitation or organic acids derived from the vegetation play in the rates of solubilization of nutrients and/or toxic trace elements from dry-deposited particles on leaf surfaces are largely unknown. Lindberg et al. (1979) found increased deposition of sulfate during periods of foliage wetness, and hypothesized that this reflected adsorption of sulfur dioxide by surface moisture followed by oxidation (possibly catalyzed by Mn) to sulfate.

SUMMARY

Dry deposition of gaseous pollutants to vegetation and the resultant effects have long been recognized as an important factor in the long-term "health," productivity, and stability of natural and agricultural ecosystems. Recognition of the important role that wet deposition and wet/dry deposition interactions play in pollutant transport, fate, and effects has only come in the last decade. As each of these is studied in greater detail, working hypotheses such as some of those suggested in this discussion will lead to a clearer understanding, and ultimately to the predictive capabilities necessary to evaluate the long-range environmental consequences of an energy-constrained industrial society.

Additional refinements in experimental design necessary for understanding plant-pollutant interactions should include studies of the following:

(1) the influence of leaf surface microstructure and leaf exudates on particle capture, retention, and gas absorption;

(2) dose-response relationships for particle deposition;

(3) possible effects of shifting to a dominant direct atmospheric source of plant nutrients and toxins rather than to a dominant soil-uptake route;

(4) episodic exposure of the canopy to high deposition rates of both wet and dry components;

(5) wet- and dry-deposited material distribution on individual leaves and within the canopy;

(6) deposited particle solubility under various moisture and chemical regimes; and

(7) interaction of dew and intercepted fog or mist with deposited particles on vegetation and subsequent effects of the resulting solution on plant growth and trace element composition.

ACKNOWLEDGMENTS

Research sponsored by the Office of Health and Environmental Research, U.S. Department of Energy, under contract W-7405-eng-26 with Union Carbide Corporation. Publication No. 1725, Environmental Sciences Division, Oak Ridge National Laboratory (ORNL), Oak Ridge, Tennessee 37830.

LITERATURE CITED

Abrahamsen, G. 180. Acid Precipitation, Plant Nutrients, and Forest Growth. In: D. Drabløs and A. Tollan, eds., "Ecological Impact of Acid Precipitation," pp. 58-63. SNSF, Norway.

Albersheim, P. 1965. Biogenesis of the Cell Wall. In: J. Bonner and J. Varner, eds., "Plant Biochemistry," pp. 151-188. Academic Press, New York, New York.

Carlson, R.W., F.A. Bassaz, J.J. Stukel, and J.B. Wedding. 1976. Physiological Effects, Wind Reentrainment, and Rainwash of Pb Aerosol Particulates Deposited on Plant Leaves. Environ. Sci. Technol. 10:1139-1142.

Chamberlain, A.C. 1975. The Movement of Particles in Plant Communities. In: J.L. Monteith, ed., "Vegetation and the Atmosphere, Vol. I, Principles," pp. 115-203. Academic Press, London.

Cowling, E.B.   1980.   Acid Precipitation and Its Effects on Terrestrial and Aquatic Ecosystems.   In: "Aerosols and Natural Sources and Transport," pp. 540-555.   Annals N.Y. Acad. Sci. 338.

Dana, M.T.   1980.   Overview of Wet Deposition and Scavenging. In: D.S. Shriner, C.R. Richmond, and S.E. Lindberg, eds., "Atmospheric Sulfur Deposition: Environmental Impact and Health Effects," pp. 263-274.   Ann Arbor Science Publishers, Ann Arbor, Michigan.

Eaton, J.A., G.E. Likens, and F.H. Bormann.   1973.   Throughfall and Stemflow-Chemistry in a Northern Hardwood Forest.   J. Ecol. 61:465-508.

Evans, L.S. and D.M. Bozzone.   1977.   Effect of Buffered Solutions and Sulfate on Vegetative and Sexual Development in Gametophytes of Pteridium aquilinum.   Amer. J. Bot. 64:897-902.

Evans, L.S., N.F. Gmur, and F. DaCosta.   1977.   Leaf Surfaces and Histological Perturbations of Leaves of Phaseolus vulgaris and Helianthus annuus After Exposure to Simulated Acid Rain. Amer. J. Bot. 64:903-913.

Evans, L.S. and T.M. Curry.   1979.   Differential Responses of Plant Foliage to Simulated Acid Rain.   Amer. J. Bot. 66:953-962.

Evans, J.S. and G.R. Hendrey (eds.).   1979.   Proc. International Workshop on the Effects of Acid Precipitation on Vegetation, Soils, and Terrestrial Ecosystems.   BNL-51195.   Brookhaven National Laboratory, Upton, New York.

Ferenbaugh, R.W.   1976.   Effects of Simulated Acid Rain on Phaseolus vulgaris L. (Fabaceae).   Amer. J. Bot. 63:283-288.

Galloway, J.N., G.E. Likens, and E. Edgerton.   1976.   Acid Precipitation in the Northeastern U.S. - pH and Acidity. Science 194:722-723.

Garrells, R.M. and C.L. Christ. 1965. Solutions, Minerals, and Equilibria. Harper and Row, New York, New York. 450 pp.

Harris, W.F. 1977. Walker Branch Watershed: Site Description and Research Scope. In: D.L. Correll, ed., "Watershed Research in Eastern North America," Smithsonian Institute Press, Edgewater, Maryland. 924 pp.

Hoffman, W.A., Jr., S.E. Lindberg, and R.R. Turner.   1980a. Precipitation Acidity: The Role of the Forest Canopy in Acid Exchange. J. Environ. Qual. 9:95-100.

Hoffman, W.A., Jr., S.E. Lindberg, and R.R. Turner. 1980b. Some Observations of Organic Constituents of Rain Above and Below the Forest Canopy. *Environ. Sci. Technol.* 14:999-1002.

Hosker, R.P. and S.E. Lindberg (eds.). 1981. Atmospheric Transport, Deposition, and Plant Assimilation of Airborne Gases and Particles. Proceedings of the Second Ecology - Meteorology Workshop, Pellston, Michigan, Aug. 16-20, 1976 (in press).

Irving, P.M. 1979. Response of Field-Grown Soybeans to Acid Precipitation Alone and In Combination With Sulfur Dioxide. Ph.D. Dissertation, University of Wisconsin, Milwaukee, Wisconsin.

Jacobson, J.S., J. Troiano, L.J. Colavito, L.I. Heller, and D.C. McCune. 1981. Polluted Rain and Plant Growth. In: "Polluted Rain," 12th Rochester International Conference on Environmental Toxicity. Plenum Publishing Co., New York, New York.

Jacobson, J.S. and P. Van Leuken. 1977. Effects of Acid Precipitation on Vegetation. In: "Proc. Fourth International Clean Air Congress, Tokyo," pp.124-127. Japanese Union of Air Pollution and Prevention Assoc., Tokyo, Japan.

Jacobson, J.S. 1980. Experimental Studies on the Phytotoxicity of Acid Precipitation: The United States Experience. In: T.C. Hutchinson and M. Havas, eds., "Effects of Acid Precipitation on Terrestrial Ecosystems," pp. 151-160. Plenum Press, New York, New York.

Krause, G.H.M. and H. Kaiser. 1977. Plant Response to Heavy Metals and SO_2. *Environ. Pollut.* 12:63-71.

Likens, G.E., F.H. Bormann, R.S. Pierce, J.A. Eaton, and N.M. Johnson. 1977. Biogeochemistry of a Forested Ecosystem. Springer-Verlag, New York.

Lindberg, S.E. Factors Influencing Trace Metal, Sulfate, and Hydrogen Ion Concentrations in Rain. *Atmos. Environ.* (in press).

Lindberg, S.E. and R.C. Harriss. 1981. The Role of Atmospheric Deposition in an Eastern United States Deciduous Forest. *Water Air Soil Pollut.* 16:13-31.

Lindberg, S.E., R.C. Harriss, R.R. Turner, D.S. Shriner, and D.D. Huff. 1979. Mechanisms and Rates of Atmospheric Deposition of Selected Trace Elements and Sulfate to a Deciduous Forest Watershed. ORNL/TM-6674. Oak Ridge National Laboratory, Oak Ridge, Tennessee.

Lindberg, S.E., R.R. Turner, N.M. Ferguson, and D. Matt. 1977. Walker Branch Watershed Element Cycling Studies: Collection and Analysis of Wetfall for Trace Elements and Sulfate. In: D.L. Correll, ed., "Watershed Research in Eastern North America," Smithsonian Institute Press, Edgewater, Maryland. 924 pp.

Martin, J.T. and B.E. Juniper. 1970. The Cuticles of Plants. St. Martins Press, New York, New York. 247 pp.

McLaughlin, S.B., D.S. Shriner, R.K. McConathy, and L.K. Mann. 1979. The Effects of SO_2 Dosage Kinetics and Exposure Frequency on Photosynthesis and Transpiration of Kidney Beans (Phaseolus vulgaris L.). Environ. Exp. Bot. 19:179-191.

Mudd, J.B. and T. Kozlowski. 1975. Response of Plants to Air Pollutants. Academic Press, New York, New York. 383 pp.

Noggle, J.C. 1980. Sulfur Accumulation by Plants: The Role of Gaseous Sulfur in Crop Nutrition. In: D.S. Shriner, C.R. Richmond, and S.E. Lindberg, eds., "Atmospheric Sulfur Deposition: Environmental Impact and Health Effects," pp. 289-298. Ann Arbor Science Publishers, Ann Arbor, Michigan.

Shriner, D.S. 1978a. Effects of Simulated Acidic Rain on Host-Parasite Interactions in Plant Diseases. Phytopathology 68:213-218.

Shriner, D.S. 1978b. Interactions Between Acidic Precipitation and SO_2 or O_3: Effects on Plant Response. Phytopathol. News 12:153.

Shriner, D.S. 1981. Terrestrial Vegetation - Air Pollution Interactions: Non-Gaseous Pollutants, Wet Deposition. In: S.V. Krupa and A. Legge, eds., "Air Pollutants and Their Effects on Terrestrial Ecosystems," Wiley Interscience, New York, New York (in press).

Shriner, D.S. and G.S. Henderson. 1978. Sulfur Distribution and Cycling in a Deciduous Forest Watershed. J. Environ. Qual. 7:392-397.

Tukey, H.B., Jr. 1975. Regulation of Plant Growth by Rain and Mist. Proc. Int. Plant Prop. Soc. 25:403-406.

Turner, J. and M.J. Lambert. 1980. Effects of Atmospheric Sulfur Deposition on Forest Growth. In: D.S. Shriner, C.R. Richmond, and S.E. Lindberg, eds., "Atmospheric Sulfur Deposition: Environmental Impact and Health Effects," pp. 321-334. Ann Arbor Science Publishers, Ann Arbor, Michigan.

Unsworth, M.N. and D. Fowler. 1976. Field Measurements of Sulfur Dioxide Fluxes to Wheat. In: "Atmosphere - Surface Exchange of Particulate and Gaseous Pollutants, 1974," pp. 342-353. ERDA Symposium Series 38, CONF-740921. National Technical Information Services, Springfield, Virginia.

Vaughan, B.E. 1976. Suspended Particle Interactions and Uptake in Terrestrial Plants. In: "Atmosphere - Surface Exchange of Particulate and Gaseous Pollutants, 1974," pp. 228-243. ERDA Symposium Series 38, CONF-740921. National Technical Information Services, Springfield, Virginia.

Wood, T. and F.H. Bormann. 1974. The Effects of an Artificial Acid Mist Upon the Growth of Betula alleghaniensis Britt. Environ. Pollut. 7:259-268.

Zimdahl, R. 1976. Entry and Movement in Vegetation of Lead. J. Air Pollut. Control Assoc. 26:655-660.

CHAPTER 18

POTENTIAL EFFECTS OF ACID PRECIPITATION ON SOIL
NITROGEN AND PRODUCTIVITY OF FOREST ECOSYSTEMS

John D. Aber
Department of Forestry
University of Wisconsin
Madison, Wisconsin 53706

George R. Hendrey and A.J. Francis
Department of Energy and Environment
Brookhaven National Laboratory
Upton, New York 11973

Daniel B. Botkin
Environmental Studies Program
University of California
Santa Barbara, California 93016

Jerry M. Melillo
Ecosystems Center
Marine Biological Laboratory
Woods Hole, Massachusetts 02543

INTRODUCTION

A number of effects of increased acidity of precipitation
on productivity of forest ecosystems has been proposed. These
include direct damage to living plants (Tamm and Cowling, 1977)
and indirect effects through altered soil chemistry and biology
(Norton et al., 1980; McFee et al., 1977; Francis et al., 1980).
The impact of direct damage effects has been evaluated by field
(Abrahamsen et al., 1980) and computer simulation (Botkin and
Aber, 1979) methods using current and forseeable levels of
precipitation acidification. Results indicate no reduction in
forest productivity. Effects on soils can take many forms
(Table 1). A number of these processes have shown significant
sensitivity to changes in pH, including carbon dioxide evolution
from humus (Tamm et al., 1977; Abrahamsen et al., 1980; Francis
et al., 1980) and leaching of aluminum from soils (Cronan and
Schofield, 1979; Meyer and Ulrich, 1977; Norton et al., 1980;
Johnson, 1979). Base saturation and soil pH appear less

411

Table 1

Some Potential Effects of Acid Precipitation on
Soil Chemistry and Nutrient Availability

A.   Direct Effects of Increased Hydrogen Ion Concentration
     1.   Decreased base saturation
     2.   Reduced availability of cations
     3.   Increased solubility of aluminum, iron, heavy metals,
          precipitation of phosphate
     4.   Increased foliar leaching

B.   Indirect Effects of Increased Hydrogen Ion Concentration
     1.   Reduced decomposition rate, and alterations in soil
          microbial populations and soil chemistry
     2.   Reduced root activity

B.   Effects of Increased Inputs of Nitrogen Oxides
     1.   Increased availability of nitrogen
     2.   Effect on decomposition rate

---

sensitive to acid precipitation effects (Abrahamsen et al.,
1980; Singh et al., 1980; Sposito et al., 1980).

Of the soil processes listed in Table 1, we have
concentrated on those affecting nitrogen availability because
it is the only nutrient which can be shown, by fertilizer trials,
to be growth-limiting in the northern hardwood forests of the
eastern United States which are the subject of this chapter.
We feel that alterations in productivity, if they occur, will
result from changes in nitrogen availability which in turn can
be affected by any modification in the composition of the soil
solution (e.g., concentrations of hydrogen, aluminum, and
calcium ions, heavy metals, etc.).

The purpose of this chapter is to evaluate the potential
effects of acid precipitation on nitrogen availability in forest
ecosystems and to test the sensitivity of northern hardwood
forests to these potential changes through a series of computer
simulations.

ACID PRECIPITATION, NITROGEN AVAILABILITY, AND TREE GROWTH

To date, no changes in forest production have been reported
even under severe, although short-term, acidifications in both
laboratory (Woods and Bormann, 1977) and field trials

(Abrahamsen et al.; Tamm et al., 1977).   Actual increases are occasionally reported (Woods and Bormann, 1977; Tviete and Abrahamsen, 1980).   Attempts to measure changes in tree growth through time under natural conditions have also shown little effect or no effect (Jonsson, 1979; Cogbill, 1977).   However, it has been shown for northern hardwoods that any change in the total amount of nitrogen becoming available in a year will cause a nearly proportional change in productivity (Aber et al., 1980, this paper).   Matching the latter result with laboratory incubation studies on the effect of pH on organic matter decomposition (Tamm et al., 1977; Francis et al., 1980), by which nitrogen is made available, indicates that changes in soil pH should reduce production.

Five factors which could delay or counteract the appearance of a growth response are:

(1)   Acid precipitation may not cause acidification of soils. Peterson (1980), Frink and Voigt (1977), and McKee et al. (1977) all suggest that proton inputs in precipitation may be minimal compared with the generation of organic acids and protons from soil processes, particularly litter decay, nitrification, and plant uptake.   However, Mayer and Ulrich (1977) emphasize that pH changes in the top few centimeters would required smaller hydrogen ion inputs before reducing process rates. Tamm (1977) also states that effects at colloid surfaces or in microsites might be important well before changes in bulk properties occur.   Other workers stress that acid precipitation will essentially increase the rate of podzolization, a long-term process of soil horizon development and surface acidification (Peterson, 1980; Norton, 1977).   Thus, long term measurement may be required to document these changes. In the first mention of at least middle-term results, Norton et al. (1980) report measurable differences in pH and base saturation over 8- and 16-yr periods.   They also report a spatial sequence representing a time-intensity sequence showing foliar changes in Ca:Al and Mn:Al ratios as well as the lead and zinc concentrations. They conclude that effects can now be seen in forest litter and shallow inorganic soils. Finally, extreme, short-term field trials have shown increased leaching of cations and decreased soil pH (Abrahamsen, 1980), with the former more extreme than the latter. The question of long-term acidification remains crucial and unanswered.

(2)   pH changes in soils may not cause changes in total nitrogen mineralization.   This would be counter to

effects observed in lab incubations and generally accepted notions of the effects of pH on microbial activity. However, in situ incubations under undisturbed conditions over a wide range of natural soil pH conditions will be required to answer this question completely.

(3)   Decreases in nitrogen mineralization may be offset by increases in nitrogen inputs in acid precipitation. Such inputs can total over 20 kg/ha/yr in a heavily impacted area (Heinrich and Mayer, 1977; Likens et al., 1978). This may be half of the total annual requirement for nitrogen in conifer stands such as those studied by Abrahamsen (1980) and Tamm et al. (1977). An important failing of many acid precipitation forest studies has been the exclusion of the nitrogen component. Controlling pH alone with sulfuric acid is misleading.

(4)   Short-term changes in nitrogen availability may be buffered by drawing on an internal pool of this nutrient in the plant. Nitrogen content of leaf/needle litter fall is roughly half that of green tissues on the tree. This retranslocated nitrogen is stored until the next growing season. Forty percent of total aboveground nitrogen demand can be met in this way in hardwood stands, more in conifer stands. This would act as a buffer, delaying the appearance of deficiency symptoms.

(5)   The vegetaton could be tolerant of low nitrogen availability. This is the case for the conifer stands studied by Abrahamsen (1980) and Tamm et al. (1977). No experimental acidification of soils in hardwood stands has been reported. This vegetation type is more sensitive to changes in nitrogen availability and should, therefore, be more susceptible to acid precipitation-induced changes in its availability (Aber et al., 1980). Hardwood forests also occupy most of the heavily impacted areas in the United States.

There is a substantial difference between these five processes in terms of long-term impact on forest production. The second and third would suggest that no such effects will occur as no long-term changes in nitrogen availability will occur. The first, fourth, and fifth would represent delays in the expression of symptoms which would eventually appear. Thus, it is crucial to sort out the relative importance of each.

The following is our working hypothesis of the interaction and relative importance of these and other factors in determining acid precipitation effects on nitrogen availability in hardwood forests in the northeastern United States.

Changes in soil pH will generally depress rates of soil organic carbon and nitrogen mineralization (rejecting number 2 above).   All   laboratory   studies   associated   with   acid precipitation research have reported decreased carbon dioxide evolution at lower pH (Abrahamsen et al., 1980; Tamm et al., 1977; Francis et al., 1980). Apparent increases in nitrogen availability (measured as leachate from soil incubations, e.g., Tamm et al., 1977) must be a short-term phenomenon resulting from the death and lysing of microbial cells and displacement of ammonium ions from exchange sites. Long-term increases in nitrogen mineralization and decreases in carbon mineralization could occur only through a drastic increase in the C:N ratio in soils. This is very unlikely; but if it did occur, increased C:N ratio would almost certainly result in a further decrease in mineralization rate. Likewise, number 5 has been shown by field   and   simulation   studies   to   be   inapplicable   to   these hardwood forests (Auchmoody and Filip, 1973; Mitchell and Chandler, 1939; Aber et al., 1980). Nitrogen is an important limiting nutrient in most northeastern hardwood forests (see analysis by Mitchell and Chandler, 1939). Internal storage through retranslocation (number 4) should provide only short-term buffering. Fertilizer studies show a 5-yr carry-over effect on growth, at least partially due to this mechanism. This could help to explain a lack of growth response in short-term, extreme acidification studies such as those carried out in Scandinavia, but would not provide significant long-term protection.

We are left, then, with the interaction of increased nitrogen inputs and potentially decreased mineralization rates as important acid precipitation effects. The effect of the former is fairly easy to predict. Most of these inputs are as mineral nitrogen directly available for plant uptake.

The rate and effect of soil acidification remain the most important   unknowns.   These   are   affected   not   only   by precipitation pH but also by initial soil properties (e.g., CEC, base saturation, organic matter content, texture) and rates of plant uptake and recycling of nitrogen and other non-limiting nutrients (calcium, magnesium, potassium, phosphorus, and iron) and elements (aluminum). All of these affect the content of the soil solution and hence microbial dynamics. Interactions of the soil solution with primary and secondary minerals, and chemical precipitates (e.g., oxides and phosphates of iron and aluminum) will also affect this solution. The nature of the soil solution determines the environment of microbial activity and hence organic matter catabolism and nitrogen mineralization. We need to know much more about the interactions of these processes with precipitation inputs of different pH before soil acidification estimates can be accurately made.

Within the limitations of this discussion, we have carried out a series of computer simulations on the sensitivity of northern hardwood forest ecosystems to potential acid precipitation-induced changes in nitrogen availability.

METHODS

Our simulation trials were carried out in two stages using two different models. The first stage used the JABOWA forest growth simulator of Botkin et al. (1972) modified to include the effects of nitrogen availability on tree growth (Aber et al., 1979). Northern hardwood and boreal species are assigned to one of three classes expressing their tolerance of low nitrogen availability (Table 2). Tolerance of low nitrogen is not necessarily correlated with tolerance of shade. A growth multiplier for each class is developed, as shown in Figure 1, from the data of Mitchell and Chandler (1939). The response of a forest to changes in nitrogen availability can be expected to be strongly related to the response of its dominant species. This set of simulations contains no feedbacks from changes in production to litter fall quantities and future site conditions. Thus, a series of nitrogen availabilities are imposed as initial conditions in the model and are not altered by it.

In the JABOWA simulations, a pattern of nitrogen availability predicted following clearcutting (Aber et al., 1978; Figure 2) was altered in different ways. From the literature review above, we could estimate up to a 50 to 60 percent reduction in soil organic matter decomposition (Tamm et al., 1977) or up to a 20 kg/ha/yr addition of nitrogen due to acidified precipitation (May and Ulrich, 1977). Thus, the values for nitrogen availability in Figure 2 were multiplied by 0.5, 0.75, and 0.9, and increased by 10 and 20 kg/ha/yr in five different trials. Simulations were run at two "elevations" described by the number of growing degree days (DEGD). At DEGD = 3000, the forest is pure northern hardwoods with boreal components (spruce and fir) absent. At DEGD = 2500, the forest is dominated by spruce and fir with paper birch an important early successional species and a minor beech and sugar maple component.

Table 2

List of Species and Their Nitrogen Tolerance Classes
Represented in the JABOWA Forest Growth Model - Species
Need Not Fall Into the Same Tolerance Class as They Would
With Regard to Light - Tolerance Class Assignments Were
Made on the Basis of Mitchell and Chandler's Study (1939)
and Information on Foliar Nitrogen Levels and Growth
Obtained from the Hubbard Brook Ecosystem Study - Species
With a Dash (-) in the Reference Column Are Very Minor
Stand Components Under All Conditions Tested - Nomenclature
After Fernald (1950) Unless Otherwise Noted

Species and Tolerance Class	Reference
Tolerants:	
Paper birch (Betula papyrifera)	Aber, 1976
Red maple (Acer rubrum)	Mitchell and Chandler, 1939
Red spruce (Picea rubens)	Likens and Bormann, 1970
Balsam fir (Abies balsamea)	Likens and Bormann, 1970
Intermediates:	
Sugar maple (Acer saccharum)	Mitchell and Chandler, 1939
Beech (Fagus grandifolia)	Mitchell and Chandler, 1939
Yellow birch (Betula alleghaniensis Britt.)	Aber, 1976
Striped maple (Acer pennsylvanicum)	--
Mountain maple (Acer spicatum)	--
Mountain ash (Pyrus americana)	--
Choke cherry (Prunus virginiana)	--
Intolerants:	
White ash (Fraxinus americana)	Mitchell and Chandler, 1939
Pin cherry (Prunus pennsylvanica)	Marks, 1974

In the second stage, the feedbacks between production,
litter fall, and future decomposition and nitrogen availabilities are added by linking a much modified version of the
forest growth model to litter production and decomposition
models. These allow changes in primary productivity to alter
the quantity and type of litter produced which will affect soil

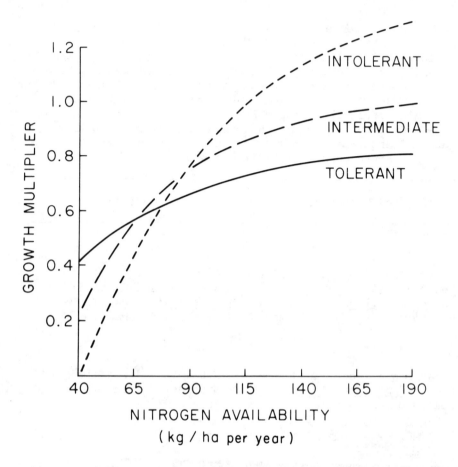

Figure 1. *Relationship between nitrogen availability and a
relative growth factor used in the JABOWA and
FORTNITE models (see Aber et al., 1979, for
derivation).*

stocks of organic matter and nitrogen and allow secondary
effects of acid rain on nitrogen availability.   The
decomposition model uses data from Melillo et al. (1981) and
Aber and Melillo (1980) to assign a decomposition rate and
nitrogen dynamics parameters to each of 19 types of litter which
are generated annually and followed individually until they
become forest floor organic matter (F+H layer or 02 horizon).
This F+H material all decomposes at a single rate and
mineralizes nitrogen in proportion to its concentration in this
material. Nitrogen is immobilized by decomposing litter. This

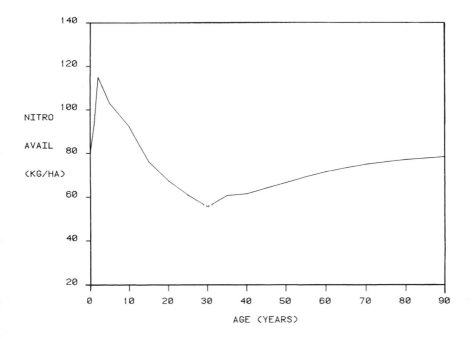

Figure 2. *Patterns of nitrogen availability predicted*
following clearcutting in northern hardwood
forests (from Aber et al., 1978).

combined model is called FORTNITE (FOResT NITrogEn model) and
is described in detail elsewhere (Aber and Melillo, 1981).

Due to a lack of data on litter fall and decomposition in
the higher elevation boreal forest, this set of simulations was
run for the lower elevation hardwood forest only. A similar
range of conditions was tested, except that instead of altering
nitrogen availability directly, decomposition rates were
increased or decreased, resulting in a changed nitrogen
availability. This alters production rates, which also changes
the litter production rates, which further affect nitrogen
cycling and availability. It is clear that acid precipitation
effects should be considered in this context of total ecosystem
function. However, this demands a larger data base and a much
more complex model whose large number of feedback loops requires
substantially increased accuracy to keep the state variables
(e.g., nitrogen availability or live tree biomass) within
bounds.

The experimental conditions tested included 10 and 20 percent increases in decomposition in litter and forest floor and 10, 25, and 50 percent decreases. All trials were run from identical initial conditions beginning with a clearcut in year 0.

RESULTS

The JABOWA simulations yield changes in equilibrium live biomass and biomass accumulation rate for the two forest types at different elevations. Figures 3 and 4 show that patterns of biomass accumulation are quite different in the two forest types, with the early peak in year 160 in Figure 3 absent in Figure 4. This is a function of different growth rates and

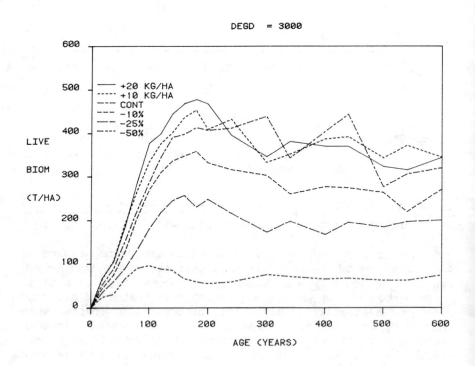

Figure 3.   *Successional changes in total live biomass at two elevations (as simulated by DEGD) as a function of simulated changes in nitrogen availability (see text for explanation of treatments).*

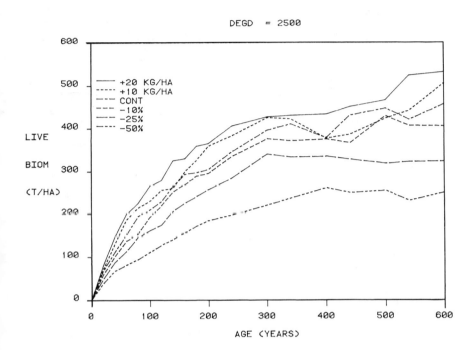

Figure 4. Successional changes in total live biomass at two elevations (as simulated by DEGD) as a function of simulated changes in nitrogen availability (see text for explanation of treatments).

life histories of the dominants in each type, with the higher elevation forest (DEGD = 2500) dominated by slower-growing and longer-lived species (especially spruce).

Error limits in both forest types are extremely low (less than 10% of the means) for most of the first 200 yr. Figures 5 and 6 show the results for this time period in greater detail and include significant differences between treatments. The range of responses is much smaller in the boreal zone forest, as would be expected from a forest dominated by nitrogen-tolerant species (see Table 2).

Figures 7 and 8 summarize these trends by comparing total accumulated live biomass in the two forest types expressed as averages for the period 500 to 600 yr (a "climax" or equilibrium forest, Figure 7) and the period from 120 to 220 yr (the peak

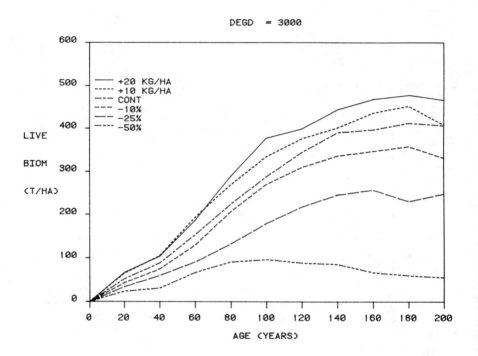

*Figure 5.   Successional changes in total live biomass at two
             elevations (as simulated by DEGD) as a function of
             simulated changes in nitrogen availability (see
             text for explanation of treatments).*

biomass period in the hardwood stands, Figure 8). The first
reflects changes in long-term equilibrium biomass and the second
changes in rate of biomass accumulation. In each case, the
abscissa is the absolute availability of nitrogen during the
period after the successional changes shown in Figure 2.
Interestingly, both forest types show similar absolute
reductions in equilibrium biomass, although the hardwoods show
a greater percentage reduction. For the 120 and 220 yr period,
reduced nitrogen availability causes a much larger decline in
the lower elevation (higher DEGD) hardwood stand. In either
case, it is apparent that even small changes in nitrogen
availability due to acid precipitation will cause a calculable
reduction in forest biomass accumulation when concurrent
changes in litter production and its decomposition are not
included.

DEGD = 2500

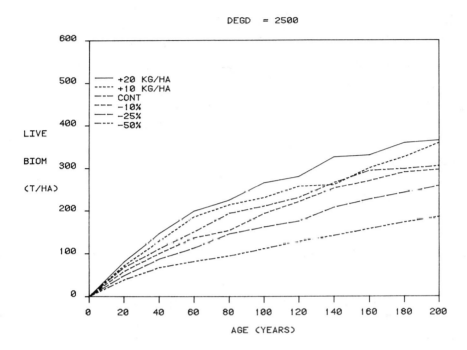

Figure 6. Successional changes in total live biomass at two elevations (as simulated by DEGD) as a function of simulated changes in nitrogen availability (see text for explanation of treatments).

The FORTNITE trials yield changes in live tree biomass, forest floor biomass, and nitrogen availability for the lower elevation deciduous forest.

Figure 9 shows trends in forest floor organic matter through time for the different treatments. All show the initial decline following cutting which has been measured for this forest type (Covington, 1981) but the magnitude of the decline is very different as is the rapidity of recovery and the value of type asymptote. The relative position, of the lines are what would generally be expected, with the highest decomposition rate yielding the lowest steady-state forest floor biomass and the lowest rate yielding the highest.

Figure 10 shows predicted patterns in nitrogen availability. The initial peaks following cutting are due to

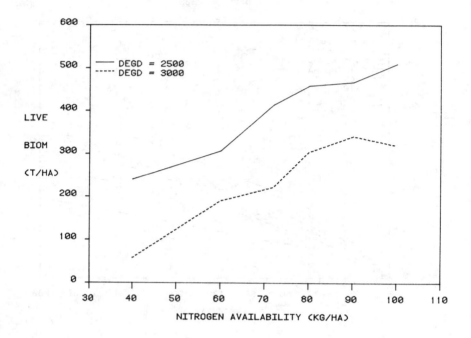

*Figure 7.   Average values for total biomass from 500 to 600
yr as a function of steady-state nitrogen availabilities.*

a reduction in leaf biomass which results in increased
decomposition in the model. For the two largest reductions in
decomposition rate, the reduction is large enough to overcome
the leaf area effect by year 5 and nitrogen availabilities are
already well below initial values.

In all cases, nitrogen availability declines between year
5 and 20 with only a small range in values present by year 20.
This decline is due to the decreasing amount of organic matter
in the forest floor from which the nitrogen is mineralized.
After year 20, nitrogen availability increases with increasing
forest floor biomass.   The curves in Figure 10 beyond year 20
are basically identical to Figure 9 scaled by the reduction or
increase in decomposition rate specified by the treatment.

Of particular interest here is that nitrogen availability
in the 50 percent decomposition reduction treatment eventually

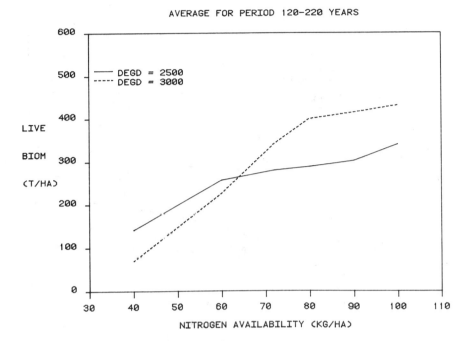

AVERAGE FOR PERIOD 120-220 YEARS

*Figure 8. Average values for total biomass from 120 to 220
yr as a function of steady-state nitrogen availabilities.*

becomes much higher than any other due to the tremendous
accumulation of organic matter. There is no field evidence
supporting the idea of higher nitrogen availabilities in stands
with low organic matter turnover rates. Two possible factors
currently not in FORTNITE, and not well understood in general,
could explain this discrepancy.

The first is concerned with assigning a single
decomposition rate to all of the non-litter organic matter in
the forest floor. It is well known that this is a combination
of many kinds of materials in various stages of decay which are
decomposing at different rates. For example, C.A. Federer
(personal communication) has shown that nitrogen is mineralized
much more rapidly per unit organic matter from the F than from
the H horizon. Both of these are included in a single
compartment in FORTNITE and mineralize at the same rate. In
forest soils which accumulate large amounts of organic matter,

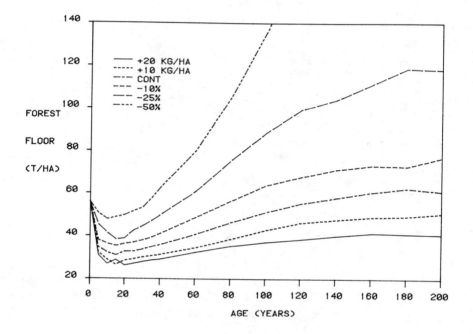

Figure 9. *Successional changes in total forest floor*
organic matter as a function of change in
decomposition rate in soil and litter as
estimated by FORTNITE.

this material may be mainly recalcitrant humic substances rather
than more labile material with faster turnover rates. Thus,
slow decomposition may be a function of the relative amounts
of the two (or more) classes of materials and may result more
from the incomplete decomposition of litter (larger portion of
litter becoming "humus") than from the decomposition rate of
stable soil organics.

A second possibility is that decomposition of soil organic
matter by microbes is generally nitrogen-limited. An increase
in nitrogen availability speeds decomposition and also allows
microbes to attack complex materials not readily decomposable
under low-nitrogen conditions. Thus, the increase in nitrogen
availability beyond yr 80 for the 50 percent reduction treatment
might, in the field, cause an increase in the decomposition
rate in a positive feedback relationship, thus reducing the
forest floor organic matter content. This would represent a

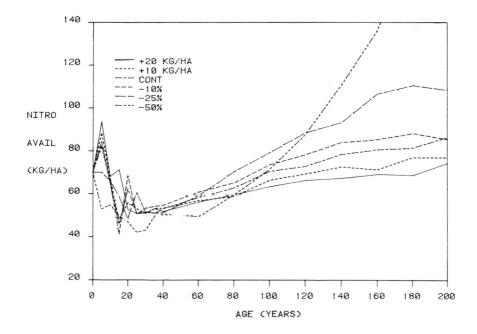

Figure 10. Successional changes in nitrogen availability as a function of change in decomposition rate in soil and litter as estimated by FORTNITE.

strong homeostatic control on total forest floor organic matter which might, if documented, buffer the ecosystem responses presented in this report.

In general, it is interesting that, after year 20, higher decomposition rates mean reduced nitrogen availabilites due to reductions in the amount of organic matter and hence nitrogen present for mineralization. This is a counter-intuitive result.

Figure 11 shows the rate of live tree biomass accumulation under each treatment. These are generally in the same relation as the nitrogen availability curves except that the degree of difference has been transformed by the relationship between nitrogen availability and tree growth used in FORTNITE (Figure 1). This curvilinear relationship makes differences toward low availabilities more important to productivity than high values. Thus, the growth reductions associated with the low nitrogen

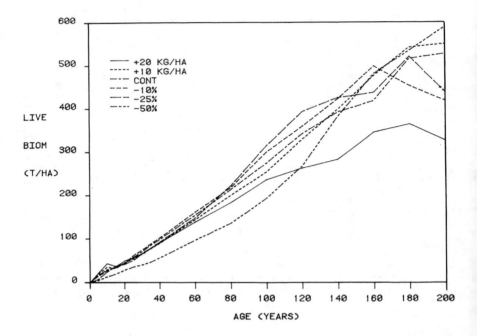

Figure 11. *Successional changes in total live biomass as*
            a function of changes in decomposition rates
            in soil and litter as estimated by FORTNITE.

availability periods (year 20 to 80, 50% reduction; after year
80, 20% increase) are greater than the growth increases
coincident with high availabilities in the 25 and 50 decrease
treatments.

    What are the implications of these trials for the value
of forests as wood producers? FORTNITE can be used to test
different management practices and can estimate yields of wood
products under different conditions. We ran a series of sample
clearcuttings for each treatment, removing all stem wood after
90 yr of growth. Table 3 reveals that yields are highest in
the control and are reduced under either increased or decreased
decomposition rates (the difference between control and 10 and
25 percent reductions are not significant). These values result
from changes in total productivity and also the pattern of tree
mortality associated with each treatment, which alters the size
distribution of stems.

Table 3

Estimated Yield From 90-Yr Clearcutting Rotations
for Changes in Decomposition Rate

Treatment	Yield (tonne/ha)
Increased 20 percent	126.3
Increased 10 percent	141.1
Control	163.6
Decreased 10 percent	158.9
Decreased 25 percent	153.0
Decreased 50 percent	109.0

CONCLUSIONS

Our first set of simulations (JABOWA) indicate that even small changes in nitrogen availablity due to acid rain will result in calculable short-term changes in production for these nitrogen-limited forests. However, when the further effects of altered productivity are included in the second set of simulations (FORTNITE), the results are complicated by temporal changes in the production/decomposition balance which determines nitrogen mineralization. The conclusion that small increases or large decreases in decomposition rate would significantly reduce harvestable yields should be considered tentative. More information on the types of material represented in the F+H layer, and their decomposition rates, as well as the relationship between nitrogen availability and decomposition of soil organic matter, might improve the accuracy of the FORTNITE projections and alter them significantly.

LITERATURE CITED

Aber, J.D., D.B. Botkin, and J.M. Melillo. 1979. Predicting the Effects of Different Harvesting Regimes on Productivity and Yield in Northern Hardwoods. Can. J. For. Res. 9:10-14.

Aber, J.D., G.R. Hendrey, D.B. Botkin, A.J. Francis, and J.M. Melillo. 1980. Simulation of Acid Precipitation Effects on Soil Nitrogen and Productivity in Forest Ecosystems. Brookhaven Nat. Lab. Inf. Rept. BNL-28658.

Aber, J.D. and J.M. Melillo. 1981. FORTNITE: A Computer Model of Organic Matter and Nitrogen Dynamics in Forest Ecosystems. University of Wisconsin Res. Bull., Madison, Wisconsin 53706 (in press).

Abrahamsen, G. 1980. Impact of Atmospheric Sulphur Deposition on Forest Ecosystems. In: D.S. Shriner, C.R. Richmond, and S.E. Lindberg, eds., "Atmospheric Sulphur Deposition: Environmental Impacts and Health Effects," pp. 397-415. Ann Arbor Science Publishers, Ann Arbor, Michigan.

Abrahamsen, G., R. Horntvedt, and B. Tviete. 1977. Impacts of Acid Precipitation on Coniferous Forest Ecosystems. Water Air Soil Pollut. 8:57-73.

Abrahamsen, G., J. Hovland, and S. Hagvar. 1980. Effects of Artificial Acid Rain and Liming on Soil Organisms and the Decomposition of Organic Matter. In: T.C. Hutchinson and M. Havas, eds., "Effects of Acid Precipitation on Terrestrial Ecosystems." Plenum Press, New York. 654 pp.

Auchmoody, L.R. and S.M. Filip. 1973. Forest Fertilization in the Eastern United States: Hardwoods. In: "Forest Fertilization," USFS USDA Gen. Tech. Rept. NE-3.

Bääth, E., B. Berg, U. Lohm, B. Lundgren, H. Lundkvist, T. Rosswall, B. Söderström, and A. Wiren. 1980. Effects of Experimental Acidification and Liming on Soil Organisms and Decomposition in a Scots Pine Forest. Pedobiologia 20:85-100.

Botkin, D.B. and J.D. Aber. 1979. Some Potential Impacts of Acid Rain on Forest Ecosystems: Implications of a Computer Simulation. Brookhaven Natl. Lab. Rept. BNL-50889.

Botkin, D.B., J.F. Janak, and J.R. Wallis. 1972. Some Ecological Consequences of a Computer Model of Forest Growth. J. Ecol. 60:849-872.

Cogbill, C. 1976. The History and Character of Acid Precipitation in Eastern North America. Water Air Soil Pollut. 6:407-413.

Cogbill, C. 1977. The Effect of Acid Precipitation on Tree Growth in Eastern North America. Water Air Soil Pollut. 8:89-93.

Covington, W.W. 1981. Secondary Succession and Forest Floor Dynamics in Northern Hardwood Forests. Ecology 62:41-48.

Cronan, C.S., R.C. Reynolds, and G.E. Lang. 1978. Forest Floor Leaching: Contributions From Mineral, Organic, and Carbonic Acids in New Hampshire Subalpine Forests. Science 200:309-311.

Cronan, C.S. and C.L. Schofield. 1979. Acid Precipitation: Effects on High-Elevation Watershed in the Northeast. Science 204:304-306.

Dennison, R., B. Caldwell, B. Bormann, L. Eldred, C. Swanberg, and S. Anderson. 1977. The Effects of Acid Rain on Nitrogen Fixation in Western Washington Coniferous Forests. Water Air Soil Pollut. 8:21-34.

Edwards, N.T. and W.F. Harris. 1977. Carbon Cycling in a Mixed Deciduous Forest Floor. Ecology 58:431-437.

Federer, C.A. and D. Lash. 1978. BROOK: A Hydrologic Simulation Model for Eastern Forests. University of New Hampshire Water Resource Res. Cent. Rept. 19, Durham, New Hampshire.

Fernalo, M.L. 1950. Gray's Manual of Botany. Rheinhold Co., New York.

Francis, A.J., D. Olson, and R. Bernatsky. 1980. Effects of Acidity on Microbial Processes. In: Proc. Int. Conf. on the Ecol. Impact of Acid Precip. Sandefjord, Norway.

Frink, C.R. and G.K. Voigt. 1977. Potential Effects of Acid Precipitation on Soils in the Humid Temperate Zone. Water Air Soil Pollut. 7:371-388.

Gosz, J.R., G.E. Likens, and F.H. Bormann. 1972. Nutrient Content of Litter Fall on the Hubbard Brook Experimental Forest, New Hampshire. Ecology 53:769-784.

Gosz, J.R., G.E. Likens, and F.H. Bormann. 1973. Nutrient Release From Decomposing Leaf and Branch Litter in the Hubbard Brook Forest, New Hampshire. Ecol. Mono. 43:173-191.

Heinrichs, H. and R. Mayer. 1977. Distribution and Cycling of Major and Trace Elements in Two Central European Forest Ecosystems. J. Environ. Qual. 6:402-407.

Hutchinson, T.C. and L.M. Whitby. 1977. The Effects of Acid Rainfall and Heavy Metal Particulates on a Boreal Forest Ecosystem Near the Sudbury Smelting Region of Canada. Water Air Soil Pollut. 7:421-438.

Johnson, N.M. 1979. Acid Rain: Neutralization Within the Hubbard Brook Ecosystem and Region Implications. Science 204:497-499.

Jonsson, N.M.   1979.   Soil Acidification by Atmospheric
Pollution and Forest Growth. Water Air Soil Pollut. 7:497-501.

Likens, G.E., F.H. Bormann, R.S. Pierce, J.S. Eaton, and N.M.
Johnson.   1978.   Biogeochemistry of a Forest Ecosystem.
Springer-Verlag. 145 pp.

Mayer, M. and B. Ulrich. 1977. Acidity of Precipitation as
Influenced by the Filtering of Atmospheric Sulphur and Nitrogen
Compounds - Its Role in the Element Balance and Effect on Soil.
Water Air Soil Pollut. 7:409-416.

McFee, W.W., J.M. Kelly, and R.H. Beck. 1977. Acid Precipitation Effects on Soil pH and Base Saturation of Exchange Sites.
Water Air Soil Pollut. 7:401-408.

Melillo, J.M., J.D. Aber, and J.M. Muratore. 1981. Nitrogen
and Lignin Control of Hardwood Leaf Litter Decomposition
Dynamics. Ecology (in press).

Mitchell, H.L. and R.F. Chandler. 1939. The Nitrogen Nutrition
and Growth of Certain Deciduous Trees of Northeastern United
States. Black Rock Forest Bull. 11.

Norton, S.A.   1977.   Changes in Chemical Processes in Soils
Caused by Acid Precipitation. Water Air Soil Pollut. 7:389-400.

Norton, S.A., D.W. Hanson, and R.J. Campana. 1980. The Impact
of Acidic Precipitation and Heavy Metals on Soils in Relation
to Forest Ecosystems. In: Miller, ed., "Int. Symp. on Effects
of Air Poll. on Mediterranean and Temperate Forest Ecosystems,"
pp. 152-157. U.S Forest Service, Berkeley, California.

Nyborg, M., J. Crepin, D. Hocking, and J. Baker. 1977. Effects
of Sulphur Dioxide on Precipitation and on the Sulphur Content
and Acidity of Soils in Alberta, Canada. Water Air Soil Pollut.
7:439-448.

Petersen, L. 1980. Podzolization: Mechanisms and Possible
Effects of Acid Precipitation. In: T.C. Hutchinson and M.
Havas, eds., "Effects of Acid Precipitation on Terrestrial
Ecosystems." Plenum Press, New York. 654 pp.

Roberts, T.M., T.A. Clarke, P. Ineson, and T.R. Gray. 1980.
Effects of Sulphur Deposition on Litter Decomposition and
Nutrient Leaching in Coniferous Forests. In: T.C. Hutchinson
and M. Havas, eds., "Effects of Acid Precipitation on
Terrestrial Ecosystems." Plenum Press, New York. 654 pp.

Runge, M. 1971. Investigations of the Content and Production of Mineral Nitrogen in Soils. <u>In</u>: "Integrated Experimental Ecology: Methods and Results of the German Solling Project," pp. 191-202. Springer-Verlag.

Singh, B.R., G. Abrahamsen, and A. Stuanes. 1980. Effects of Simulated Acid Rain on Sulfate Movement in Acid Forest Soils. <u>Soil Sci. Soc. Amer. J.</u> 44:75-80.

Sposito, G., A.L. Page, and M.E. Frink. 1980. Effects of Acid Precipitation on Soil Leachate Quality: Computer Calculations. EPA-600/3-80-015. Riverside, California. 47 pp.

Tamm, C.O. 1977. Acid Precipitation and Forest Soils. <u>Water Air Soil Pollut.</u> 7:367-369.

Tamm, C.O. and E.B. Cowling. 1977. Acidic Precipitation and Forest Vegetation. <u>Water Air Soil Pollut.</u> 7:503-511.

Tamm, C.O., L. Wiklander, and B. Popovic. 1977. Effects of Application of Sulphuric Acid to Poor Pine Forests. <u>Water Air Soil Pollut.</u> 8:75-87.

Tveite, B. and G. Abrahamsen. 1980. Effects of Artificial Acid Rain on the Growth and Nutrient Status of Trees. <u>In</u>: T.C. Hutchinson and M. Havas, eds., "Effects of Acid Precipitation on Terrestrial Ecosystems." Plenum Press, New York. 654 pp.

Vitousek, P.M., J.R. Gosz, C.C. Grier, J.M. Melillo, W.A. Reiners, and R.L. Todd. 1979. Nitrate Losses From Disturbed Ecosystems. <u>Science</u> 204:469-474.

Whittaker, R.H., F.H. Bormann, G.E. Likens, and T.G. Siccama. 1974. The Hubbard Brook Ecosystem Study: Forest Biomass and Production. <u>Ecol. Mono.</u> 44:233-254.

Wood, T. and F.H. Bormann. 1977. Short-Term Effects of a Simulated Acid Rain Upon the Growth and Nutrient Relations of <u>Pinus strobus</u>. <u>Water Air Soil Pollut.</u> 7:479-488.

CHAPTER 19

THE ACTION OF WET AND DRY DEPOSITION COMPONENTS
OF ACID PRECIPITATION ON LITTER AND SOIL

William W. McFee
Agronomy Department
Purdue University
West Lafayette, Indiana 47907

Christopher S. Cronan
Department of Botany and
Plant Pathology
Deering Hall
University of Maine
Orono, Maine 04469

INTRODUCTION

Soils and litter of the Great Lakes region are receiving acid inputs from atmospheric deposition. The net effects of these inputs are not yet clear. This chapter will raise several key questions about the effects of acid deposition on litter and soil, review some of the published results and some of our own findings, and discuss the importance of this information for the Great Lakes region.

DEPOSITION EXAMPLES

We examined the acidity of precipitation from National Atmospheric Deposition Program (NADP) collection stations in Minnesota, Wisconsin, and Michigan for a twelve-month period, ending April, 1980. In that year, Lamberton in southern Minnesota (Figure 1) received precipitation that was generally above pH 5.6, while Marcell in northern Minnesota (Figure 2) had much more frequent incidence of low pH (around 5.0). Three months of data from Trout Lake, Wisconsin, ranged from pH 4.3 to 5.2, a pattern that was similar to a comparable period at Marcell. The trends in Michigan were toward much lower precipitation pHs. The bulk of values recorded at the USDA Forest Service station at Wellston in the northwestern lower peninsula (Figure 3) and at the Kellogg Biological Station near

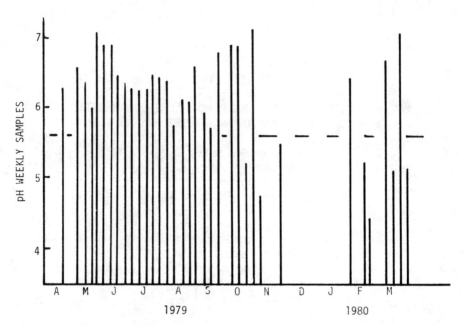

*Figure 1. Acidity of weekly samples of wet deposition at
Lamberton, Minnesota, April, 1979 to March, 1980
(NADP, 1979, 1980).*

many values near 4.0. The values from the Kellogg Station are
not presented; however, nine months of data, ending in April,
1980, appeared to be similar to data gathered at Wellston. In
another study, Adams (1980) indicated that bulk precipitation
in the Pellston area of Michigan had an average pH of 4.27 in
the summer of 1978 and pH 4.53 the following winter. Thus,
while the lake states region is not as heavily impacted as some
regions of the northeastern United States, it is receiving
significant amounts of acidic material.

     To put the quantities in perspective, we summarized the
quarterly deposition of sulfate, nitrate, and hydrogen ions at
the Wellston station from the NADP data (Table 1). Depositions
were calculated from the wet collector concentrations times the
rain gauge volumes. For that reason, and since this is only
one year's data, these values should be interpreted with caution
and it must be assumed that the error involved could be large.
The values reported in Table 1 for wet deposition may represent
only about one-half of total deposition. Several workers have
suggested that dry deposition may be roughly equal to wet
deposition in the eastern United States (e.g., Galloway and
Whelpdale, 1980).

Table 1

Wet Deposition at Wellston, Michigan, April, 1979, Through March, 1980

Period	Precipitation	NO_3^-	SO_4^{-2}	H^+	K^+	Ca^{+2}
	-----cm-------	-------------------------------------kg/ha--------------------------------				
April to June, 1979	20	4.7	7.0	0.077	0.085	0.70
July to September, 1979	13	2.5	4.3	0.054	0.030	0.29
October to December, 1979	23	4.9	5.7	0.064	0.069	0.69
January to March, 1980	13	4.2	3.9	0.051	0.059	0.81
Total	68	16.3	20.9	0.246	0.243	2.49
Total (keq/ha)		0.26	0.44	0.25	0.006	0.12

Data from NADP (1979, 1980)

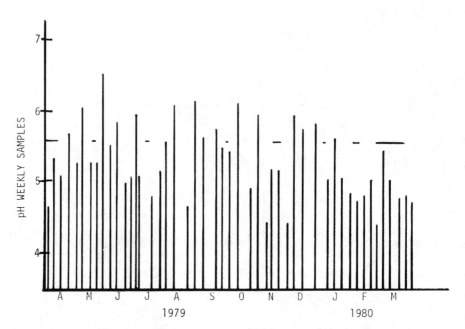

Figure 2.   Acidity of weekly samples of wet deposition at
*   Marcell, Minnesota, April, 1979 to March, 1980*
*   (NADP, 1979, 1980).*

How does this magnitude of input compare with other fluxes
and components of a forest ecosystem? Sixteen kg/ha of nitrate
is a small input in an agricultural system, ranking well below
that expected from microbial fixation. It is nearly as high
as the average nitrate input in bulk precipitation at Hubbard
Brook, New Hampshire (19.7 kg/ha) which exceeded the stream
losses only slightly (Likens et al., 1977). Thus, in an
unmanaged system, it may be an important addition, although it
is still a small fraction of the amount found in the forest
floor.

Sulfate input is roughly 50 percent of the average
deposition at Hubbard Brook, and is a significant portion of
the 24.5 kg S/ha/yr that Likens et al. (1977) estimated for
root uptake and the 124 kg S/ha they estimated for the forest
floor. It is sufficiently large to balance the sulfur removal
in many annual agricultural harvests.

The potassium and calcium inputs indicated in Table 1 are
quite small compared to the expected annual plant uptake or
that contained in the standing biomass of a forested system
(Likens et al., 1977).

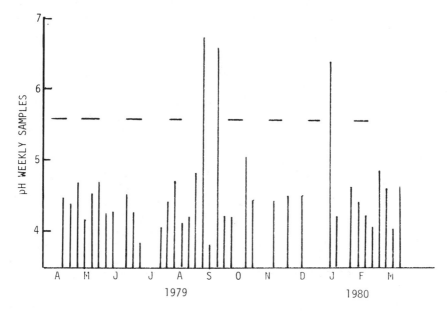

Figure 3. Acidity of weekly samples of wet deposition at Wellston, Michigan, April, 1979 to March, 1980 (NADP, 1979, 1980).

It is interesting to note that when the inputs are expressed in terms of chemical equivalents (Table 1), sulfate constitutes about two-thirds of the anion input while roughly two-thirds of the cations are basic cations rather than hydrogen. When considering soil leaching, soil pH changes, and related effects, the hydrogen ion input that would result from 100 cm of precipitation at pH 4.0 (1 keq/ha), can be compared with the size of the soil's exchange capacity. For every 1 meq/100 gm cation exchange capacity in the mineral soil matrix, the exchange capacity would equal 26 keq/ha to a depth of 20 cm (assuming a soil bulk density of 1.3 gm/cc). In most of the lake states soils, the exchange capacity is likely to range upward from a low of 2 to 3 meq/100 gm (about 52-78 keq/ha to 20 cm depth), and is already partially saturated with acidic cations, hydrogen, and aluminum. An annual input of 0.246 keq of hydrogen is not likely to cause rapid changes in the base saturation of the soils.

In spite of these indications that the atmospheric strong acid inputs are small compared to the pools present in the system, the question of long-term impacts still remains. If, in fact, these current inputs of acid, sulfate, and nitrate are replacing centuries of input that were less acid and lower in sulfates and nitrates, then new trends will develop and a new balance between gains, losses, and productivity will be established.

POTENTIAL IMPACTS OF ACID DEPOSITION ON SOILS

The concern regarding atmospheric deposition and soils centers on both the long-term ability of the soils to support a vigorous plant community and the interaction of soils, through runoff and leaching, with aquatic systems. There are a large number of potential impacts of acidic deposition on litter and soils which relate to these two areas of concern. Several of these potential effects are listed in Table 2, along with references which support or refute the hypothesized impacts.

These impacts, taken together, present the possibility of serious damage to the soil's ability to support vigorous plant growth and to protect the quality of water. However, this has not yet been confirmed as a general conclusion; there is still much research that is needed to permit a holistic assessment of this problem. Some of the evidence for these effects is based on experimental rates of acidic deposition far exceeding current loadings. The same amount of input spread over a much longer term may have a different net effect. Although many of the potential effects could result in decreased site productivity over the long term, no one has yet been able to demonstrate this conclusively. Offsetting effects may even occur in some instances. For example, an increased weathering rate of primary minerals may leave the supply of cations available for production unchanged even in the presence of increased leaching. Likewise, compensating shifts in microbial communities may result in a new dynamic equilibrium just as desirable as the previous one.

One of the difficulties with much of the writing concerning acidic deposition effects on soils has been our failure to differentiate clearly between effects of "acidic deposition" and "soil acidification." The latter is an almost ubiquitous process in humid climates of the world. The effects of soil acidity on plant nutrition, soil chemical processes, and soil organisms have been studied for decades. Many of the undesirable effects of soil acidity have occasionally been attributed to acidic deposition without regard to the significance of soil acidification in bringing about the acid condition. Many natural phenomena influence this process of acidification as much or more than the phenomenon of acid precipitation.

The reference we make to Jackson's (1963) work (Table 2) is an example. He was examining the influence of acid weathering on clay minerals, but not in the context of acid deposition.

The conditions he created may or may not be created by acidic deposition.   Likewise, Sawhney's (1968) work illuminates changes in an acid soil environment without regard to the origin of the acidity.   The experiments conducted by Francis et al. (1980) utilized soil that had been adjusted to a pH of 3.0 for comparison with the native soil pH of 4.6.   Their results indicate significant changes in microbial behavior at that very low pH; however, there is some difficulty in extrapolating from such experiments to real world impacts of acidic deposition. A reduction in soil pH to 3.0 is not likely to occur as a result of acidic precipitation.   It has been pointed out by several authors (Wiklander, 1973/74; Reuss, 1975; McFee et al., 1977) that reduction in soil base saturation and increases in soil acidity due to acidic precipitation would take quite a while, on the order of decades, for a noticeable shift to occur from that cause alone.

In spite of these qualifications, most of the processes indicated in Table 2 should be viewed with concern.   A significant change in carbon mineralization rate from any cause could bring about a disruption in nutrient flow to growing plants in a forest ecosystem.   Likewise, if the acidic inputs associated with atmospheric deposition change something as basic to plant nutrition as the soil cation exchange capacity, then serious consequences could result.

SOIL ALUMINUM LEACHING

One of the potential ecological effects of acidic deposition is the increased mobilization of soil aluminum that may result. Changes in soil aluminum leaching may be indicative of changes in weathering rates, changes in base saturation, and changes in clay mineralogy and morphology; furthermore, increased aluminum concentrations in natural waters may be harmful to plant and animal communities. We shall examine what is known about soil aluminum leaching in more detail.

As shown by Lind and Hem (1975) and others, the solubility of aluminum is very pH-dependent and declines dramatically in the pH range between 5 and 7.   However, in the presence of organic ligands, the solubility of aluminum can be greatly enhanced by the formation of soluble organic-aluminum complexes. In some Spodozols, the solution acidification and organometallic complexation provided by plant- and soil-derived organic acids result in the dissolution and transport of aluminum and other metals from upper into lower soil horizons (Malcolm and McCracken, 1968). During water percolation through the B horizon of such soils, aluminum and other metals leave solution in response to:

Table 2

The Potential Impacts of Atmospheric Deposition on Litter and Soils – For Each Major Soil
Process or Property, There Are One or More Hypotheses
Describing the Possible Consequences of Acidic Deposition

Process or Property	Hypothetical Impact of Acidic Deposition	Evidence (+) Confirmation (−) Rejection
I. Decomposition-Mineralization		
--Organic Matter Turnover	Decreased rate of C mineralization as a result of soil acidification and/or associated trace metal toxicity	+Francis et al., 1980 +Lohm, 1980
--Microbial Community Dynamics	Shift from bacteria toward more acid-tolerant fungi	+Lohm, 1980
--Nitrogen Mineralization	Decreased ammonification Decreased nitrification	+Francis et al., 1980 +Francis et al., 1980 +Alexander, 1980 +Abrahamsen, 1980
--Root Uptake	Increased N availability Trace metal toxicity associated with soil acidification	+Mayer and Ulrich, 1977 +Ulrich et al., 1980
II. Soil Exchange Complex		
--Clay Mineral Morphology	Increased formation of hydroxy-Al interlayers under acid weathering	+Jackson, 1963
--Exchange Capacity	Decrease in CEC as a result of clay alumination	+Sawhney, 1963

Category	Effect	References
--Base Saturation and Exchangeable Acidity	Increase in the CEC of Ultisols as a result of sulfate adsorption	+Johnson, 1980
	Decrease in base saturation and increase in soil acidity	+Abrahamsen, 1980 +Farrell et al., 1980 +Stuanes, 1980 +Bjor and Teigen, 1980 -Linzon and Temple, 1980
III. Element Leaching		
--Aluminum	Increased mobilization and leaching	+Ulrich et al., 1980 +Cronan, 1980a +Abrahamsen et al., 1976 +Mayer and Ulrich, 1977 +Baker et al., 1977
--Manganese	Increased leaching	+Mayer and Ulrich, 1977 -Tyler, 1978
--Heavy Metals	Increased leaching	+Abrahamsen, 1980 +Overrein, 1972
--Nutrient Cations	Increased leaching loss	+Mayer and Ulrich, 1977 +Stuanes, 1980 +Cronan, 1980b -Johnson and Cole, 1977
--Sulfur	Retained, thus reducing S loss and cation loss	+Farrell et al., 1980 +Johnson, 1979
IV. Primary Mineral Weathering	Increased weathering loss	+Gjessing et al., 1976 +Johnson, 1979

(1)   the removal of organic ligands by immobilization or decomposition, and
(2)   the associated rise in the soil solution pH.

Consequently, water percolating below the B2 horizon of such soils will typically exhibit very low concentrations of dissolved aluminum. This pattern has been described for several podzolized soils away from the influence of acid precipitation (Ugolini et al., 1977; Johnson et al., 1977; Granstein, 1976), and is presumed to represent the historical aluminum leaching pattern for podzolized soils, including those of the eastern and Great Lakes regions of the United States.

In contrast to that pattern, podzolized soils that are now subject to leaching by atmospheric sulfuric acid may show a very different pattern of aluminum transport (Cronan and Schofield, 1979; Cronan, 1980c).   This aluminum leaching response to acidic deposition can be illustrated with soil solution chemistry data obtained from Panther Lake watershed in the central Adirondack Park, New York (Cronan and Schofield, unpublished data).   The soils at that site are classified as Typic Fragiorthods and Haplorthods and exhibit the classic B2ir horizon accumulation of organic matter, aluminum, and iron that one would expect in a soil that has developed from aluminosilicate parent materials under the influence of podzolization and organic acid leaching. However, as shown in Table 3, the soil solution chemistry data from this site imply that the historical podzolization pattern of aluminum transport has been altered. The data illustrate that dissolved organic carbon (DOC), solution acidity, and organically complexed monomeric aluminum concentrations are greatest at the 02/A2 horizon interface (20 cm depth), just as one would predict for a soil undergoing podzolization. Yet, at the lower B horizon depth (50 cm), there is an intriguing divergence of leaching patterns. On the one hand, there is a dramatic decline in DOC, solution acidity, and dissolved monomeric organic aluminum- -just the pattern of solution chemistry that one would expect in a Spodosol. However, at the same time that the organic acidity and organic aluminum have declined, there is a major peak in the concentrations of inorganic monomeric aluminum and total dissolved monomeric aluminum. These soil solution chemistry patterns are interpreted to indicate the following:

(1)   In these and other podzolized forest soils, the mobilization and transport of soil aluminum by plant- or soil-derived organic acids is primarily limited to the upper soil profile. The major constraint on the movement of organic-aluminum is imposed by the limited mobility of the organic molecules in the clay- rich B horizon. Under this type of leaching regime,

Table 3

The Solution Chemistry Patterns Associated With
Aluminum Mobilization and Leaching in the Panther
Lake Watershed, Adirondack Park, New York (Aluminum
Data Courtesy of Dr. C. L. Schofield)

Water Sample Source	pH	DOC (mg/l)	Al_o[a] (μg/l)	Al_i[b]	SO_4 (μeq/l)
20 cm soil solution	3.9	23.6	302	368	172
50 cm soil solution	4.6	5.5	161	1,111	151

[a]Al_o corresponds to organically complexed monomeric aluminum.
[b]Al_i corresponds to inorganic monomeric aluminum.

one would expect very little transport of dissolved inorganic aluminum to groundwater and aquatic systems.

(2) With atmospheric inputs of strong acids to these soils, the historical trend of aluminum accumulation in the B2 horizon is apparently altered. Under conditions where the mobility of the atmospheric sulfate (or nitrate) is not limited by adsorption or absorption processes, the anthropogenic strong acids may contribute to increased leaching of free inorganic aluminum from the B horizon to groundwater and surface waters. The explanation for this modern increase in soil aluminum leaching may involve two factors: atmospheric sulfuric acid may produce a small but significant decrease in the soil solution pH throughout the soil profile, thereby increasing the solubility and hence mobility of aluminum; and the mineral acidity introduced from the atmosphere may be less susceptible to removal from solution by adsorption and chemical precipitation reactions than the natural organic acids (Cronan, 1980b).

The extent to which soils in the Great Lakes region might be expected to exhibit this type of soil aluminum leaching response to acidic deposition will depend on several important factors. The first of these is the seasonal and long-term pattern and magnitude of strong acid loading to a given soil. The next most important determinant is the availability of "labile aluminum" in the soil. Table 4 shows the soil aluminum fractions that may potentially be affected by either organic or mineral acid leaching and which may contribute to dissolved aluminum in the soil solution. As indicated, each of these aluminum pools will exhibit a separate thermodynamic and kinetic leaching response to changes in hydrogen ion activity. Thus, the aluminum transport that one might observe in a given soil profile would be very much dependent upon the abundances of each aluminum fraction in the soil and upon the rate of water movement through the soil. These different aluminum fractions can be characterized and quantified by means of the progressive acid dissolution technique of Grandquist and Sumner (1957), while the general soil hydrologic properties can be characterized with hydraulic conductivity measurements. For ecosystems which do exhibit increased aluminum mobilization and leaching, there may be effects upon pedogenesis, clay morphology and ion exchange capacity, nutrient cycling, and the health of aquatic and terrestrial communities.

The only research we are aware of that examines acid precipitation effects on soils of the Great Lakes region is that of James Boyle and his students at the University of Michigan. One of his students, Adams (1980), reports mixed indications of increased leaching of copper, magnesium, potassium, and sodium measured at the bottom of the O and A2 horizons in the field. His leaching columns in the laboratory did not respond significantly to acid inputs. In both studies,

Table 4

Soil Aluminum Fractions Which May Release Inorganic Aluminum to the Soil Solution in Great Lakes Forest Ecosystems Exposed to Acidic Deposition - The Fractions are Ranked According to Their Hypothesized Ability to Release Dissolved Aluminum to a Sulfuric Acid-Dominated Soil Solution

Aluminum Fraction	Reaction Rate With Sulfuric Acid
Exchangeable aluminum	Rapid
Amorphous aluminum hydroxide	Moderately rapid
Organic-aluminum complexes	Moderately rapid
$Al_x(OH)_y$ interlayers in expansible clay minerals (e.g., vermiculite)	Relatively slower
Aluminum in clay lattices	Relatively slower
Undecomposed aluminum silicates	Relatively slower

the aluminum values were below his detection limits, approximately 1 mg/l.

Both Norton's (Hendrey *et al.*, 1980) and McFee's (1980) analyses of regional sensitivity indicate that a large portion of the Great Lakes region may be somewhat sensitive. Norton's classification places the emphasis on the nature of the bedrock and determines a region's sensitivity by the percent of the region underlain by rocks of high alkalinity. Admittedly, it is not a powerful system in regions of deep glacial deposits that are unrelated to the bedrock. McFee's system is primarily concerned with the surface soil cation exchange capacity, but fails to consider the sulfate adsorption capacity of the soil, the presence of easily weatherable materials, or differences among soils of low cation exchange capacity. In spite of their limitations, they do give us some indication that compared to much of the United States, there are numerous areas within the Great Lakes region that should be watched for evidences of acid precipitation effects.

LITERATURE CITED

Abrahamsen, G. 1980. Impact of Atmospheric Sulfur Deposition on Forest Ecosystems. *In*: D.S. Shriner *et al.* (eds.), "Atmospheric Sulfur Deposition - Environmental Impact and Health Effects," Chapter 40, pp. 397-415. Ann Arbor Science Publishers, Ann Arbor, Michigan 48106.

Abrahamsen, G., K. Bjor, R. Horntvedt, and B. Tveite. 1976. Effects of Acid Precipitation on Coniferous Forest. *In*: F.H. Braekke, ed., "Impact of Acid Precipitation on Forest and Freshwater Ecosystems in Norway," pp. 38-63. SNSF-Project.

Adams, P.W. 1980. Forest Soil and Site Productivity Considerations in Intensive Wood Harvest Systems in Northern Lower Michigan. Ph.D. Thesis (pp. 43-88), University of Michigan, Ann Arbor, Michigan 48109.

Alexander, M. 1980. Effects of Acid Precipitation on Biochemical Activities in Soil. *In*: D. Drabløs and A. Tollan, eds., "Ecological Impact of Acid Precipitation - Proceedings of an International Conference," pp. 47-52. SNSF, Sandefjord, Norway.

Baker, J., D. Hocking, and M. Nyborg. 1977. Acidity of Open and Intercepted Precipitation in Forests and Effects on Forest Soils in Alberta, Canada. *Water Air Soil Pollut.* 7:449-460.

Bjor, K. and O. Teigen. 1980. Lysimeter Experiment in Greenhouse. In: D. Drabløs and A. Tollan, eds., "Ecological Impact of Acid Precipitation - Proceedings of an International Conference," pp. 200-201. SNSF, Sandefjord, Norway.

Cronan, C.S. 1980a. Solution Chemistry of a New Hampshire Subalpine Ecosystem: A Biogeochemical Analysis. Oikos 34:272-281.

Cronan, C.S. 1980b. Consequences of Sulfuric Acid Inputs to a Forest Soil. In: D.S. Shriner, C.R. Richmond, and S.E. Lindberg, eds., "Atmospheric Sulfur Deposition - Environmental Impact and Health Effects," pp. 335-343. Ann Arbor Science Publishers, Ann Arbor, Michigan 48106.

Cronan, C.S. 1980c. Controls on Leaching From Coniferous Forest Floor Microcosms. Plant Soil 56:301-322.

Cronan, C.S. and C.L. Schofield. 1979. Aluminum Leaching Response to Acid Precipitation: Effects on High-Elevation Watersheds in the Northeast. Science 204:304-306.

Farrell, E.P., I. Nilsson, C.O. Tamm, and G. Wiklander. 1980. Effects of Artificial Acidification With Sulfuric Acid on Soil Chemistry in a Scots Pine Forest. In: D. Drabløs and A. Tollan, eds., "Ecological Impact of Acid Precipitation - Proceedings of an International Conference," pp. 186-187. SNSF, Sandefjord, Norway.

Francis, A.J., D. Olson, and R. Bernatsky. 1980. Effect of Acidity on Microbial Processes in a Forest Soil. In: D. Drabløs and A. Tollan, eds., "Ecological Impact of Acid Precipitation - Proceedings of an International Conference," pp. 166-167. SNSF, Sandefjord, Norway.

Galloway, J. and D. Whelpdale. 1980. Atmospheric Sulfur Budget for North America. Atmos. Environ. 14:409-417.

Gjessing, E.T., A. Henriksen, M. Johannessen, and R.F. Wright. 1976. Effects of Acid Precipitation on Freshwater Chemistry. In: F.H. Braekke, ed., "Impact of Acid Precipitation on Forest and Freshwater Ecosystems in Norway," pp. 64-85. SNSF-Project.

Grandquist, W.T. and G.G. Sumner. 1957. Acid Dissolution of a Texas Bentonite. Clays Clay Min. 6:292-301.

Granstein, W.C. 1976. Organic Complexes and the Mobility of Iron and Aluminum in Soil Profiles. Geol. Soc. Amer. Abstr. Programs 8:891.

Hendrey, G.R., J.N. Galloway, S.A. Norton, C.L. Schofield, P.W. Shaffer, and D.A. Burns. 1980. Geological and Hydrochemical Sensitivity of the Eastern United States to Acid Precipitation. USEPA-600/3-80-024. U.S. Environmental Protection Agency, Environmental Research Laboratory, Corvallis, Oregon. 99 pp.

Jackson, M.L. 1963. Interlayering of Expansible Layer Silicates in Soils by Chemical Weathering. Clays Clay Min. 11:29-46.

Johnson, N.M. 1979. Acid Rain: Neutralization Within the Hubbard Brook Ecosystem and Regional Implications. Science 204:497-499.

Johnson, D.W. 1980. Site Susceptibility to Leaching by H_2SO_4 in Acid Rainfall. In: T.C. Hutchinson and M. Havas, eds., "Effects of Acid Precipitation on Terrestrial Ecosystems," pp. 525-535. Plenum Publishing Corp., New York.

Johnson, D.W. and D.W. Cole. 1977. Sulfate Mobility in an Outwash Soil in Western Washington. Water Air Soil Pollut. 7:489-495.

Johnson, D.W., D.W. Cole, S.P. Gessel, M.J. Singer, and R. Minden. 1977. Carbonic Acid Leaching in a Tropical, Temperate, Subalpine, and Northern Forest Soil. Arc. Alp. Res. 9:329-343.

Lind, C.J. and J.D. Hem. 1975. Effects of Organic Solutes on Chemical Reactions of Aluminum. USGS Water Suppl. Paper 1827-G. 83 pp.

Likens, G.E., H.H. Bormann, R.S. Pierce, J.S. Eaton, and N.M. Johnson. 1977. Biogeochemistry of a Forestry Ecosystem. Springer-Verlag, New York. 146 pp.

Linzon, S.N. and P.J. Temple. 1980. Soil Resampling and pH Measurements After an 18 Year Period in Ontario. In: D. Drabløs and A. Tollan, eds., "Ecological Impact of Acid Precipitation - Proceedings of an International Conference," pp. 176-177. SNSF, Sandefjord, Norway.

Lohm, U. 1980. Effects of Experimental Acidification on Soil Organisms. In: D. Drabløs and A. Tollan, eds., "Ecological Impact of Acid Precipitation - Proceedings of an International Conference," pp. 178-179. SNSF, Sandefjord, Norway.

Malcolm, R.L. and R.J. McCracken. 1968. Canopy Drips: A Source of Mobile Soil Organic Matter for Mobilization of Iron and Aluminum. Soil Sci. Soc. Amer. Proc. 32:834-838.

Mayer, R. and B. Ulrich. 1977. Acidity of Precipitation as Influenced by the Filtering of Atmospheric Sulfur and Nitrogen Compounds - Its Role in the Element Balance and Effect on Soil. Water Air Soil Pollut. 7:409-416.

McFee, W.W. 1980. Sensitivity of Soil Regions to Acid Precipitation. USEPA-600/3-80-013. U.S. Environmental Protection Agency, Environmental Research Laboratory, Corvallis, Oregon. 179 pp.

McFee, W.W., J.M. Kelly, and R. Beck. 1977. Acid Precipitation Effects on Soil pH and Base Saturation of Exchange Sites. Water Air Soil Pollut. 7:401-408.

NADP. 1979. NADP Data Report of Precipitation Chemistry. Volume II, No. 2-4. National Atmospheric Deposition Program, National Resource Ecology Laboratory, Fort Collins, Colorado. 180 pp.

NADP. 1980. NADP Data Report of Precipitation Chemistry. Volume III, No. 1. National Atmospheric Deposition Program, National Resource Ecology Laboratory, Fort Collins, Colorado. 184 pp.

Overrein, L.N. 1972. Sulfur Pollution Patterns Observed: Leaching of Calcium in Forest Soil Determined. Ambio 1:145-147.

Reuss, J.O. 1977. Chemical and Biological Relationships Relevant to the Effect of Acid Rainfall on the Soil-Plant System. Water Air Soil Pollut. 7:461-478.

Sawhney, B.L. 1968. Al Interlayers in Layer Silicates. Effect of OH/Al Ratio of Al Solution, Time of Reaction, and Type of Structure. Clays Clay Min. 16:157-163.

Stuanes, A.O. 1980. Release and Loss of Nutrients From a Norwegian Forest Soil Due to Artificial Rain of Varying Acidity. In: D. Dabløs and A. Tollan, eds., "Ecological Impact of Acid Precipitation - Proceedings of an International Conference," pp. 198-199. SNSF, Sandefjord, Norway.

Tyler, G. 1978. Leaching Rates of Heavy Metal Ions in Forest Soil. Water Air Soil Pollut. 9:137-148.

Ugolini, F.C., R. Minden, H. Dawson, and J. Zachara. 1977. An Example of Soil Processes in the Abies amabilis Zone of Central Cascades, Washington. Soil Sci. 124:291-302.

Ulrich, B., R. Mayer, and P.K. Khanna. 1980. Chemical Changes Due to Acid Precipitation in a Loess-Derived Soil in Central Europe. <u>Soil Sci.</u> <u>130</u>:193-199.

Wiklander, L. 1973/74. The Acidification of Soil by Acid Precipitation. <u>Grund Forbattring</u> <u>26</u>:155-164.

Wiklander, L. 1975. The Role of Neutral Salts in Ion Exchange Between Precipitation and Soil. <u>Geoderma</u> <u>14</u>:92-105.

CHAPTER 20

THE EFFECTS OF ACID PRECIPITATION ON CROPS

Jeffrey J. Lee
Corvallis Environmental Research Laboratory
U.S. Environmental Protection Agency
Corvallis, Oregon 97333

INTRODUCTION

The regions of North America impacted by acid precipitation encompass vast acreages of fertile farmland. In the United States, agricultural crops had a farm value of 55 billion dollars in 1977. Crops grown in states bordering the Great Lakes had a 1977 farm value of more than 16 billion dollars (Table 1); crops grown in the Great Lakes region of Canada added significantly to this total. Thus, the potential impact of acid precipitation on crops of this region is a major concern.

Acid precipitation is rain, snow, or fog whch contains an excess of strong acid anions (primarily sulfate and nitrate) over basic cations (primarily calcium). The pH value dividing "acid precipitation" from "clean precipitation" is arbitrary. Pure rainwater in equilibrium with atmospheric carbon dioxide would have a pH of 5.65 due to carbonic acid. However, contamination by soil dust, organic matter, or other natural material could lower the pH value (higher pH values are also possible). For crop effects, a value of 5.0 might provide a suitable, although arbitrary, upper boundary on the pH of acid precipitation. Monitoring data indicate that at a location with average pH of 4.0, a few percent of the events may be at 3.0. Thus, pH 3.0 to pH 5.0 will be taken as the range of ambient concentrations. Since research on crop effects has overwhelmingly emphasized acid rain, this chapter will only deal with this type of acid precipitation.

From the point of view of crops, acid rain exposes foliar surfaces to an aqueous solution containing primarily sulfate, nitrate, and hydrogen ions. Other ions (calcium, magnesium, ammonium, potassium, sodium, chloride) are also present, but generally in lesser amounts. Foliar contact is of greatest

Table 1

Farm Value (1977) of Crops Grown in States Bordering
the Great Lakes - Source: Statistical Abstract
of the United States, 1978

State	Dollar Value $(\times 10^6)$	National Rank	Principal Crops
Minnesota	3,085	5	Corn, soybeans, hay, wheat
Wisconsin	1,575	11	Hay, corn, oats, potatoes
Michigan	1,108	17	Corn, hay, dry beans, wheat
Illinois	4,632	2	Corn, soybeans, hay, wheat
Indiana	2,359	6	Corn, soybeans, hay, wheat
Ohio	1,880	8	Corn, soybeans, hay, wheat
Pennsylvania	853	22	Corn, hay, mushrooms, potatos
New York	724	27	Hay, corn, potatoes, wheat
Region	16,216	--	Corn, soybeans, hay, wheat
United States	55,634	--	Corn, soybeans, hay, wheat

concern, since highly managed agricultural soils are not
considered sensitive to acid rain; thus, the root environment
is unlikely to be directly affected by acid rain. However,
this might not apply to alkaline soils, which could benefit
from acidification, or to relatively poorly buffered sandy
soils.

Each major constituent of acid rain can affect crop yield.
Sulfate and nitrate solutions in contact with foliage are very
effective in altering crop growth. Thus, the question is not
whether acid rain could, at some concentration, affect crop
productivity, but rather what is the nature of the response at
ambient levels and does this response have practical
significance. The general nature of this response and some
results from experiments with particular crops are discussed
in this chapter.

NATURE OF THE DOSE-RESPONSE CURVE

Acid rain is just one of several environmental factors which can influence crop yield. A common modeling approach is to represent the effect of several variables as a product. For example, we might have:

$$\text{Yield} = Y = S \times W \times P \times AR \tag{1}$$

where S and W represent the effects of soil and weather, and P and AR represent the effects of gaseous pollutants and acid rain. If only the acid rain factor, AR, is being investigated, we can write:

$$\text{Yield} = (\text{Control Yield}) \times (\text{Acid Rain Factor})$$

$$Y - C \times AR$$

so that

$$AR = Y/C \tag{2}$$

The factor AR will, in general, be a function of a vector of variables describing the characteristics of acid rain (e.g., average concentrations, temporal variability, peak concentrations, etc.). This function defines the <u>dose-response curve</u>.

This formulation assumes that there are no interactions among the various environmental variables, so that the yield equation is separable by products. In this overly simplified model, variations in factors other than AR contribute to the variability of the control, but not to the acid rain effect. Thus, the acid rain factor is independent of the other environmental factors. Note, however, that for some ranges of variables, Equation 1 can be approximated as a sum with, perhaps, cross-terms. In this case, a statistical test for interaction (i.e., a test for cross-terms) based on yield (i.e., on an additive model) would give positive results, even if the underlying processes are independent. These false positives can be avoided by log-transforming Equation 1:

$$\ln Y = \ln S + \ln W + \ln P + \ln AR \tag{3}$$

Positive results for interactions based on Equation 3 (i.e., log-transformed data) indicate that the factors are not separable (i.e., terms containing more than one variable must be added to Equation 3). This could indicate that the underlying mechanisms depend on the joint action of more than one variable. Thus, intrinsic, biologically significant, interactions among

variables are likely. In these cases, the variable defined in
Equation 2 will still be useful if:

(1)   the interaction terms are small;
(2)   the range of variation of environmental variables is
      small; or
(3)   the specific purpose is to detect interactions by
      inspection.

In any case, AR provides a basis for comparing results from
different experiments.

Three idealized examples of commonly found dose-response
curves are shown in Figure 1. These differ primarily in the
placement of certain features with respect to ambient doses.
These features result from a shifting balance between
stimulatory and inhibitory effects. In this figure, "zero dose"
does not necessarily refer to absolute zero dose, but rather
to zero increment above some reference (e.g., background) dose.

A toxic response curve (T) generally falls below the
reference yield (yield at zero dose). Although there might be
a stimulatory response at small doses (observed for air
pollutants, trace elements, herbicides), for this type response
random processes such as atmospheric deposition are likely to
produce doses in the inhibitory range. A nutrient response
curve (N) flattens and reaches a maximum at high doses. Although
very high doses (e.g., over-fertilization) can cause smaller
yields, random fluctuations are likely to produce doses in the
stimulatory range. A third curve is intermediate (I) to the
toxic and nutrient curves. The peak of the curve and the
crossover point from stimulatory to inhibitory response fall
well within the range of frequently encountered ambient doses.
A fourth curve, not shown, is the zero-response curve, falling
on the reference yield for common ambient doses.

The terms "stimulation" and "inhibition" cannot be equated
with "desirable" or "undesirable." For example, inhibition of
a weed species would be desirable; stimulation of a non-
harvested portion of a crop would be undesirable if it occurs
by diversion of energy from the harvested portion. In less-
managed systems such as grass and forests, total system response
to a change in any component would be difficult to predict.

In experiments conducted collaboratively by the Corvallis
Environmental Research Laboratory (CERL) and Oregon State
University (OSU), we have observed intermediate responses (I)
to acid rain in forest tree seedling emergence and growth, leaf
litter decomposition, and crop growth and yield. Results from
one experiment on crops are given in Table 2. For crops,

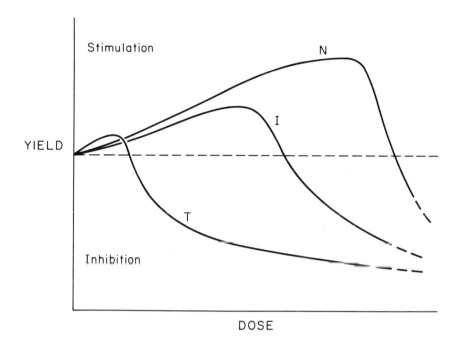

Figure 1.   *Three idealized dose-response curves. T: Toxic*
response, generally below reference yield at zero
(added) dose. N: Nutrient response, generally
above reference yield. I: Intermediate response,
with peak and crossover point from stimulatory to
inhibitory response well within range of common
ambient doses.

deviations from control responses were as large as ± 70 percent
and deviations of ± 20 percent were frequent. Results from
this and other studies suggest that this might be a common
response curve for crops which do not follow the zero-response
curve, with the locations of the peak and crossover point
dependent on other environmental factors. This would be a type
of interaction, as discussed above. However, for approximately
half the crops tested, no statistically significant responses
have been documented for acid rain at pH 3.0 or above.   In
addition, type T and N responses have also been observed.

     For doses near the crossover point, small changes in either
dose or in the location of the crossover point can result in
a switch between stimulatory and inhibitory responses. If the
curve is steep in this region, fluctuations in yield due to
shifts across the crossover point could be large. In a research

Table 2

Yield of Marketable Portion of Crops Grown in 1979 (from Lee et al., 1981)

Crop	Control Mean[a] g/pot	Ratio of Treatment Mean[a] to Control Mean (Fresh Weight of Yield Per Pot)			SE[c]	F-Test Significance Level[d]
		pH 3.0[b]	pH 3.5[b]	pH 4.0[b]		
Radish 1	43.23	0.44[j]	0.83[j]	0.92[h]	0.04	0.000[j]
Radish 2	42.12	0.40[j]	0.81[i]	0.84[h]	0.06	0.000[j]
Radish 3	47.74	0.24[j]	0.73[j]	1.14[h]	0.06	0.000[j]
Radish 4	26.79	0.38[j]	1.03	0.86	0.09	0.000[j]
Radish 5	18.07[g]	0.59	1.41	1.56	--[g]	--[g]
Beet	55.07	0.57[j]	1.02[j]	1.09[i]	0.11	0.011[i]
Carrot	138.54	0.56[j]	0.55[j]	0.73[i]	0.08	0.001[j]
Mustard green	59.28	0.70[j]	0.87[h]	0.83[i]	0.05	0.003[j]
Spinach	32.33	0.85	0.99	0.90	0.07	0.388
Swiss chard	99.72	0.90	1.04	0.94	0.07	0.561
Bibb lettuce	129.97	1.01	1.02	1.03	0.04	0.932
Tobacco	--	--	--	--	--	--
Cabbage	240.81	0.91	1.47	1.01	0.17	0.131
Broccoli	44.63	0.75[j]	0.92	0.89	0.07	0.063[h]
Cauliflower	69.62	1.03	1.46	1.20	0.15	0.185
Potato	691.79	0.92[j]	1.11[i]	1.07[h]	0.03	0.001[j]

Crop						
Green pea	21.55	1.04	0.98	1.05	0.04	0.674
Alfalfa	--	--	--	--	--	--
Red clover[e]	--	--	--	--	--	--
Tomato[e]	302.88	1.31[j]	1.01	0.95	0.07	0.001[j]
Green pepper[e]	193.12	1.05	1.20[i]	1.05	0.06	0.103[h]
Strawberry[e]	113.04	1.72[j]	1.72[j]	1.51[j]	0.13	0.001[j]
Oats	--	--	--	--	--	--
Wheat	--	--	--	--	--	--
Barley	--	--	--	--	--	--
Corn[f]	--	--	--	--	--	--
Onion	410.11	1.01	1.12	1.04	0.06	0.426
Fescue[e]	--	--	--	--	--	--
Orchardgrass	--	--	--	--	--	--
Bluegrass	--	--	--	--	--	--
Ryegrass	--	--	--	--	--	--
Timothy	--	--	--	--	--	--
Dry Weight of Yield Per Pct						
Radish 1	2.66	0.45[j]	0.79[j]	0.86[i]	0.04	0.000[j]
Radish 2	2.51	0.47[j]	0.83[i]	0.86[h]	0.05	0.001[j]
Radish 3	2.53	0.31[j]	0.77[j]	1.15[h]	0.05	0.000[j]
Radish 4	1.71	0.42[j]	1.01	0.87	0.03	0.000[j]
Radish 5	1.08[g]	0.64	1.40	1.52	--[g]	--[g]
Beet	10.38	0.55[i]	1.03[j]	1.10[j]	0.11	0.012[j]
Carrot	13.36	0.53[j]	0.57[j]	0.69[j]	0.08	0.000[j]

Table 2 (continued)

Crop	Control Mean[a] g/pot	Ratio of Treatment Mean[a] to Control Mean			SE[c]	F-Test Significance Level[d]
		pH 3.0[b]	pH 3.5[b]	pH 4.0[b]		
Mustard green	7.30	0.69[j]	0.90	0.86[h]	0.06	0.002[j]
Spinach	3.58	0.93	1.03	0.98	0.08	0.871
Swiss chard	16.66	0.98	1.04	1.03	0.06	0.827[h]
Bibb lettuce	6.13	1.05	0.97	1.07	0.03	0.087[h]
Tobacco	27.64	0.97	0.97	1.03	0.03	0.443
Cabbage	29.89	0.87	1.19	0.92	0.13	0.378[j]
Broccoli	6.07	0.75[j]	0.88	0.91	0.06	0.078[i]
Cauliflower	6.36	1.01	1.39	1.27	0.13	0.164
Potato	149.53	0.86[j]	1.05	1.05	0.03	0.000[j]
Green pea	4.21	1.06	0.97	1.06	0.06	0.547
Alfalfa[e]	28.72	0.94	1.31[j]	1.17[j]	0.05	0.000[j]
Red clover[e]	31.05	0.99	1.03	1.02	0.04	0.911
Tomato[e]	--	--	--	--	--	--
Green pepper	12.72	1.13	1.17[i]	1.06	0.06	0.207
Strawberry[e]	--	--	--	--	--	--
Oats	31.41	0.92	1.00	1.00	0.05	0.500
Wheat	29.30	0.97	0.98	0.98	0.06	0.976
Barley	34.71	1.05	1.06	1.00	0.05	0.727

Corn[f]	35.56	1.13[h]	0.95	0.99	0.05	0.085[h]
Onion	29.11	1.10	1.14	1.09	0.06	0.295
Fescue[e]	25.25	0.96[i]	1.07	0.92	0.04	0.018[i]
Orchardgrass	22.47	1.23[i]	1.10	1.00	0.07	0.097[h]
Bluegrass	12.81	0.98	0.94	1.00	0.05	0.725
Ryegrass[e]	20.24	0.99[i]	0.98	0.96	0.03	0.787
Timothy	21.07	1.24[i]	1.09	0.86	0.07	0.003[j]

[a] Sample size per mean: for SR and LR chambers, 14 pots; for SQ chambers, 25 pots (see Table 5 for chamber types).

[b] Significance of difference between acid rain treatment mean and control mean determined by two-sided t-test.

[c] Standard error of the mean (computed using error mean square from analysis of variance), divided by mean control yield.

[d] Significance level of F-test from one-way analysis of variance among four experimental groups.

[e] Seasonal total of multiple harvests.

[f] For corn, data refer to total aboveground (stem plus leaves) weight.

[g] Unreliable data for control; see text.

[h] Significant with $p \leq 0.10$.

[i] Significant with $p \leq 0.05$.

[j] Significant with $p \leq 0.01$.

situation, this would result in an increase in variability and a lack of reproducibility which could only be understood by considering the curve as a whole. For example, we have found both inhibitory and stimulatory responses of radish yield to pH 4.0 rain; at pH 3.5, inhibition dominates (examples are given in Table 2). This indicates that for radish the crossover point is near pH 4.0.

Fluctuations due to shifts across the crossover point could also be of practical importance. The increased uncertainty in productivity would be economically detrimental to farmers even if the fluctuations averaged to zero over several growing seasons.

The curves shown in Figure 1 assume a single peak for each chemical species. However, the constituents of acid rain cannot vary independently; any change in hydrogen ion concentration (pH) must be matched by a change in at least one anion concentration. In general, both sulfate and nitrate will vary with hydrogen ion. Thus, in the simplest case, a curve of yield versus hydrogen ion concentration could have three peaks (i.e., one each for the hydrogen, sulfate, and nitrate ions). The situation is further complicated by the fact that plants are multi-compartment systems, with the response of any compartment potentially influenced by the response of the other compartments. The portion identified as "marketable yield" may encompass several components. Thus, the curves in Figure 1 might best be thought of as forming a basis for constructing the yield-dose curve. The resultant curves might have complicated structures with multiple peaks. Although most experimental results are consistent with a single peak within the ambient range of concentrations, some data suggest multiple peaks.

A fundamental question is the definition of dose (i.e., which acid rain variables are most relevant to response). Possible definitions include:

(1)   total loading (concentration times volume);
(2)   average concentration;
(3)   maximum concentration sustained for some minimum time interval; and
(4)   concentration at a particular time.

Soil-mediated effects might be most sensitive to total loading (1). Foliar leaching and injury might respond mainly to maximum concentration sustained for some minimum time interval (3). Interactions of acid rain with growth stage or temperature could depend on seasonal components of temporal variability (4). If highly correlated with other measures, average concentration

(2) could be a useful index of dose; it can also be used to define a reference situation. The utility of the various measures of dose is an experimental question which remains to be answered.

EXPERIMENTAL APPROACHES AND RESULTS

In any experiment on the effects of acid rain on crops, there is a trade-off between control and realism. Control is needed to reduce variability, to separate the influences of various environmental variables, and to determine the underlying mechanism of response. Realistic growing conditions and rain application are necessary for the results to be transferable to agricultural situations. Greenhouse and field exposure chamber experiments emphasize control; field experiments emphasize realism at a cost of greater variability of response and uncertainty of cause. Various approaches have been used to study the effects of acid rain on crops. The advantages and disadvantages of several approaches have been reviewed by Jacobson (1980). Selection of an approach must be carefully matched to the scientific question being addressed.

Experimental apparatus can alter environmental conditions without their becoming unrealistic. In Figure 2, X_1 and X_2 are two environmental variables which affect yield. In the central shaded portion, yields are high; in the outer shaded portion, yields are lower, but still acceptable. Response curves such as shown in Figure 1 imply that these regions have no "holes" or "islands"; these can only exist for multiple-peaked response curves. Significant perturbations of growing conditions are those which result in movement from one region to another, or in a distortion of the regions. Small changes (e.g., $1^{\circ}C$, typical of many chamber designs) would not substantially affect the acid rain dose-response curve (AR, Equation 2) unless the interactions were large. However, potted plants grown in field-exposure chambers might experience large environmental distortions, such as unrealistic wind speeds and soil temperatures.

In the CERL-OSU crop studies, we have used two approaches: field-exposure chambers (potted plants) and field-grown plants; we have also used large lysimeter tanks to study forest response (Lee and Weber, 1980). Our field experiments would be classified by Jacobson (1980) as "application of simulated acid rain to crops that receive ambient rain." However, since National Acid Deposition Program (NADP) data from the site suggest average rain pH of 5.6 (occasionally down to 5.0), with very low sulfate and nitrate concentrations, the results are not confounded by the exposure of the control plants to ambient

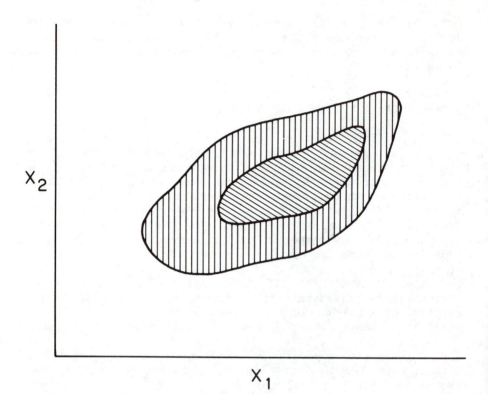

Figure 2.   Crop response regions. X_1 and X_2 are two environ-
mental variables.   Central shaded region repre-
sents superior crop yields; in other shaded
region, crops are lower, but acceptable.

acid rain, even though no rain exclusion system is used. Also, since summer rainfall is minimal (e.g., approximately 1 cm average for August), simulated rain can be applied without over-watering the crops. However, crops grown and harvested during spring (such as radish, lettuce) receive about equal amounts of ambient and simulated rain; thus, average concentrations of major acid rain constituents may be effectively half those of the simulated rain. Low levels of ambient air pollutants present at our site reduce the level of potential concern in experiments for possible acid rain-air pollutant interactions, and thus eliminate the need to exclude ambient gaseous pollutants.

In 1979, CERL-OSU used field exposure chambers (potted plants) to conduct a study of the relative sensitivities of 28 major crops in the United States to sulfur acid rain, applied at constant concentration (pH 3.0, 3.5, 4.0, or pH 5.6 control; Lee et al., 1981). Results (Table 2) indicated that, within the range of ambient concentration, sulfuric acid rain can:

(1) decrease the yield of root crops such as radish, carrot, and beet by as much as 76 percent;

(2) decrease the yield (as much as 30%) and marketability of leaf crops such as spinach, Swiss chard, and mustard greens by causing severe foliar injury;

(3) cause visible injury to tomatoes;

(4) increase the yield of fruit crops such as tomato, green pepper, and strawberry (up to 72%);

(5) increase the productivity of forage crops such as alfalfa, orchard grass, and timothy (up to 31%).

Extreme caution should be used when considering the results of this (or any other) experiment on the effects of acid rain on plant growth. Specifically, it would be wrong to associate an effect on yield at a particular experimental pH with effects on yield from ambient rain at the same average pH. For example, a location with average pH 4.0 might experience a few pH 3.0 events, with most events above pH 4.0. If response is most sensitive to peak concentrations, then yield effects might be more similar to the experimental results at pH 3.0 than to those at pH 4.0. However, if frequency of acid rain events is more important, then the experimental effects at pH 4.0 would tend to overestimate actual effects. Thus, at best, the data of Table 2 might define a potential range of effects, and might tentatively define potentially sensitive classes of crops.

Results of a 1980 field experiment tended to support the general conclusions of the 1979 study for both sulfuric-only and sulfuric-nitric acid rain. However, results of a different experiment (field-exposure chamber, potted plants) using alkaline soil (pH 7.5-8.2) rather than a slightly acidic soil (pH 5.8-7.0) indicated that acid rain can, for some crops, partially offset the adverse effects of overly alkaline soil. Although the economic importance of such crop-soil combinations is minimal, these results do suggest that, under some conditions, acid rain-soil interaction can be more important than any adverse effects of acid rain.

Experimental results for the four economically most important crops in states bordering the Great Lakes (Table 1) are discussed below. Because of the scarcity of data, and for the reasons given above, none of these results should be considered definitive.

Corn

In a 1980 field study, we found that sulfuric-nitric acid rain (2:1 by equivalents) at pH 4.0 decreased corn yield by 9 percent (statistically significant at p = 0.05). Decreases at pH 3.5 and 3.0 were 4 percent and 0 percent; however, these were not statistically significant. Corn seems to be an example of a crop for which the components of yield respond differently to acid rain. Acid rain tended to decrease the number of ears per plant, decrease the number of kernels per ear, and increase the weight per kernel; however, these differences were not statistically significant. Since yield per plant is the product of these three factors, the result was a minimum in yield at pH 4.0; at pH 3.0 the increase in kernel weight compensated for decreases in the other components. Presumably, at lower pH values the dose-response curve would start to decrease, resulting in a multi-peaked function.

Soybean

In studies at Brookhaven and Argonne National Laboratories, simulated sulfuric acid rain was applied to field-grown soybeans which also received ambient acid rain (approximately pH 4.1). Evans et al. (1980) found that the supplemental acid rain caused incremental decreases (i.e., in addition to any effect of ambient rain) of 3 percent (pH 4.0) to 11.5 percent (pH 2.7). However, Irving and Miller (1980) observed no differences related to the supplemental simulated acid rain. These studies differed in cultivars used, experimental design, amount and manner of acid rain application, ambient acid rain volume and acidity, soil type, and other environmental factors, probably including ambient ozone and sulfur dioxide concentrations. Jacobson (1980) reported that simulated acid rain increased field-grown soybean yield at low ozone levels, but decreased yield at higher levels; this indicates that the acid rain response and the ozone response of soybean interact.

Hay

We found that alfalfa and grass productivities were stimulated by sulfuric acid rain in both field-exposure chamber (potted plants, Table 2) and field experiments. However, in field experiments with 2:1 sulfuric-nitric (by equivalent) rain, stimulation occurred only during the early part of the growing season; total annual production was unaffected by 2:1 acid rain. Thus, response depended on the specific anions present, and not just on pH.

Wheat

No field experiment data are available for wheat. In a field-exposure chamber (potted plants, Table 2) experiment, wheat was not significantly affected by simulated sulfuric acid rain.

CONCLUSION

Experimental results have shown that the chemicals in acid rain applied under ambient conditions can affect crop growth and yield at concentrations within the range of ambient concentrations. The dose-response curve for each crop probably is non-monotonic, and, for some crops, may have several peaks. Some peaks may fall within the range of ambient concentrations. The locations of the peaks and the crossover points from stimulatory to inhibitory response may be dependent on other environmental factors. Plant parts are affected differently, suggesting that acid rain might change the allocation of energy within plants. For example, because the limed agricultural soils used in most studies are insensitive to acid rain, the observed effects on roots probably are not due to changes in root environment.

Important questions remain regarding the influence of temporal variability of concentrations on dose-response curves, and the transferability of experimental curves to ambient conditions. Interactions between acid rain and other environmntal factors (such as air pollutants, water stress, soil characteristics, temperature) have scarcely been studied. Before a credible assessment of the economic impact of acid rain on crops can be done, the mechanisms of response have to be studied and the predictive capability enhanced and validated.

ACKNOWLEDGMENTS

In addition to the author, the CERL-OSU acid rain team consists of Grady Neely (CERL) and Louis Grothaus, Shelton Perrigan, and Cynthia Cohen (OSU).

LITERATURE CITED

Evans, L.S., C.A. Conway, and K.F. Lewin. 1980. Yield Responses of Field-Grown Soybeans Exposed to Simulated Acid Rain. In: D. Drabløs and A. Tollan, eds., "Ecological Impact of Acid Precipitation," pp. 162-163. Proceedings of an International Conference, Sandefjord, Norway, March 11-14. SNSF Project, Ås-NLH, Norway.

Irving, P.M. and J.E. Miller. 1980. Response of Field-Grown Soybeans to Acid Precipitation Alone and In Combinaton With Sulfur Dioxide. In: D. Drabløs and A. Tollan, eds., "Ecological Impact of Acid Precipitation," pp. 170-171. Proceedings of an International Conference, Sandefjord, Norway, March 11-14. SNSF Project, Ås-NLH, Norway.

Jacobson, J.S. 1980. The Influence of Rainfall Composition on the Yield and Quality of Agricultural Crops. In: D. Drabløs and A. Tollan, eds., "Ecological Impact of Acid Precipitation," pp. 41-46. Proceedings of an International Conference, Sandefjord, Norway, March 11-14. SNSF Project, Ås-NLH, Norway.

Lee, J.J., G.E. Neely, S.C. Perrigan, and L.C. Grothaus. 1981. Effect of Simulated Sulfuric Acid Rain on Yield, Growth, and Foliar Injury of Several Crops. Environ. Exp. Bot. 21:171-185.

Lee, J.J. and D.E. Weber. 1980. Effects of Sulfuric Acid Rain on Two Model Hardwood Forests: Throughfall, Litter Leachate, and Soil Solution. EPA-600/3-80-014. U.S. Environmental Protection Agency, Corvallis Environmental Research Laboratory, Corvallis, Oregon 97333. 38 pp.

Statistical Abstract of the United States. 1978. 99th Edition, U.S. Bureau of the Census, Washington, D.C.

CHAPTER 21

EFFECTS OF DRY DEPOSITION COMPONENTS OF
ACIDIC PRECIPITATION ON VEGETATION

Sagar V. Krupa
Department of Plant Pathology
University of Minnesota
St. Paul, Minnesota 55108

INTRODUCTION

Molecular forms of sulfur and nitrogen are of major concern
in the production of atmospheric acidity. On a comparative
basis, significantly more progress has been achieved in our
understanding of the atmospheric chemistry of sulfur compounds
(Atmos. Environ., Volume 12, 1978), their removal processes and
field effects on terrestrial vegetation. Only recently have
research efforts been directed to the nitrogen compounds.
Therefore, in the following sections, the discussion is
restricted to atmospheric sulfur species.

Sulfur in the atmosphere exists in both gaseous and
particulate (solid and liquid) forms, their individual
concentrations varying in time and space. Sulfur dioxide (SO_2)
and particulate sulfate (SO_4) constitute the most abundant
sulfur species in the atmosphere. Reduced forms of sulfur such
as hydrogen sulfide (H_2S) are most likely deposited to the
ground after oxidation to sulfur dioxide and particulate sulfate
(Atmos. Environ., Volume 12, 1978).

On an annual basis, estimates of the importance of dry vs
wet processes of atmospheric sulfur removal to the earth range
from a ratio of 60:40 percent to 40:60 percent (NAS, 1978).
The relative magnitude of atmospheric sulfur removal by the two
processes varies with the season and the geographic location.

Results of investigations on the dry deposition of sulfur
dioxide and particulate sulfate on vegetation have been
summarized by several authors (Chamberlain, 1981; Garland,
1978; Lindberg et al., 1979; Sehmel, 1980). The objective of
this chapter is to review the results of a multi-year study of
atmospheric sulfur inputs into the terrestrial ecosystem in the

vicinity of a coal-fired power plant and its potential consequences to vegetation. Emphasis is directed to dryfall inputs of sulfur as sulfur dioxide, as an addition over the wetfall sulfate ion (SO_4^{-2}) inputs.

POINT SOURCE CHARACTERISTICS AND SULFUR CHEMISTRY

Northern States Power Company's (NSP) SHERCO coal-fired power plant is located in Sherburne County, near Becker, Minnesota. The facility is surrounded by a highly agricultural and irrigated area with flat terrain, approximately 65 km northwest and downwind (summer months) from an urban complex (Minneapolis-St. Paul). The power plant became operational in 1976.

During summer months, meteorological conditions have produced periodic ozone episodes in the study area due to transport of urban plumes (Laurence et al., 1977). The power plant is equipped with a 200 m stack and employs advanced pollutant scrubber technology (Table 1), resulting in the control of 99.9 percent of particulate matter and about 60 percent of the sulfur dioxide in the flue gas. Net sulfur dioxide emissions from the point source during 1978 were approximately 59 tonnes per day.

Hegg and Hobbs (1980) investigated the sulfur chemistry of the plume using aircraft monitoring (Table 2). Sulfate concentrations in the center of the plume ranged from 0.7 to 4.96 percent (0.88 ± 0.30 to 7.86 ± 0.35 $\mu g/m^3$) relative to the sulfur dioxide levels (55 to 1100 $\mu g/m^3$). The conversion rate of sulfur dioxide to sulfate in the plume was calculated to be 0 to 2.2 ± 1.3 percent/hr (Hegg and Hobbs, 1980). Limited analyses of ground level fine particulate aerosols for sulfate showed low values (~2.0 $\mu g/m^3$) for the study area (Krupa, unpublished data).

Ground level sulfur dioxide concentrations were continually measured over a 3 yr period at 8 locations in the vicinity of the point source based upon predicted isopleths. These data were modeled according to Larsen (1977), and a summary is presented in Table 3. Highest measured average ground level sulfur dioxide concentrations were 150 $\mu g/m^3$/0.5 hr, 136 $\mu g/m^3$/hr, 121 $\mu g/m^3$/3 hr, and 107 $\mu g/m^3$/8 hr, during 0.01 percent of the time annually.

ATMOSPHERIC SULFUR INPUTS AND VEGETATION RESPONSE

During the 1977-1980 (summertime) crop season, in the vicinity of the point source, on an average, wetfall inputs

Table 1

Northern States Power Company Coal-Fired Power Plant – Source Description and Characteristics –
Location: Becker, Sherburne County, Minnesota (93° 55' W; 45° 20' N)

Characteristics of Surrounding Area	Power-Generating Capacity (MW)	Sulfur Content of Coal (%)	Ash Content of Coal (%)	Flue Gas Concentration of SO_2 (ppm)	Flue Gas Concentration of NO_x (ppm)	Scrubber Characteristics
Summer months: downwind from urban plume; high relative humidity; flat terrain, agricultural (irrigated)	1400 (stack height 200 m), plume exit velocity 90 m/sec	0.8	8.4	250 $(6.55 \times 10^5$ $\mu g/m^3)$	400	Wet scrubber followed by limestone ($CaCO_3$) scrubber, heated plume (82°C)

Adapted from Hegg, D.A. and P.V. Hobbs, 1980.

Table 2

Sulfur Dioxide and Sulfate Concentrations in the Plume[a]

Date	Range (km)	Concentration of SO_2 ($\mu g/m^3$)[b]	Concentration of SO_4 ($\mu g/m^3$)[c]
June 17, 1978	3.2	1000	1.90+0.40
	17.6	346	6.01+0.77
	20.0	275	4.70+0.67
June 21, 1978	3.2	686	5.16+0.34
	9.6	320	2.38+0.30
	32.0	369	2.72+0.31
	64.0	55	2.73+0.31
	Ambient	26	0.88+0.30
June 22, 1978	3.2	1079	7.87+0.35
	19.2	1108	7.77+0.30
	32.0	482	3.70+0.36
	48.0	79	1.56+0.37
	Ambient	34	1.37+0.36

[a]Adapted from D.A. Hegg and P.V. Hobbs, 1980.
[b]These are "point" measurements (obtained from bag sampler) near the center of the plume. The error in these measurements is $\pm$ 1.3 $\mu g/m^3$. Actual measured values in parts per billion were converted to micrograms per cubic meter using reference conditions: 25°C; 270 mm Hg.
[c]These values were derived from ion chromatography of Teflon[R*] filters.

of sulfur as sulfate ions were an order of magnitude greater (10:1 kg/ha) than the dryfall sulfur dioxide inputs. However, sulfate ions and nitrate ions in these rainfalls appeared to be predominantly in a non-acidic form [H^+ on (SO_4^{-2} + NO_3^-) = 0.04 and 0.3 R^2, depending on the year]. Air parcel trajectory analysis suggests long-range transport of the sulfate and nitrate and removal by rainfall (Krupa, unpublished data). Additionally, ambient sulfur dioxide and sulfate concentrations appeared to be too low in the study area to account for much of the high sulfate ion concentrations measured in rain (up to 30.0 mg/l).

*Registered trademark of E.I. duPont de Nemours and Company, Inc., Wilmington, DE.

Table 3

Average Ground Level Sulfur Dioxide Concentrations
(μg/m^3) in the Vicinity of the Point Source[a]

Averaging Time (hr)	Percent Frequency of Occurrence (hr/year)			
	0.01	0.1	1.0	10
0.5	121.8+16.5 (107.2-145.7)[b]	73.9+11.3 (63.1-93.5)	25.2+6.0 (18.9-36.2)	0
1.0	112.9+15.5 (112.9-136.2)	67.9+10.7 (57.6-86.5)	21.2+5.8 (15.7-31.4)	0
3.0	99.0+13.9 (86.7-120.8)	57.9+9.4 (48.7-74.7)	15.2+5.6 (9.4-23.6)	0
8.0	86.5+12.6 (74.9-106.6)	50.0+8.4 (40.6-63.9)	9.7+5.8 (3.4-19.4)	0
24.0	72.0+10.7 (61.8-90.1)	38.5+7.3 (31.4-51.4)	3.9+5.5 (0-14.9)	0

From Krupa et al., 1980.
[a]Values were derived from three years of measured SO$_2$ concentrations at 8 locations based on isopleths. Model used was according to R.I. Larsen, 1977.
[b]Values in parentheses represent concentration range.

Ground level increments in sulfur dioxide inputs in the vicinity of the source during its operation were modeled using U.S. Environmental Protection Agency's Climatological Dispersion Model (CDM) applied to a single source (Figure 1)(Krupa et al., 1980). Predicted isolines of average ground level sulfur dioxide concentrations ranged from 0.25 to 2.0 μg/m^3 during June to September.

During each summer foliar sulfur concentrations were determined in several plant species, in a total of 48 permanent field plots located in 8 different directions from the point source at distances of 4 to 80 km. Significant increases in foliar sulfur concentrations were found in several plant species without a corresponding measurable increment in the soil sulfur concentrations (Krupa et al., 1980). For example, foliar sulfur in corn increased significantly during 1977 (operational at one half generating capacity). The geographic locations of these

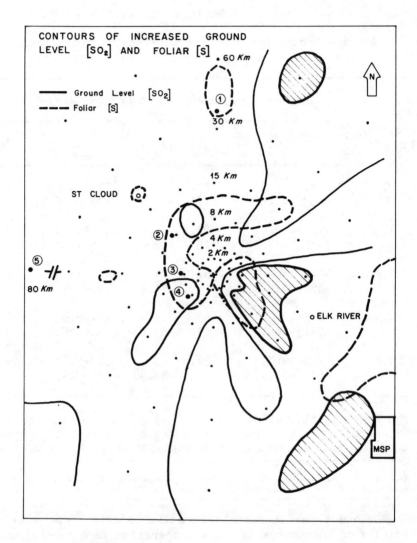

Figure 1.   *Modeled contours of increased ground level sulfur
dioxide concentrations (μg/m³) relative to the
point source (+).   Contours depict increments of
sulfur dioxide concentrations (μg/m³) during 1977
or 1978 compared with 1976.   Contours enclosing the
hatched areas depict increments in sulfur dioxide
levels both during 1977 and 1978, with the highest
increment in 1978.   Dotted lines represent geographic areas with increased foliar total sulfur concentrations (% dry wt.) in vegetation in 1978 compared with 1975.   Numbers 1 through 5 in the Figure
indicate sites where corn was sampled for [34]S
analyses.*

corn plots (1 through 4) were consistent with the modeled isopleths of increased ground level sulfur dioxide concentrations (Figure 1). Plot number 5 in Figure 1 was the background site.

The point source uses sub-bituminous Montana coal. This coal has a higher ^{34}S value (+8.7) compared to the background soil (+2 to +3) for the study area.

$$^{34}S \% = \frac{^{34}S/^{32}S \text{ Sample}}{^{34}S/^{32}S \text{ Trolite Standard}} - 1 \times 10^3$$

Therefore, corn foliar samples were analyzed for ^{34}S to substantiate whether the tissue increments in sulfur were due to the dryfall inputs of sulfur as sulfur dioxide from the source. The foliar values of $^{34}S \% \times \Delta\% S$ (1975 to 1977) were 5 to 11 times higher in the corn samples at locations 1 through 4 compared to location 5 (Figure 1 and Table 4).

In order to evaluate the effects of dryfall sulfur dioxide inputs of sulfur on crop productivity, soybean Hodgson 78 (an indeterminate cultivar) was exposed in a computerized open-top field fumigation system to sulfur dioxide regimes roughly comparable to the measured ambient concentrations (Figure 2 and Table 5). These exposures occurred during periods of no rain and, therefore, any observed effects should be viewed as contributions of the dryfall sulfur as sulfur dioxide inputs. The effects of sulfur dioxide on the yield parameters were evaluated relative to plants grown in charcoal-filtered air, with all other conditions identical to the sulfur dioxide exposed (Table 6). At the pollutant concentration exposure durations used, comparatively greater effects were observed on seeds/plant, pods/plant, and seed weight/plant relative to the other characteristics evaluated.

The pollutant exposure-plant response data are being modeled according to Blackie and Dent (1974). The effect of any stress factor at one point in a crop's growth may be expressed as a proportion of response relative to a check. The check may be the crop's response under background or zero stress. The general empirical model is expressed as:

$$y = f(x_{t_i}) \qquad\qquad (1)$$

Table 4

Changes in Total Foliar Sulfur in Corn Sampled During 1975 and 1977 at Different Locations in the Vicinity of the Point Source and δ^{34}S During 1977 (1975- Pre-Operational, 1976- 700 MW Output and 1977- Fully Operational, 1400 MW Output)

(a)	(b)	(c)	(d) % Foliar S		(e)	(f)	(g)	(h)
Site #[a]	Direction	Distance km	1975	1977	Δ % S 75-77	%δ^{34}S	Δ% S x % δ^{34}S	Relative Proportions of (g)
1	N	32.0	0.17	0.22	29.0	7.0	203	8.46
2	NW	8.0	0.20	0.23	15.0[b]	8.4	126	5.25
3	W	4.0	0.18[b]	0.22	22.0[b]	9.0	198	8.25
4	SW	4.0	0.16	0.22	37.5[c]	7.3[c]	274	11.42
5	W	80.5	0.22	0.23	4.5[c]	5.3[c]	24[c]	1.00[c]
	(Background)							

[a] For site locations, refer to Figure 1.

[b] 1976 data, %ΔS 1976-1977.

[c] Other values in each vertical column are signficantly different, $0.010 < p < 0.025$.

Krupa (1981, unpublished).

Table 5

Open-Top Field Chamber Artificial Hourly Exposure
Regime of Soybean to Sulfur Dioxide, Roughly in
Comparison to Hourly Ambient Ground Level Sulfur
Dioxide Concentrations in the Vicinity of the
Point Source During 1980 Summer (An Example)

SO$_2$ Concentration Range µg/m^3/hr	Percent Hr Frequency of Occurrence	
	Open-top Chamber[a]	Ambient
0-23	97.87	98.80
26-50	0.40	0.88
52-76	0.76	0.12
79-102	0.30	0.03
105-128	0.07	--
131-155	0.13	--
157-181	0.07	--
183-207	0.09	--
209-233	0.06	--
262-286	0.04	0.05
288-311	0.04	--
314-338	0.01	--
341-364	0.02	--
367-390	0.01	--
393-417	0.01	--
455-469	0.01	--
472-495	0.01	--
550-574	0.01	--
576-599	0.04	--
602-626	0.01	0.12
655-678	0.02	(629-655
969-992	0.01	µg/m^3/hr)

[a]The exposure regimes represent the cumulative frequency of
occurrence of average hourly sulfur dioxide concentrations
derived from continuous sulfur dioxide measurements at 8
isopleths over three summers.

Pratt, Hendrickson, and Krupa (1981 unpublished data)

Table 6

Effects of SO_2 on Soybean Yield Parameters - Plants Were Exposed in Open-Top Chambers to SO_2 Regimes as Specified in Table 5

Component	x̄ Control	x̄ SO_2 Treated	x̄ Yield Loss	S.D. (Pooled)	% Yield Loss
Pods/row	815.9	804.2	11.73	69.17	1.44
Seed wt/row	292.0	295.0	--	25.10	--
Pods/plant	41.92	37.20	4.73	3.16	11.27*
Seed wt/plant	15.05	13.64	1.41	1.29	9.35*
Seeds/row	1798.3	1736.7	61.58	141.75	3.43
Seeds/plant	92.43	80.53	11.90	6.30	12.88**
TSW	161.7	169.9	--	7.13	--
Seeds/pod	2.20	2.16	0.033	0.026	1.50*

TSW = Thousand seed weight; all seed weight in g.

Yield loss = $(1 - \frac{Treated}{Control})$

Statistical significance, *p = 0.10; **p = 0.05 by t test
Pratt and Krupa (1981 unpublished)

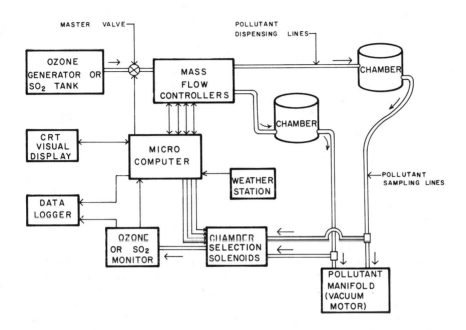

Figure 2. *Schematic diagram showing computer-controlled pollutant dispensing and monitoring systems to the open-top field fumigation chambers. This system can be programmed to roughly mimic ambient pollutant conditions as single pollutants or as multiple-pollutant mixtures.*

y = proportion of reduction in the response parameter (e.g., yield)

x_{t_i} = stress dosage parameter at time t_i

With linear least squares approximation, an example of a functional relationship may be:

$$y = b_0 + b_1 x + b_2 x^2 \qquad (2)$$

y = proportion of yield loss

x = dose at a crop development stage, quantified as one
    of three summary statistics, h_i (total hourly dose/24
    hr), $h_i/24$ (average hourly dose), and h_n (average
    hourly dose for hours when dose exceeded 0 µg/l)

The approach adapted is a dynamic matrix model, where for each
specified dt, a functional relationship will be developed, as
in equation (2).   A proportion of yield reduction Δy will be
incurred during each time interval dt.   The net crop response
over the growing season will be:

$$_1\int^n y \cdot dt$$

n = maximum number of growing days.

The time step for integration, dt, will depend on the
specific crop and the extent of available, actually measured
raw air quality data.   For example, with alfalfa, where
pollutants have a direct effect on the economic biomass, dt =
1 day.   With a cereal, where the pollutants have an indirect
loss effect (i.e., the plant part directly affected by the
pollutant) that is different from the part with economic value
(the grain), dt is treated at a bigger time step.   This model
allows a 95 percent confidence limit to be assigned to each
loss contribution over each dt, and to the total season loss
estimate. Using this modeling approach and the measured ambient
sulfur dioxide concentrations at different geographic locations
in a given state, crop loss predictions can be developed on a
county basis.

SUMMARY

In conclusion, in Minnesota, while wetfall inputs of sulfur
ions appear to be far greater than the dryfall sulfur dioxide
inputs during the summertime, the dryfall sulfur as sulfur
dioxide effects on terrestrial vegetation can be separated and
identified. With new point sources such as the one discussed
in this chapter, dryfall inputs of sulfur as sulfur dioxide in
the vicinity of the source will be low and any effects identified
will be subtle. Evaluation of these effects requires long-term
intensive studies.   Additionally, in these studies, joint
effects of non-sulfur pollutants with the sulfur inputs must
be considered.

ACKNOWLEDGMENT

The author is most grateful to Dr. H.R. Krouse of the University of Calgary, Canada, for the stable sulfur isotope analyses. Thanks are also extended to Dr. P.S. Teng, Dr. B.I. Chevone, Mr. Riley Hendrickson, and Mr. G.C. Pratt for their cooperation and help.

LITERATURE CITED

Blackie, M.J. and J.B. Dent. 1974. The Concept and Application of Skeleton Models in Farm Business Analysis and Planning. J. Ag. Econ. 25:165-175.

Chamberlain, A.C. 1982. Deposition of Gases and Particles to Vegetation and Soil. In: S.V. Krupa and A.H. Legge, eds., "Air Pollutants and Their Effects on the Terrestrial Ecosystem." John Wiley and Sons, Inc., New York (in press).

Garland, J.A. 1978. Dry and Wet Removal of Sulphur From the Atmosphere. Atmos. Environ. 12:349-362.

Hegg, D.A. and P.V. Hobbs. 1980. Measurements of Gas to Particle Conversion in the Plumes From Five Coal-Fired Electric Power Plants. Atmos. Environ. 14:99-116.

Krupa, S.V., B.I. Chevone, J.L. Bechthold, and J.L. Wolf. 1980. Vegetation: Effects of Sulfur Deposition by Dry-Fall Processes. In: D.S. Shriner, C.R. Richmond, and S.E. Lindberg, eds., "Atmospheric Sulfur Deposition. Environmental Impacts and Health Effects." Ann Arbor Science Publishers, Ann Arbor, Michigan 48106. 568 pp.

Larsen, R.I. 1977. An Air Quality Data Analysis System for Interrelating Effects, Standards, and Needed Source Reductions: Part 4. A Three-Parameter Averaging-Time Model. J. Air Pollut. Control Assoc. 27(5):454-459.

Laurence, J.A., F.A. Wood, and S.V. Krupa. 1977. Possible Transport of Ozone and Ozone Precursors in Minnesota. Ann. Amer. Phytopathol. Soc. (Abstr.)4:89.

Lindberg, S.E., R.C. Harris, R.R. Turner, D.S. Shriner, and D.D. Huff. 1979. Mechanisms and Rates of Atmospheric Deposition of Selected Trace Elements and Sulfate to a Deciduous Forest Watershed. Oak Ridge National Laboratory, Environmental Sciences Division Publ. No. 1299.

NAS. 1978. Sulfur Oxides. U.S. National Academy of Sciences, Washington, DC. pp. 8-50.

Sehmel, G.A.   1980.   Model Predictions and a Summary of Dry Deposition Velocity Data.   <u>In</u>:   D.S. Shriner, C.R. Richmond, and S.E. Lindberg, eds., "Atmospheric Sulfur Deposition. Environmental Impacts and Health Effects."   Ann Arbor Science Publishers, Ann Arbor, Michigan 48106.   568 pp.

INDEX